Alexander W. Koch (Ed.)

Optomechatronics

MDPI

This book is a reprint of the special issue that appeared in the online open access journal *Sensors* (ISSN 1424-8220) in 2013 (available at: http://www.mdpi.com/journal/sensors/special_issues/optomechatronics).

Guest Editor
Alexander W. Koch
Technische Universität München
Institute for Measurement Systems and Sensor Technology (MST)
Munich, Germany

Editorial Office
MDPI AG
Klybeckstrasse 64
Basel, Switzerland

Publisher
Shu-Kun Lin

Managing Editor
Lucy Lu

1. Edition 2014

MDPI • Basel • Beijing • Wuhan

ISBN 978-3-03842-008-8 (PDF)
ISBN 978-3-03842-001-9 (Hbk)

Table of Contents

Preface

The field of optomechatronics provides synergistic effects of optics, mechanics and electronics for efficient sensor development. Optical sensors for the measurement of mechanical quantities, equipped with appropriate electronic signal (pre)processing means have a wide range of applications, from surface testing, stress monitoring, thin film analysis to biochemical sensing. The aim of this special issue is to provide an overview of actual research and innovative applications of optomechatronics in sensors. Papers addressing, inter alia, optical sensor principles, fiber-optic sensors, electronic speckle pattern interferometry, surface analysis, thin film measurement, FGB sensors, and biochemical sensors are provided.

Prof. Dr.-Ing. Dr. h.c. Alexander W. Koch
Guest Editor

1

Reprinted from *Sensors*. Cite as: Levi, A.; Piovanelli, M.; Furlan, S.; Mazzolai, B.; Beccai, L. Soft, Transparent, Electronic Skin for Distributed and Multiple Pressure Sensing. *Sensors* **2013**, *13*, 6578–6604.

Article

Soft, Transparent, Electronic Skin for Distributed and Multiple Pressure Sensing

Alessandro Levi [1,2]**, Matteo Piovanelli** [1,2]**, Silvano Furlan** [3]**, Barbara Mazzolai** [1] **and Lucia Beccai** [1,]*

[1] Center for Micro-BioRobotics@SSSA, Istituto Italiano di Tecnologia, Viale Rinaldo Piaggio 34, Pontedera 56025, PI, Italy; E-Mails: alessandro.levi@iit.it (A.L.); matteo.piovanelli@iit.it (M.P.); barbara.mazzolai@iit.it (B.M.)

[2] The BioRobotics Institute, Scuola Superiore Sant'Anna, Polo Sant'Anna Valdera, Viale Rinaldo Piaggio 34, Pontedera 56025, PI, Italy

[3] Department of Applied Mathematics and Theoretical Physics, Centre for Mathematical Sciences, University of Cambridge, Wilberforce Road, Cambridge CB3 0WA, UK; E-Mail: s.furlan@damtp.cam.ac.uk

* Author to whom correspondence should be addressed; E-Mail: lucia.beccai@iit.it; Tel.: +39-050-883-079; Fax: +39-050-883-402.

Received: 23 March 2013; in revised form: 19 April 2013 / Accepted: 3 May 2013 / Published: 17 May 2013

Abstract: In this paper we present a new optical, flexible pressure sensor that can be applied as smart skin to a robot or to consumer electronic devices. We describe a mechano-optical transduction principle that can allow the encoding of information related to an externally applied mechanical stimulus, e.g., contact, pressure and shape of contact. The physical embodiment that we present in this work is an electronic skin consisting of eight infrared emitters and eight photo-detectors coupled together and embedded in a planar PDMS waveguide of 5.5 cm diameter. When a contact occurs on the sensing area, the optical signals reaching the peripheral detectors experience a loss because of the Frustrated Total Internal Reflection and deformation of the material. The light signal is converted to electrical signal through an electronic system and a reconstruction algorithm running on a computer reconstructs the pressure map. Pilot experiments are performed to validate the tactile sensing principle by applying external pressures up to 160 kPa. Moreover, the capabilities of the electronic skin to detect contact pressure at multiple subsequent positions, as well as its function on curved surfaces, are validated. A weight sensitivity of 0.193 gr^{-1} was recorded, thus making the electronic skin suitable to detect pressures in the order of few grams.

Keywords: optical; artificial skin; electronic skin; tactile sensor; pressure sensor; pressure distribution; soft; flexible

---

1. Introduction

In recent years, touch screens have represented one of the major drivers for new technological developments in the field of flexible touch sensors suitable for extended surfaces [1,2]. Resistive and capacitive methods represent the leading approaches in the field. In this context, resistive sensors show advantages of low cost and low power consumption, but their drawbacks include a reduction of the light transmittance of the screen, as a result of the overlaying sensitive layer on the display, and the existence of a minimum pressure threshold required for touch detection. These sensors consist of a layer of conductive elastomer or foam and they suffer of a highly non-linear force-resistance characteristic that requires the use of signal processing algorithms and causes poor long term stability [3,4]. Capacitive sensing gained importance, especially in consumer electronics, but its drawbacks are the high costs and complexity of fabrication, power consumption, stray capacitance and lack of pressure detection, especially when this transduction method is applied to extended areas in consumer electronic devices. Moreover, the materials these sensors are made of cause a reduced light transmittance through the screen [1,2,5].

Looking more closely at actual touch panels, they detect only the position of the contacts and use this input for human-machine interactions. This kind of interface is suitable for most applications but others would benefit from a further degree of interaction. Possible 3D user interfaces include, for example: graphic applications, games, 3D virtual object manipulation tasks etc. Though 2D gestures could emulate the interaction with a third dimension, they can't provide a mapping as intuitive and direct as it is for the two dimensions on a surface, causing a complex and not natural interaction. Indeed, it has been reported that pressure based interactions improve usability. For example, pressure based keyboards can improve key click performance on touch screens [6]. Therefore, pressure detection can add the third touch dimension to user interfaces but some pressure sensitive technologies are not suitable for application to touch screens. As a matter of fact, in order to be integrated over the screen, touch sensors need to be not only pressure sensitive but also transparent. The features of the sensor (mechanical, optical) as well as its function (pressure detection) are both fundamental for the feasibility of a new technology in practical touch screen application.

Besides pressure sensing, it is desirable that touch sensors be flexible, thin and bendable so that they can be applied over curved surfaces or flexible displays. These features would enable new form factors for devices and innovative products: the electronic industry is working on the next generation of mobile devices that have pocket size but have wider screens that can be rolled. The dramatic push towards having flexible systems (including all related electronics) by the consumer electronics market is rapidly improving the available technologies for developing new tactile sensing systems in modern robotics, where additional stringent mechanical characteristics are required. In particular, the emulation of the mechanical characteristics of the biological skin model is one of the major goals for humanoid and rehabilitation robotics [7,8]. In parallel, it is noteworthy to highlighting that new classes of robots are being investigated [9,10] that will find uses in applications where conventional hard robots are unsuitable. They represent the emerging field of soft robotics, which will highly benefit from the development of soft and flexible smart skins, since these will endow the soft artifacts with the capability to interact with the environment. In the mentioned robotic research areas, in addition to flexibility, features like softness and stretchability represent the bottleneck towards real skin-like devices that can both be integrated in 3D systems and imitate nature when interfacing with the outside world, *i.e.*, have a suitable compliance at sensor/environment interface.

Among more consolidated transduction methods for touch sensors, like resistive and capacitive ones, optical approaches have been investigated and developed to the extent that new sensors have been commercialized [11]. Main important reasons in support of optical sensors include: they are immune to electromagnetic and electrostatic fields that are common in industrial environments; they are not affected by humidity; their signals can be easily multiplexed and integrated using light emitting sources and demultiplexed using photodetectors, making them a good candidate for potential large area electronics [12–18].

Optical sensors can be divided in two main categories: fiber Bragg based optical sensors and microbending optical sensors. The first ones are composed of optical fibers with internal Bragg grating that reflects narrow spectral components of the light emitted by a broad spectrum source. Strain/pressure and temperature can be detected by analyzing the Bragg wavelength shift of the reflected light. An example of these kinds of sensors is reported in Reference [12] where it presented a sensitivity of 2.1×10^{-3} MPa^{-1}.

The second type of optical sensors consists of waveguides in which their microbending can alter the transmitted light. The determination of the light loss is used to detect the pressure. The use of flexible optical fibers allowed force discrimination below 1 N with a resolution of 0.1 N, and up to 30 N with 1 N resolution [13]. A resolution of 0.05 N for loads up to 15 N has also been registered with the same sensor approach [14]. Another optical waveguide sensor exploited a different configuration of crosslinked optical fibers embedded in a silicone elastomer allowing discrimination of pressure in the medium regime range 20–30 kPa [15]. A high sensitivity of 1 kPa^{-1} was obtained in the optical sensor with two plastic fibers separated by a compressible optical cavity made of Polydimethylsiloxane (PDMS) [16]. The deformation of the cavity with pressure changes the transmissivity of the device and therefore the pressure can be determined from the light intensity at the output. Like in the previously commented approaches, here as well there are some fabrication complexities, mostly due to the alignment requirements for the optical components. Therefore, these sensors are not easily integrated in an array, thus limiting their application over extended areas. An interesting method was to use a tapered optical fiber embedded into a PDMS-gold composite [17] that resulted in excellent pressure detection corresponding to a weight ~5 grams. The limits of this work are the poor optical transparency, the complexity of fabrication and the fact that for covering a large area the taxels need to cover the entire surface and have to be read one by one.

In this work we present and apply a novel optically-based approach by using a soft, flexible and transparent PDMS waveguide having the two-fold function of mechanical substrate and waveguide material, with air as cladding. We address the design and fabrication of the full tactile sensing system embodying the applied principle. As a result the fabricated device is stretchable, bendable, and rugged, while the principle on which it is based is potentially extendable to large areas. The mechanical characteristics of the electronic skin are due to the properties of PDMS [19], which is a soft, conformable and compliant material that can conform to large areas and surfaces of complex and not planar shapes. It is a homogeneous and optically transparent material for wavelengths ranging from 235 nm to the near-infrared, and that makes optical detection possible over the entire visible region. Moreover, it has an attenuation as low as 0.4 dB/cm. While the air-PDMS-air configuration has been previously investigated [18,20] and flexibility and stretchability proposed [18], to the authors' knowledge the possibility to provide information about multiple contacts, pressure distribution and the related shape of occurred contact, by means of a complete smart skin was not addressed. Few artificial skins are more advanced since they can detect and reconstruct the shape of the contact, but they lack stretchability and are only partially flexible [21,22].

In the present investigation, we attempt to go beyond the state of the art by proposing a new concept of smart electronic skin (e-skin) that has: (1) intrinsic mechanical compliance, stretchability and flexibility; (2) the capability to provide information about multiple contacts and the distribution of pressure externally applied; and (3) the capability to retrieve information about the shape of the contact. Although this last point is not fully investigated in the present work, a preliminary validation of such aspect is provided.

The paper is organized as follows: in Section 2 we present the electronic skin concept, we illustrate the design and physics of the tactile sensing mechanism, its embodiment in a prototype, the electronics and the reconstruction method, as well as the experimental set-ups and protocols used in the characterization. In Section 3 the results of the preliminary experimental analysis to validate the concept are reported. Finally, in Section 4 and Section 5 the discussion and the conclusions are reported, respectively.

2. Materials and Methods

The electronic skin comprises a tactile sensing mechanism, conditioning and acquisition electronics, and reconstruction software running on a PC.

2.1. Tactile Sensing Mechanism: Design and Physics

The tactile sensing mechanism is based on a mechano-optical transduction principle. The intensity of an electromagnetic wave traveling in a waveguide is modulated by mechanical deformations of the waveguide itself. From the intensity measured at the boundaries of the waveguide, it is possible to reconstruct the location and entity of mechanical deformations, by solving an inverse problem by a process that is similar to tomographic backprojection [23].

The device in this work consists of a thin flexible elastomeric transparent layer embedding along its periphery electromagnetic emitters and detectors that are positioned in a known configuration. Because of the total internal reflection phenomenon [24], signals from the emitters are bound in the elastomeric layer and they reach the detectors. Given that $n_1 > n_2$, where n_1 is the refractive index of the elastomeric layer and n_2 is the refractive index of air ($n_2 = 1$), the electromagnetic radiation emitted in the waveguide and incident on the boundary at an angle larger than or equal to a critical angle θ_c, (Equation (1)) is completely reflected, and thus results bound in the guiding layer:

$$\theta_c = \arcsin\left(\frac{n_2}{n_1}\right) \tag{1}$$

Figure 1 provides a 2D schematic representation of the tactile sensing device showing the *sensing area* bounded by emitters and detectors with specific relative positions. In principle, the *sensing area* can have various shapes without any detriment to the operation of the sensor: however, this aspect is not addressed in the present investigation.

Figure 1. Schematic of the tactile sensing device layout.

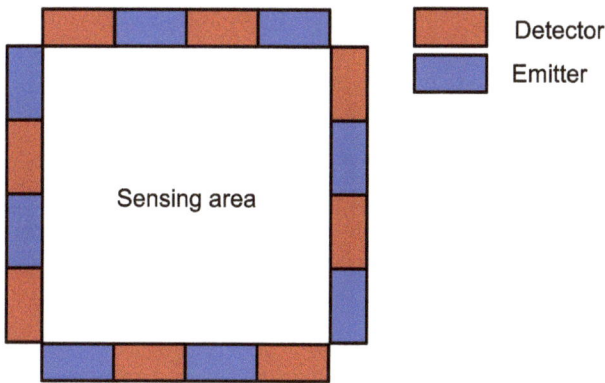

Figure 2 exemplify the operating mechano-optical transduction principle of the sensor. When no mechanical stimulus is externally applied, electromagnetic waves with a known intensity J_0 propagate to each detector from the emitters. The signal is then converted to a current I_0 which is then read and elaborated. When a mechanical contact event occurs, for example if an indentation is performed with an object on the surface of the *sensing area*, there is a variation in the output current I of the detectors, associated to both the contact area and the applied pressure of the applied mechanical contact.

Figure 2. Schematic of sensor working principle in case an external mechanical stimulus: (**a**) is not applied; (**b**) is presented at the top of the sensor. Red lines show a travelling wave direction.

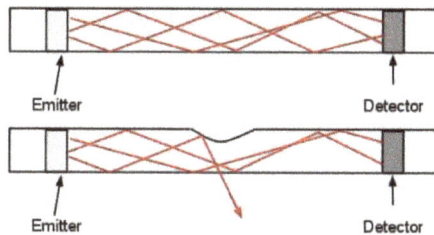

The reason for this is that the mechanical contact causes concurrent effects that lead to a change in the intensity of the light reaching the detectors, since the electromagnetic waves are partially deflected out of the waveguide. This behavior happens because of two main mechanisms: (1) the *Frustrated Total Internal Reflection* effect [25–27] which is due to a variation of the refraction index caused by the contact; and, (2) the deformation of the compliant waveguide, which causes a loss of light intensity, similarly to the bending of optical fibers [28].

The quantitative determination of the pressure applied to all points of the waveguide surface is beyond the scope of this work. We assume, as a first approximation, that we are in a regime of small deformations, where the mechanical behavior of the waveguide can be considered linear. These deformations, whose amplitude then depends linearly on the applied pressure, cause changes in the curvature of the waveguide interface. The losses of electromagnetic intensity caused by the changes in curvature can be expected to be non-linear, similarly to what is observed when deforming optical fibers [28].

Starting from the variations of signal intensity recorded at the periphery of the waveguide, the determination of the deformations that cause the electromagnetic losses is an inverse problem. Conceptually, this problem is similar to those found in tomographic imaging. Therefore the reconstruction process we used is inspired by tomographic back-projection procedures [23]: however the formal treatments developed for those applications cannot transfer directly to our case, since geometry and conditions are too different. A back-projection process allows obtaining an image mapping the deformations on the whole surface: on this map it is possible to observe the position and intensity of the deformations.

A further aspect concerns the possibility to reconstruct the shape of the contact area. Although this aspect was not thoroughly investigated in this work, some preliminary results will be provided in Section 3.

In general, we can state that the relation between the light intensity collected by the detectors, J, and the pressure applied on the waveguide, P, is given by:

$$P(i,j) = F\left(\sum_1^M \sum_1^N J_K\right) \tag{2}$$

where F is a generic function that can be used in a reconstruction algorithm in order to take into account the sum of light intensities emitted by all M light emitters and reaching each of the N detectors in the boundary. $P(i,j)$ represents the pressure calculated in the specific pixel point (i,j) resulting from a discretization of the sensing surface that will be described in Section 2.4.

2.2. Sensor Fabrication

An electronic skin prototype was built embodying the tactile sensing mechanism described above. In such a system a polydimethylsiloxane (PDMS) waveguide structure is used with embedded emitters and detectors. The emitters are infrared (IR) LEDs, while the detectors are phototransistors (*i.e.*, photodetectors, PDs). The components were chosen based on the wavelength at which their performances peak match, in either case being 950 nm. PDMS was chosen since its optical and mechanical properties are well characterized [19]. However, we verified the refractive index of PDMS for different wavelengths performed by means of reflectometry. From the data, we could extrapolate a value of $n_1 = 1.428$. Like explained in the previous section n_2 is the refractive index of air, thus $n_2 = 1$. Considering Equation (1) we obtain a critical angle $\theta_c = 44.45°$. Therefore the LEDs chosen to build the electronic skin have a narrow cone of emission, mostly included between ±30°, hence a large part of their emitted power is transmitted in the planar waveguide.

The sensor was fabricated by embedding eight LEDs (TSKS5400S, 950 nm, Vishay, Malvern, PA, USA) and eight PDs (TEKT5400S, 950 nm, Vishay) in a 5 mm thick layer of PDMS having a diameter of 5.5 cm. The 16 components were fixed to a plastic support frame by their connectors, thus leaving the head of each component free. The components were then placed on their heads in a Petri dish. PDMS (Dow Corning Sylgard 184) was prepared by mixing the curing agent to the base monomer with 1:10 weight ratio and degassed in vacuum for about 1 hour. A suitable quantity of the liquid PDMS mixture was poured in the petri dish to reach the 5 mm thickness required to completely embed the active part of emitters and detectors in the polymer. Finally, the sample was put in an oven at 60 °C for 3 hours and the curing phase ended after approximately 12 hrs at room temperature. The resulting prototype is shown in Figure 3, where the blue components are LEDs and the black components are PDs.

Figure 3. Electronic skin with external wiring: (**Left**) top view; (**Right**) side view when flexed.

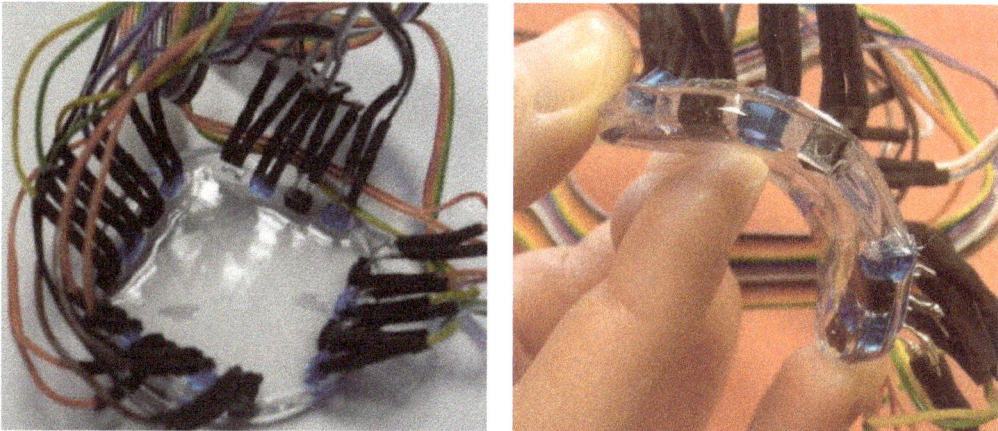

2.3. Electronics

The electronics (see Figure 4) designed for the proposed artificial skin consists of two independent parts on the same printed circuit board (PCB): driving electronics to switch on the LEDs in the desired way and readout electronics to acquire and condition the output signals from the PDs. Two power voltages were used: 3 V for both driving and readout electronics, and 5 V for $V_{control}$, which was used to polarize the LEDs.

The driving electronics comprises a 1–8 decoder/demultiplexer (Texas Instruments CD74AC138M) whose outputs activate one power switch MOSFET (Fairchild Semiconductor FDC6330L) at a time. Since the decoder is active low, inverters (Fairchild Semiconductor 74AC04MTC) are used to drive the power switch MOSFETs. When the corresponding power switch MOSFET is activated, the LED is switched on and polarized with the required current, 10 mA to be working in the linear zone of its I-V characteristic. The polarization resistance R was therefore calculated to be 390 Ω. The schematic in Figure 4 shows how an additional power switch

8

MOSFET (called Enable) is used to disconnect the common terminal of all LEDs from ground. This choice was taken to make the schematic of the electronics modular, so that it can be used with a greater number of active components. Using a specific configuration of four 1–8 decoder/demultiplexer, 32 LEDs can be sequentially activated one by one. This way it is possible to replicate the driving electronics a number of times, and sequentially activate a multiple of 32 LEDs.

Figure 4. (a) Schematic overview of the driving electronics; **(b)** Schematic overview of the readout electronics.

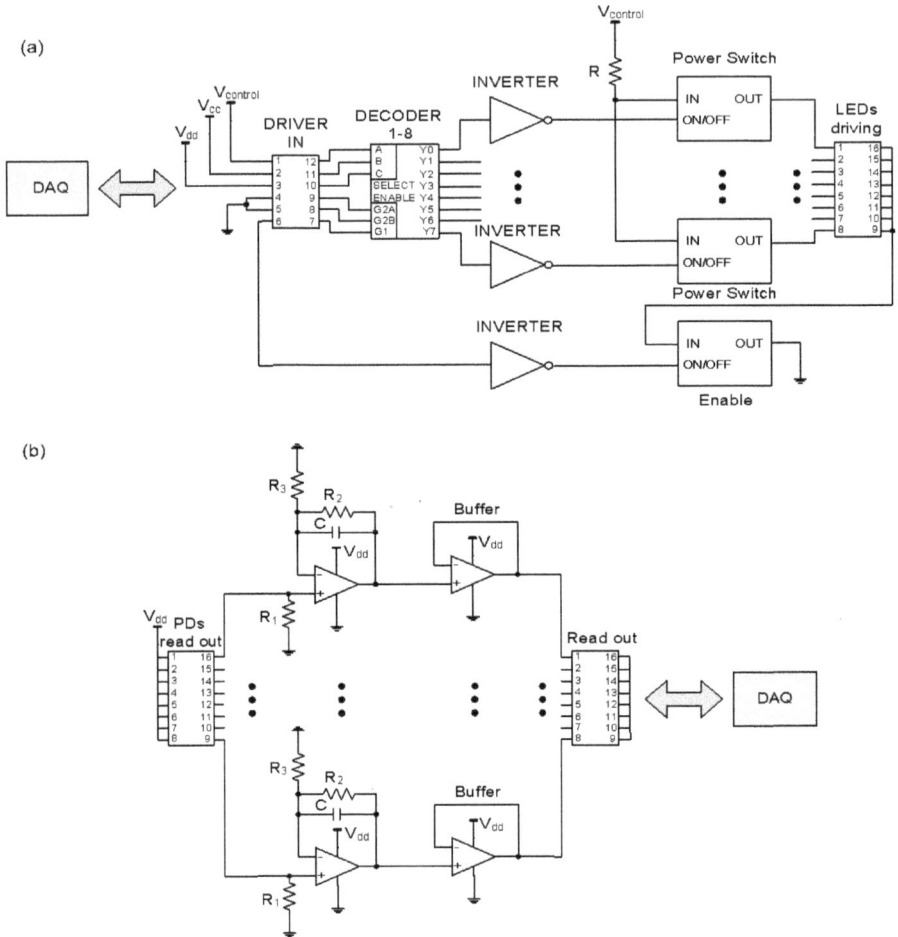

The read-out electronics comprises a polarization stage for the phototransistors, a current-voltage converter stage that converts the photocurrents in voltages and finally an amplification and filter stage.

Considering a $V_{dd} = 3$ V, a collector-emitter saturation voltage $V_{CEsat} = 0.3$ V and a required collector current for the phototransistors of 4 mA, the polarization resistor R_1 was set to 675 Ω. The gain of the amplifiers (R_2/R_3) was set to 1 by choosing for both R_2 and R_3 a value of 10 kΩ. The value for the gain should be chosen in order to avoid saturating the amplifiers' output with the highest electromagnetic signal coming from the LEDs. For future larger artificial skins the distance between LEDs and PDs would be higher, thus the gain will probably need to be set to a value larger than the unity. The refresh rate for the touch module was chosen to be 8 Hz and thus the clock frequency for the LEDs and PDs was 64 Hz. The low pass filter frequency was set to 38 Hz by choosing $C = 415$ nF, thus respecting the Nyquist theorem and avoiding back folded frequencies in the band of interest. After the amplifier/filter stage the outputs pass through a stage with op-amps in buffer mode and finally are sampled by the Data Acquisition (DAQ) system.

2.4. Reconstruction Process

The reconstruction process is inspired by tomography and consisted in the definition of three types of matrices and in addressing a backprojection procedure. The major steps in the reconstruction are described in the following.

The data acquired from the DAQ board is initially stored as a matrix, A, of double precision values. Each element in the line is the value acquired from a PD, with each line representing a period of activation for one specific LED. In the specific case of the experiments presented here and performed with the sensor described above, the resulting matrix has eight columns, one for each PD, and 8 N lines, where N is the number of times each LED has been activated, corresponding to the number of times the entire *sensing area* (see Figure 1) has been scanned. A similar matrix, B, obtained from acquisitions performed on the sensor during which there have been no contacts, is used to determine calibration values for all LED-PD pairs.

These two matrixes, A and B, are fed to a C++ algorithm alongside geometrical information about the relative position of all active components in the sensor. The geometrical information is used to generate an internal matrix representation, C, of the *sensing area* and compute correlations between each LED-PD pair and the points of that matrix. The correlations are used in the reconstruction process to determine whether the acquired values for any given LED-PD pair are related to contacts at a specific point or not. In the present work we used the simplest correlation approach: two parallel lines connect the extremities of the LED and the PD in each LED-PD pair, defining a trapezoidal section of the *sensing area*. For the reconstruction, only the points inside the trapezius are considered affected by the specific LED-PD pair.

The calibration matrix B is used to normalize the acquired matrix A, so that data is in the range between 0 and 1. This is followed by the backprojection process that consists in reading each value for each line from the normalized matrix A/B, corresponding to the measurement for a specific LED-PD pair, and adding it to the points of the matrix representation C determined in the correlation step. After having backprojected as many lines as there are LEDs in the sensor (eight in the case of the sensors used for this work), a contact intensity map of the entire surface at a given time has been computed and is saved. The images of the reconstructed *sensing area* in this paper are obtained by plotting the reconstructed maps using MatLab.

2.5. Experimental Tests

In a first phase, experimental trials were performed to test the electronic skin's working principle in order to provide significant information for the implementation and optimization of the reconstruction algorithm. This included loading tests performed on the sensor during which its output signals were analyzed without using the processing algorithm.

In a second phase, the aim of the experimental analysis was to test the capability of the electronic skin system (comprising the tactile sensing mechanism, its electronic conditioning system and the processing algorithm) to detect tactile information related to a mechanical stimulus. In particular, the position and intensity of contact were obtained. Moreover, preliminary trials for detection of one type of contact shape were addressed.

In a third phase, we performed preliminary experiments to begin validating the more advanced features of the electronic skin: detection of multi-pressure contacts and operation on curved surfaces. The experimental apparatus and protocols used in these tests are described in the next sections.

2.5.1. Experimental Setup

The experimental apparatus employed for the characterization of the smart skin system consisted of the components schematically illustrated in Figure 5. In particular, the loading system is shown more in detail in Figure 6. The force applied to the sensor was measured and recorded through a 6-axis load cell (ATI NANO 17F/T, ATI Industrial Automation, Apex, NC, USA) (A) interfaced to a loading probe (B). The vertical position of the load cell was determined with an initial rough manual positioning, by means of three orthogonal manual micrometric translation stages with crossed roller bearing (M-105.10,PI, Karlsruhe, Germany) (C), followed by an accurate controlled positioning, by means of a servo-controlled micrometric translation stage (M-111.1, PI) (D). This way the contact of the loading probe on the *sensing area* (see Figure 1) of the sensor (E) was achieved. In this work a probe with a square head (10×10 mm^2) was chosen because the limited resolution of the artificial skin imposed a constraint to the reconstruction of polygons with higher numbers of facets. Indentation experiments with probes having different head shapes will be addressed in a future work with a higher resolution skin.

The electronic and acquisition system, integrated in the experimental set-up, consisted of (see Figure 5): a NI-DAQ board (USB 6216), that was used for generating the driving signals for the LEDs and for acquiring the electronic signals from the PDs of the sensor; an ad-hoc printed circuit board (PCB) designed for amplification and filtering of such signals, and for driving the LEDs; finally, a laptop with a C++ algorithm that performs the elaboration and outputs the contact/pressure distribution. When addressing the initial testing phase of the sole optical tactile sensor without the processing algorithm, the same loading system described above (Figure 5) was used, but not the acquisition section (filtering and amplification) of the electronics: rather the sensor outputs were acquired by an oscilloscope (Agilent Technologies MSO7014A) directly connected to the sensor and analyzed using Matlab software.

For the final test on the multi pressure reconstruction capability the set-up of Figure 5 was used but the loading system was substituted by an electronic scale where the optical sensor was set for the measurements.

Figure 5. Block illustration of the experimental set-up for the artificial skin system. The loading system is depicted in detail in Figure 6.

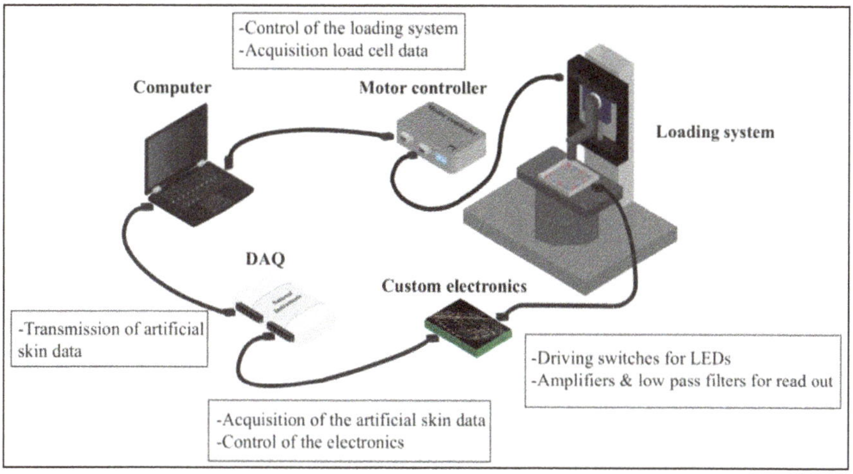

Figure 6. Image of the experimental set-up loading system, integrating: (**A**) the 3- axis load cell; (**B**) the Delrin loading probe whose 10 mm × 10 mm square shape is shown in (**B'**); (**C**) the three orthogonal manual micrometric translation stages; (**D**) the servo controlled micrometric translation stage. The 5 mm thick transparent electronic skin is shown (**E**), whose components are wired to the custom electronic system as schematized in Figure 5.

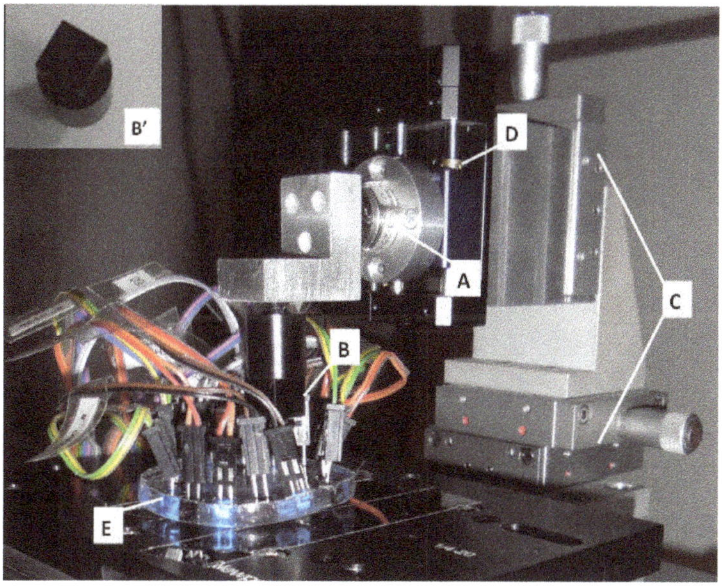

In order to address a preliminary validation of the performance of the optical sensor on curved surfaces (as it will be explained in Section 2.5.2 (5)) a half cylinder tube was positioned on an electronic scale and loaded manually.

2.5.2. Experimental Protocols

Different tests were performed to assess the overall performance of the sensor, according to the following protocols.

(1) Preliminary Experimental Analysis on the Sensor Working Principle

Acquisitions were performed without any load on the *sensing area*, with the aim to retrieve the shape of the analog signal that each PD produces when incoming electromagnetic signals, from each LED, reach them. The study of this analog signal is important because the core of the reconstruction algorithm has been developed from the analysis of variations of its shape. In particular, each column of matrix A (as defined in Section 2.4) contains a digitalization of this signal for a PD.

The output of one arbitrarily chosen PD was acquired directly with an oscilloscope during one driving cycle in which all LEDs were activated one at a time. No load was applied in this case and thereafter the measured signal could be considered as the offset state for the tested PD. The refresh rate of the sensor response was 8 Hz and thus each LED was turned on for 125 ms at a time.

(2) Indentation Tests

Loading tests were performed to assess the variation of the outputs of the PDs in function of externally applied loads and contact positions. In particular, forty different loads were applied, in the range from 0 kPa to 160 kPa. The probe chosen had a square shape of 10×10 mm^2. The two contact positions tested and the two PDs considered are shown in Figure 9 and Figure 13. For each loading condition 10 outputs for each PD were acquired by the oscilloscope and the average of the waveform values was calculated with Matlab.

(3) Pressure Map Reconstruction

The artificial skin functionality was validated by means of indentation tests with a Delrin probe having a square head (area 10×10 mm^2) as indenter, and by applying different values of normal load from 0 N to 80 KPa. The experimental set-up used is described in Section 2.5.1 and depicted in Figures 5 and 6. Four subsequent contacts, in four different positions of the *sensing area* (see Figure 16(a)), were executed for each acquisition sequence.

(4) Preliminary Multi-Pressure Tests

In order to give clues on the capability of the artificial skin to detect multiple contacts at the same time, the system was placed on an electronic scale, and four contacts were achieved on its *sensing area* by means of hand-held probes with a square section. Each contact was added sequentially, until four of them were on the surface, with the additional value measured by the scale after each new contact converted into a nominal pressure by dividing it by the known section of the probes. A multi-probe indentation setup is required to obtain a thorough characterization, but in a first validation phase a preliminary experiment was performed.

(5) Preliminary Pressure Map Reconstruction on Curved Surfaces

The response of the artificial skin positioned on a curved surface of 5 cm bending radius was tested. In this test a half cylinder structure was positioned on an electronic scale and the load was applied with a Delrin square shape probe (area 1 cm^2). A loading pressure of about 142 kPa was applied by making four subsequent contacts, each time releasing the load and waiting for 5 seconds between contacts. As in the previous case (*i.e.*, (4)) this represents a preliminary validation.

3. Results

The results of the experiments which protocols are described in the previous section are reported in the following in separate sub-sections.

3.1. Preliminary Experimental Analysis on the Sensor's Working Principle

The photovoltage output of the detectors was measured when neither physical contact nor load was applied on the *sensing area* of the artificial skin system. Therefore, Figure 7 shows the voltage acquired by the oscilloscope across the marked PD, *i.e.*, PD 7. The recorded photovoltage spans from 0.6 V to 3 V. This curve is an example of the calibration voltage (or offset) used in the process of reconstruction of contacts on the *sensing area*. Moreover, this kind of output represented the reference voltage, V_0, used to calculate the voltage variation when a load was applied on the *sensing area* surface with an indenter.

Figure 7. Photovoltage measured at PD 7 when no load is applied. The numbers in the plot on the right indicate the LED responsible for the corresponding part of detected signal.

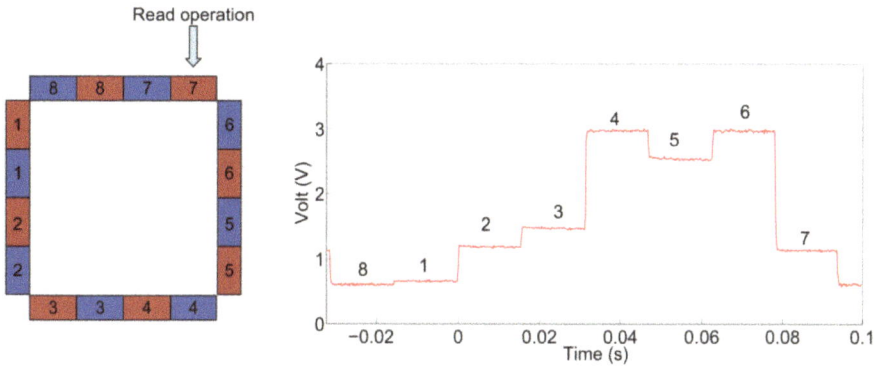

Also, Figure 8 shows the photovoltage across PD 5 in absence of any contact or load applied. Owing to the different position of this PD with respect to PD 7, this photovoltage waveform is different from the one in Figure 7 and the voltage spans between 0.4 V and 2.25 V.

Figure 8. Photovoltage measured at PD 5 when no load is applied. The numbers in the plot on the right indicate the LED responsible for the corresponding part of detected signal.

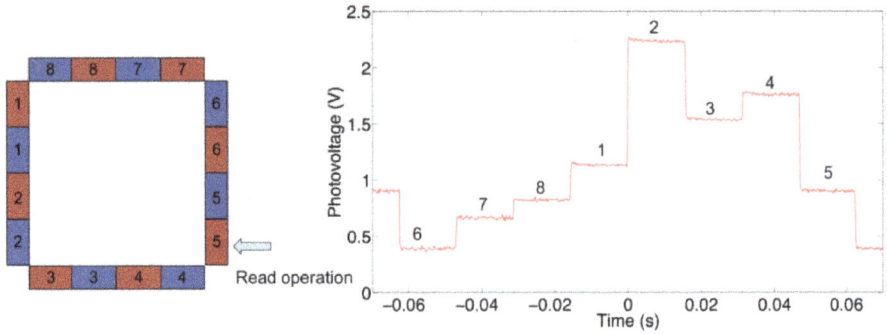

3.2. Indentation Tests

Here we show the results of the indentation experiments by considering the waveforms of the photovoltage across the same detectors of the measurements described in the previous section, *i.e.*, PD 5 and PD 7. In particular, we focus on the results obtained when the signal from a specific active emitter showed the largest variations at the detector. Moreover, the emitter-detector pair for which the results are shown in the following, highlight two typical cases of contact occurring on the *sensing area*: in one case, contact occurs in the trapezius that corresponds to the portion of waveguide directly linking the emitter and detector; in a second case, contact occurs in a region only partially in the trapezius connecting the two components.

Figure 9 shows some of the typical waveforms acquired across PD 7. Waveforms were acquired for 40 values of pressure, nevertheless in the graph four of them are reported for the sake of readability. The inset in Figure 9 contains a schematic representation of the conditions of the experiment, highlighting the relative position of indenter and detector.

Figure 9. Waveforms of photovoltage measured at PD 7 when different loads are applied. The contact area is 100 mm^2. Inset: schematic of the experiment.

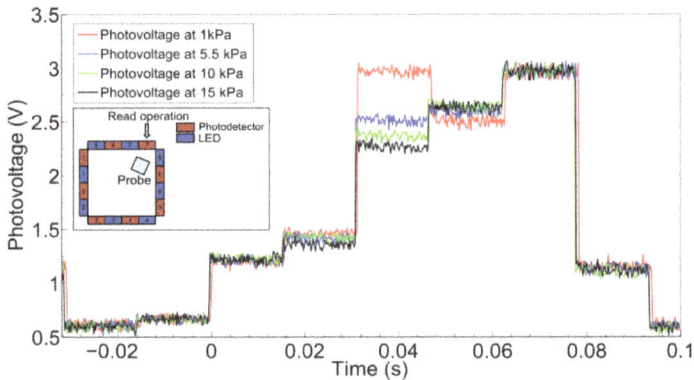

The largest voltage variation at PD 7 was observed when the active LED was number 4. Therefore this is the case in which the contact occurs on the portion of the *sensing area* directly linking the emitter (LED 4) and detector (PD 7). The absolute voltage variation ($V_{meas} - V_0$) versus the applied load for the pair defined by LED 4 and PD 7 is shown in Figure 10. As previously explained (see Section 3.1), V_0 is the voltage measured at the output of PDs when no contact or external load is applied.

Figure 11 shows a magnification of the same curve in Figure 10, focusing on the 0–40 kPa pressure range. It can be noticed that at 40 kPa the absolute photovoltage variation is 0.8 V. We can define the sensitivity as the slope of the photovoltage relative variation, $\Delta V/V_0$, versus the the applied pressure, P, as follows:

$$S = \frac{1}{\frac{\Delta V}{V_0}} \frac{d\left(\frac{\Delta V}{V_0}\right)}{dP} \tag{3}$$

Figure 10. Absolute variation of photovoltage, ($V_{meas} - V_0$), at PD 7 receiving the signal from LED 4, with error bar.

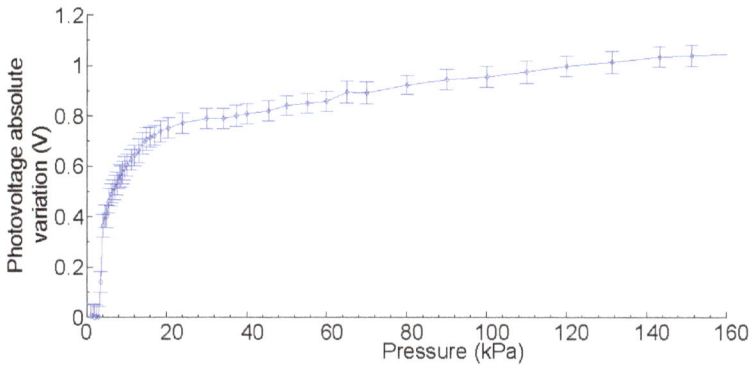

Figure 11. Absolute variation of photovoltage, ($V_{meas} - V_0$), at PD 7 receiving a signal from LED 4.

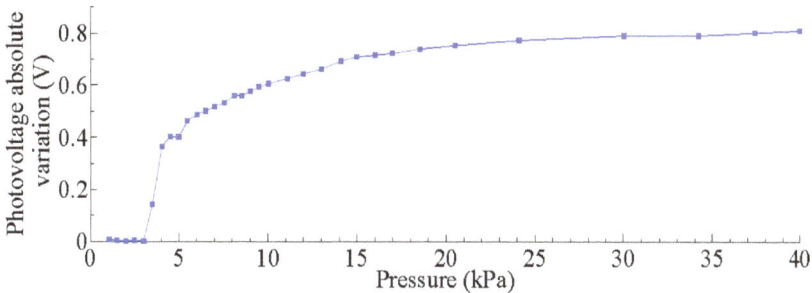

Accordingly, Figure 12 shows the sensitivity as well as the relative photovoltage variation for the pair represented by LED 4 and PD 7.

16

Figure 12. Relative variation of detected photovoltage, $(V_{meas} - V_0)/V_0$, and sensitivity for the pressure optical sensor using PDMS as bulk material. The read-out operation considers PD 7 receiving light from LED 4.

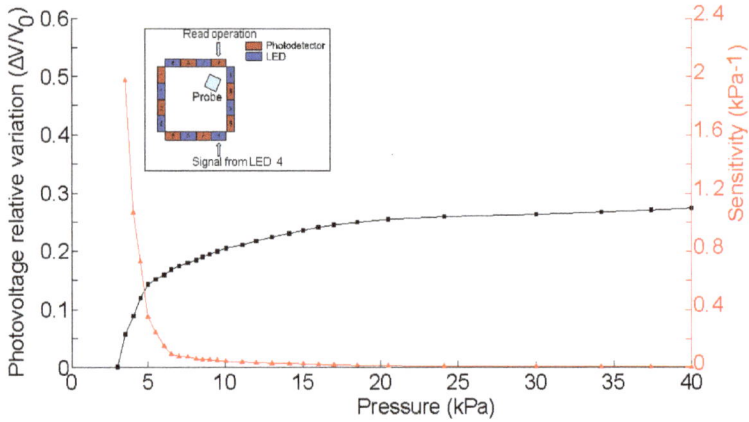

In considering the results obtained for PD 5, Figure 13 shows some of the waveforms acquired across such detector. As in the previous case, while 40 values of pressure were used, the output signals resulting from four indentations are reported in the graph for the sake of readability. The inset in Figure 13 contains a schematic representation of the conditions of the experiment, with the approximate relative position of indenter and the contact area with respect to PD 5.

Figure 13. Photovoltage measured at PD 5 when different loads are applied. The contact area is 100 mm². Inset: schematic of the experiment.

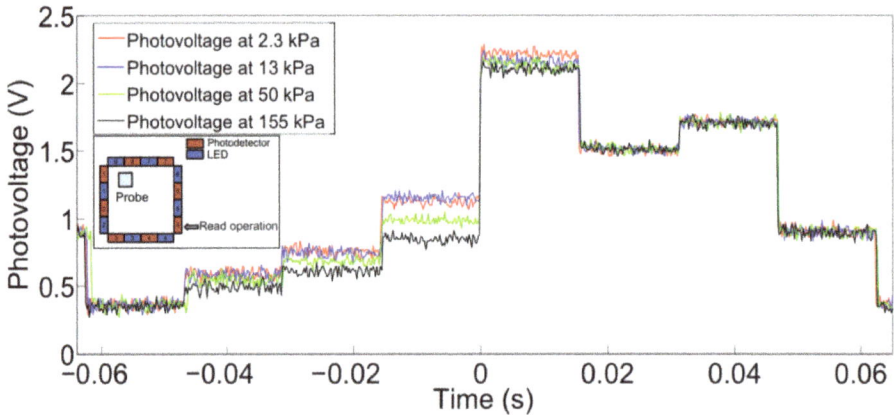

The largest photovoltage variation was observed when the active LED was number 8. Therefore, in this case contact occurs in a region only partially in the trapezius connecting emitter and detector. The absolute voltage variation $(V_{meas} - V_0)$ versus the applied load for the pair defined by LED 8 and PD 5 is shown in Figure 14.

Figure 14. Absolute variation of photovoltage, ($V_{meas} - V_0$), at PD 5 receiving signal from LED 8, with error bar.

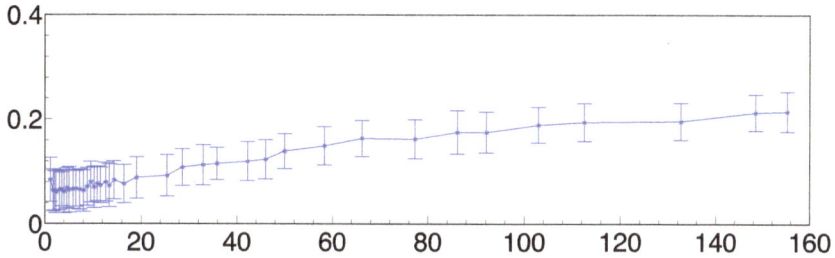

Figure 15 shows the sensitivity, defined as the slope of the photovoltage versus the applied load,in Equation (3) as well as the relative photovoltage variation, for the pair represented by LED 8 and PD 5.

Figure 15. Relative variation of detected photovoltage, ($V_{meas} - V_0$)/V_0, and sensitivity for the pressure optical sensor using PDMS as bulk material. The read-out operation considers PD 5 receiving light from LED 8.

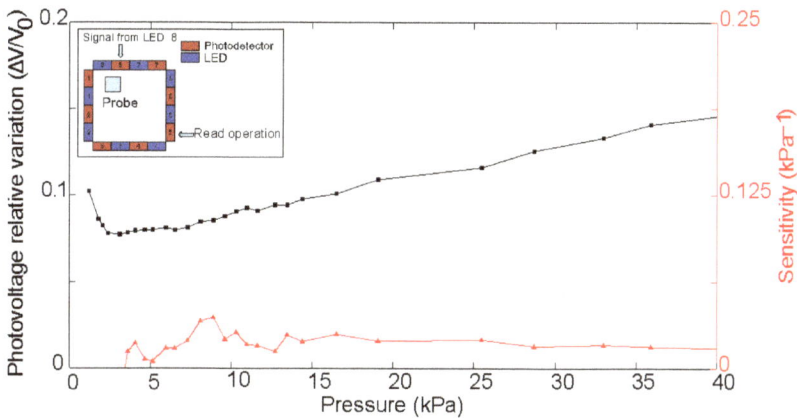

3.3. Pressure Map Reconstruction

This experiment focused on the performance of the electronic skin in mapping pressure applied to its *sensing area*. Figure 16 shows selected frames obtained through the reconstruction algorithm when the applied pressure was 70 kPa. The white stripes and black spots in the frames shown in Figure 16(b–e) are artifacts introduced by the reconstruction algorithm because of the LED-PD geometrical correlation approach chosen. To overcome these issues a coupling scheme more sophisticated than the trapezoidal approach should be used, allowing interpolation of the information on the sensing area. Alternatively, these artifacts could be reduced by using more components, possibly smaller than those used in this work, and positioned closer to each other. It is worth pointing out that no filtering of the signals was performed during or after the reconstruction.

18

Thus, the image shows the raw data obtained through simple reconstruction, without elaborating the signals to improve the quality of the results.

Figure 16. (**a**) Design layout of the sensor with the approximate position of four subsequent contacts depicted; (**b–e**) the corresponding two dimensional intensity profiles obtained from reconstruction, in which the square shape corresponding to the applied pressure profiles is highlighted, and the probe contact positions can be noticed.

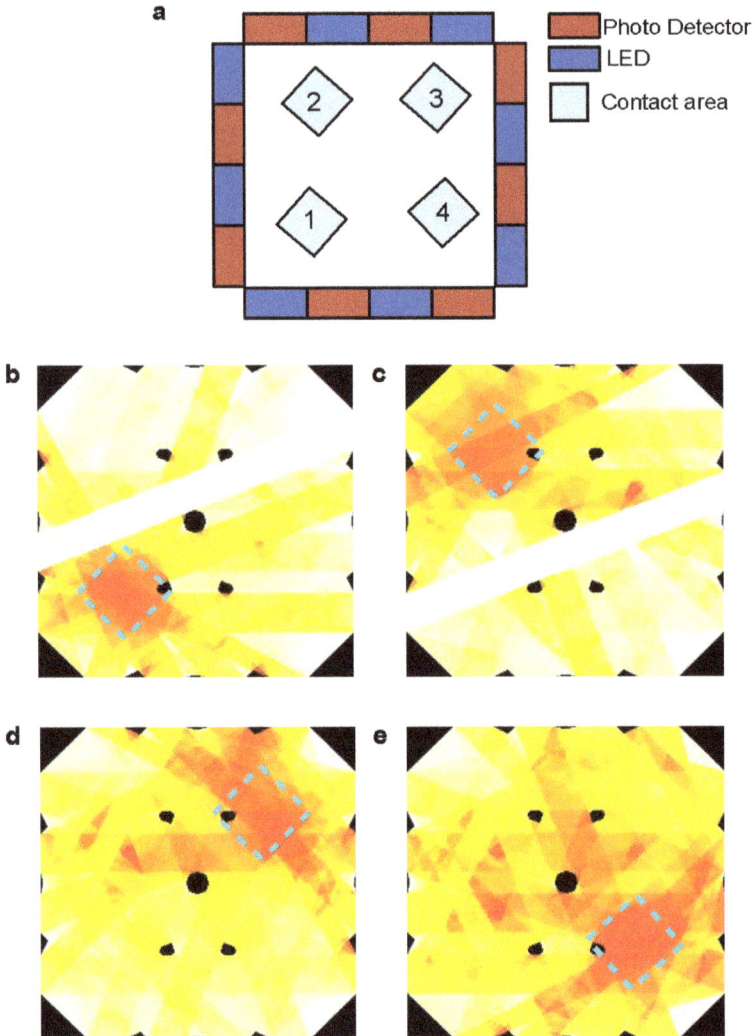

3.4. Preliminary Multi-Pressure Tests

This kind of experiment aimed at giving preliminary results for the multi-pressure contact detection of the electronic skin. Figure 17 shows selected frames obtained through the reconstruction algorithm. As explained in Section 3.3, the white stripes and black spots in the frames shown are artifacts introduced by the reconstruction algorithm, and here as well no data filtering was performed during or after the reconstruction.

Figure 17. (**a**) Design layout of the sensor with the position of 4 subsequent and multiple contacts depicted. At bottom and clockwise: (**b–e**) show the corresponding two dimensional intensity profile obtained through the reconstruction by mapping the pixels signals, in which the multiple contacts square shape, corresponding to the applied pressure profile, can be reconstructed by the e-skin, as well their different positions; (**f–i**) show the subsequent release of the square shape contacts as detected by the e-skin.

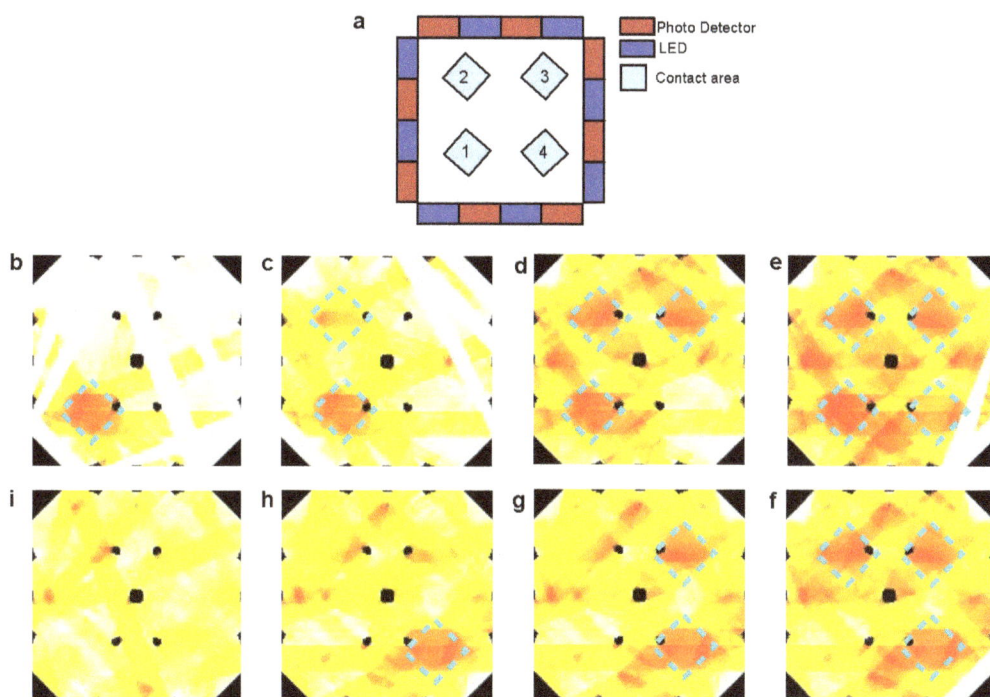

3.5. Preliminary Pressure Map Reconstruction on Curved Surfaces

This experiment was performed with the goal of showing that the electronic skin can operate even when bent on curved surfaces. Figure 18 shows selected frames obtained through the reconstruction algorithm. The same artifacts described in Section 3.3 are evident. Also in this case, no filtering was performed during or after the reconstruction.

Figure 18. (**a**) Design layout of the sensor with the position of four subsequent contacts depicted; (**b–e**) The corresponding two dimensional intensity profile obtained through the algorithm by mapping the pixels signals, in which the square shape corresponding to the applied pressure profile can be reconstructed by the e-skin positioned on a curved surface, as well its different position.

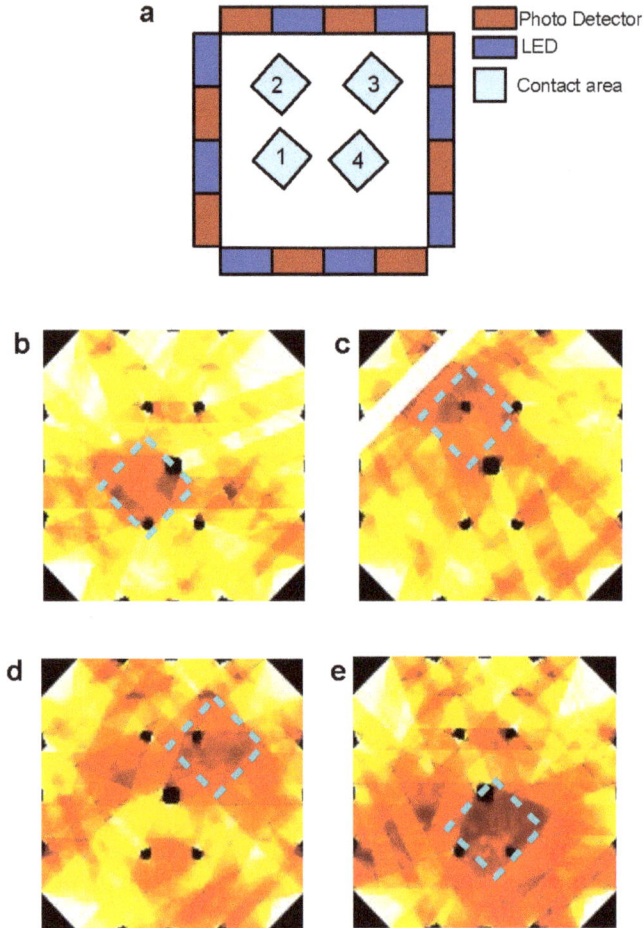

4. Discussion

The design of the system investigated in this work aimed at an electronic skin of simple fabrication overcoming some limitations of current planar pressure sensor technologies. This was achieved by the adoption of a mechano-optical principle of detection, and the use of a distributed peripheral architecture for the active components. The fabrication process is simple and repeatable as it uses PDMS as bulk material for the waveguide and through-hole components as active elements. Differently by other optical sensors approaches [17] this sensor is based only on a PDMS film functioning both as waveguide and substrate. Two concurrent phenomena are responsible for the working principle of the sensor: (1) the deformation of the flexible waveguide upon application of a pressure on the sensing area; and (2) the out-coupling of wave-guided signals at the position of the deformation. To our knowledge, only in Ramuz *et al.* [18] we can find an approach exploiting the same physical principles, hence that work will often be referred to in the present discussion. It is worth pointing out that the authors of that work considered a single emitter and a single detector, and as a consequence its results are only relevant to the first part of the present work, and not where we used the whole distributed system to reconstruct maps of the sensing area.

An immediate consequence of the used transduction method and related electronic skin architecture is the fact that no active component is in or under the sensing area. Furthermore, the latter is a planar waveguide transmitting the signals from the emitters to the detectors, and so there is no need for connections or wiring to cross it. Both these design choices allow high transparency and high flexibility of the sensing area, and are at odds with what is generally seen in planar pressure and touch sensor technologies, in which active components are sensing "pixels" in the sensing area, and thus need connections to cross it. Moreover, an important aspect related to the architecture of the tactile system is that the number of active components increases linearly with the length of the sensing area periphery, rather than with its area. This can prove to be an advantage in terms of both costs and power consumption. On the other hand, this architecture places a larger burden on the system in terms of computations required to correctly evaluate contacts on the sensing area, because this information needs to be determined indirectly.

It must be noted that in the design of the system, the thickness of the electronic skin plays a key role, since it needs to strike a balance between flexibility, sensitivity and robustness. In the present work, such parameter was mostly affected by the choice of simplifying the optical coupling between components and the waveguide by entirely embedding the former in the latter. Thus the PDMS waveguide was 5 mm thick.

Concerning the specific results that were obtained from the indentation experiments, in the following we discuss the two typical cases considered in Section 3.2. The largest variations of photovoltage with respect to increasing load were observed for emitter-detector pairs, respectively 4–7 and 8–5, for which the deformation affected the most the path of transmission of the IR signal in the planar waveguide. Conversely, we can see almost no variation in the signal from LED 6 to PD 7 when the probe is in the position specified in Figure 9, as would also be expected tracing a ray connecting the two components and observing that it does not intersect the region affected by the indentation. In the data on the LED 4 - PD 7 pair (see Figures 9–12), we see a lack of response for the initial points for which the load applied was low. This behavior probably originates from the fabrication of the electronic skin, and the chosen position of the indentation probe. Indeed, the thickness of the PDMS waveguide (*i.e.*, 5 mm) was chosen in order to completely embed the "heads" of the electronic components in the material. However, the sensitive portion of the phototransistors is actually a little smaller than the whole immersed portion, leading to the fact that a portion of the signal reaching the component, rather than being detected, is lost where it "hits" the packaging of the component. Bearing that in mind, when the indentation experiment is performed by positioning the probe just in front of the component (as in the case we are now referring to) increasing the load does indeed reduce the optical transmission channel, but it does not affect the signal portion that reaches the actual sensitive portion of the component. After the first few points in which the lack of response is apparent (see Figures 9–12), further increases in load cause a reduction in the photovoltage out of the phototransistor, PD 7, as was expected. The just described behavior is not the same when the indenting probe is positioned far from the detector, as is for example the case with the other LED-PD pair reported in Figures 12–14.

Besides the above considerations, the trend observed in Figures 9–12 for the relative photovoltage variation for the LED 4–PD 7 pair is similar to the one reported in Reference [18]. An important difference is in the amount of variation of the signal for a specific pressure range: while Ramuz *et al.* report a variation of almost 90% around 35 kPa, in our experiments we observed a 25% variation for a similar load. The most probable reason for this behavior is to be found in the thickness difference between the two systems: while the present electronic skin is 5 mm thick, in Reference [18] a waveguide of 600 μm thickness is employed. As a result, the electronic skin presented in this work suited also for higher pressure regimes: indeed, photovoltage variations could be appreciated up to the application of 160 kPa. At a first rough approximation this can be explained by invoking the fact that a thicker waveguide can support a larger number of guided optical modes. However, in increasing the pressure range a lower Signal-to-Noise Ratio (SNR) is obtained that translates in a lower sensitivity (as from Equation (3)): we see a peak of 1.93 kPa^{-1}, while in the reference they showed sensitivity up to 0.2 kPa^{-1}. These sensitivities are calculated using the pressure, and are thus dependent on the surface in contact with the probe. A possibly better parameter for comparison is the weight sensitivity, independent from the contacting surface, expressed in gr^{-1}. Our measured peak sensitivity of 1.93 kPa^{-1} corresponds then to a weight sensitivity of 0.193 gr^{-1} while in Reference [18] a higher weight sensitivity of 30 mg^{-1}was reported. This makes the electronic skin suitable to detect a measured weight in the gram range.

Furthermore, concerning the relation obtained in the photovoltage-pressure graph of Figure 12, it can be noted that it is typical of an IR light beam (bundle) crossing the contact area and consists of a power law. Such relation derives mainly from geometrical considerations of the deformation of the waveguide surface responsible of the total internal reflection phenomenon, an issue currently under investigation. Additional effects, like the proximity of the contact area to the photodetector, can affect the relation especially in the lower pressure regimes. When the light beam bundle partially crosses the contact area the sensitivity is proportionally reduced due to a reduced number of light beams sensing the deformation. The reduction of sensitivity comes with the scaling of the photovoltage-pressure graph shown in Figure 15: basically, we can argue that the curve in Figure 15 could be the first portion of a curve exhibiting the same power law shown in Figure 12, but scaled with respect to pressure. However, more experiments are required to be sure on this point.

Still regarding the LED 8-PD 5 pair, in Figure 15 we see that the sensitivity, albeit lower as was just discussed, is more or less constant on the whole pressure range. This is symptomatic of an almost linear response, most likely related to the closeness of the emitter to the contact position. There is a correlation between the spatial position of the pressure location and the sensitivity. This happens because the PDMS works as a planar waveguide and light travels in all directions. When the pressure location is far away the light source, the percentage of light intercepted is lower and this means to have a lower dampening percentage. If the pressure location is near the light source, more light can be dampened by the applied pressure and there will be a higher dampening percentage.

Concerning the experimental analysis performed to assess the capability of the system to provide a pressure map, we refer in the following to the results shown in Figures 16–18. Because of the low number of emitters and detectors used to build the electronic skin tested in this work, we did not expect a good spatial resolution for the contacts on the sensing area. The reconstruction we used to determine contact information is similar to backprojection reconstructions used in tomography, for which there is a large volume of theory showing that effective resolution, represented by number of significant reconstructed pixels, is limited to the square of the number of emitters (or detectors) [23]. While it would not be correct to directly transfer those theories on our system, because of the different geometries involved, the artifacts we find in our results are not unlike those resulting from low number of emitters (or detectors) in classical tomography. Indeed, in all the images obtained from our reconstructions (see Figures 16–18) it is easy to see stripes and black spots introduced by the particular algorithm used. A further cause of the artifacts in the reconstructed results is the particular coupling scheme chosen for emitter-detector pairs. While the use of trapezii to define the points of the sensing area affecting the signal detected for a specific pair is the simplest to implement algorithmically, its lack of any smoothing or interpolation introduces the sharp edges of the stripes in the figures. The use of a larger number of components, along with the introduction of more advanced coupling schemes for emitter-detector pairs can in the future improve on the spatial performances of the electronic skin presented in this work.

Further limits to the spatial resolution of the electronic skin are posed by the material used to fabricate the planar waveguide. The deformation of the surface cannot be expected to exactly follow the contact profile; rather, a mechanical filtering effect is introduced that tends to smear the boundaries of the detected contact. This effect can be somewhat mitigated, if the material used is well characterized, by a post-processing step in the reconstruction algorithm. However, in this work the limits to the resolution introduced by the low number of components and by the coupling scheme are such that the mechanical effect could not be appreciated.

Bearing in mind all these considerations, the results in Figures 16–18 still show quite accurately the position of the contacts applied on the sensing area. Moreover, the shape of the probes used can clearly be seen. A discussion similar to the one made for the limits of the system in this work in terms of resolution can also be made regarding the ability of the electronic skin to determine the shapes of applied contacts. In the pressure map experiments in this work we limited ourselves to square probes. Nonetheless, the obtained results are remarkable when compared to a different state-of-the-art e-skin [21] that was able to reconstruct the contact profile of a larger probe (3 cm^2) using a 19×18 sensor array: the electronic skin in this work only uses 16 components (eight emitters and eight detectors). This comparison is useful to underline the strength of the approach and architecture chosen for the electronic skin compared to planar sensor arrays [21,29–31].

Even if the results were obtained through a preliminary analysis, Figure 18 shows that the electronic skin allows the detection of multiple simultaneous contacts on its sensing area. It can be construed that using a higher number of emitters and detectors, the present electronic skin principle should ideally allow the detection of any number of contacts. Limits to that can be introduced by the materials used, because the mechanical filtering effect introduced above might prevent discrimination of contacts that are too close one to the other, by signal noise and by system constraints, like for example the maximum allowable data rate in a real time system. Moreover, a theoretical limit to performance is also present in the tomographic theories mentioned above [23], where SNR is defined proportional to $\sqrt{E}/D^{3/2}$, with E being the number of emitters and D the number of detectors.

The importance of fabricating a flexible and compliant electronic skin is strongly related to the possibility of detecting tactile cues when the device is on a curved body. This opens a plethora of applications in robotics and for consumer goods that can present this way new form factors. The preliminary experimental analysis performed so far by positioning the device on curved surfaces can thus provide an initial validation of the system, and results are shown in Figure 18. This is allowed by the fact that any condition of stress and deformation of the sensing area can be used to determine a calibration state for the system. This is simply realized by using the photovoltage values for each emitter-detector pair acquired when the sensing area is in the desired condition in the calibration matrix used by the reconstruction algorithm. As a consequence, the electronic skin does not need to be used on a flat surface.

5. Conclusions

The integrated electronic skin proposed in this work is rugged, but still has high mechanical flexibility and it is easy to fabricate. The mechano-optical sensor showed high transparency since we used a PDMS layer as both a waveguide embedding the active components and as the sensing area of the e-skin.

The artificial skin was tested by applying pressures up to 160 kPa. In such a range output voltage variations could be appreciated and this demonstrates the suitability of the e-skin for touch interfaces. In fact, normal forces in the range of 0 N to 2 N are most relevant when a fingertip makes a dynamic contact with a flat surface during a typical exploration task [32] similar to that performed when interacting with a touch screen. Therefore, if we were to consider a contact area of the fingertip of 1 cm^2, the pressure involved would range from 0 to 20 kPa, while higher pressures (e.g., up to 100 kPa) would be used in case the contact area is smaller (e.g., down to 0.2 cm^2).

We found a peak sensitivity of 1.93 kPa^{-1} which corresponds to a weight sensitivity of 0.193 gr^{-1} for an emitter- detector pair, which makes the electronic skin able to discriminate a weight as low as a few grams. Moreover, the reconstruction algorithm has been initially tested to reconstruct four simultaneous pressure contacts and to prove the correct operation of the electronic skin when bent.

We attempted to go further the state–of-the-art of optical sensors using waveguide by adopting an information processing principle inspired by tomography, that coupled with the mechanical properties of the material chosen results in an unique e-skin, able to reconstruct the intensity profile of multiple contacts on its sensing area. We expect future implementations of this system to not be limited in the number of contacts. Moreover we show preliminary results of reconstruction of the shape of contact.

The active components are positioned around the periphery of the *sensing area*, and since they are not bearing the load, the skin is rugged and doesn't have drift of electrical characteristic with time. The power consumption is low because it scales with the number of components N and not with N^2 as most artificial skins. This means that power consumption scales linearly instead of quadratically with the number of components, which is related to the desired spatial resolution.

Unlike previous approaches, that we define "point to point", since an array of taxels is used to retrieve pressure information on a sensing area, our electronic skin can "sense" in every point of its surface as it adopts a distributed reconstruction method. This way the spatial resolution dramatically increases and it can be expected it to achieve values below 1 mm; the latter representing a good reference value for a human finger [33], the most sensitive touch sensor known.

All these characteristics open a plethora of specific applications of the electronic skin in robotic devices in general, including soft robots in which the gathering of information from a soft tactile skin is fundamental for their control. In parallel, the artificial skin can find ready application as a touch interface technology for consumer electronic devices, adding flexibility and true pressure detection to the functionalities available today.

Future works will include the development of a larger electronic skin, with a higher density of active components. Moreover, smaller components (SMD), will allow integration of a thinner elastomeric waveguide. As explained in the discussions section, such a device is expected to reach higher performances, especially in terms of spatial resolution and sensitivity. Further study will be dedicated to the theoretical analysis of the system, in order to account for different non-idealities in the real systems: for example, the mechanical deformations of the layer introduce non-trivial scattering effects, which need to be modeled in order to introduce opportune filters in the information processing phase, to further improve performances. Another issue that will be addressed is the spatial filtering of the elastic materials: knowing the modeling of this effect a reverse filter in the algorithm might be implemented, allowing resolving at the maximum spatial resolution the pressure contacts.

Conflicts of Interest

The authors declare no conflict of interest.

References

1. Nishii, M.; Sakurai, R.; Sugie, K.; Masuda, Y.; Hattori, R. The use of transparent conductive polymer for electrode materials in flexible electronic paper. *SID Symp. Dig. Tech. Pap.* **2009**, *40*, 768–771.
2. Kim, H.-K.; Lee, S.; Yun, K.-S. Capacitive tactile sensor array for touch screen application. *Sens. Actuators A Phys.* **2011**, *165*, 2–7.
3. Weiss, K.; Worn, H. The Working Principle of Resistive Tactile Sensor Cells. In Proceedings of 2005 IEEE International Conference on Mechatronics and Automation, Niagara Falls, Canada, 29 July–1 August 2005; Volume 1, pp. 471–476.
4. Del Prete, Z.; Monteleone, L.; Steindler, R. A novel pressure array sensor based on contact resistance variation: Metrological properties. *Rev. Sci. Instrum.* **2001**, *72*, 1548–1553.
5. Seiichi, T.; Tomoyuki, T.; Masato, M.; Eiji, I.; Kiyoshi, M.; Isao, S. Transparent conductive-polymer strain sensors for touch input sheets of flexible displays. *J. Micromech. Microeng.* **2010**, *20*, doi:10.1088/0960-1317/20/7/075017.

6. Brewster, S.A.; Hughes, M. Pressure-Based Text Entry for Mobile Devices. In Proceedings of the 11th International Conference on Human-Computer Interaction with Mobile Devices and Services, Bonn, Germany, 15–18 September 2009.

7. Beccai, L.; Roccella, S.; Ascari, L.; Valdastri, P.; Sieber, A.; Carrozza, M.C.; Dario, P. Development and experimental analysis of a soft compliant tactile microsensor for anthropomorphic artificial hand. *IEEE/ASME Trans.Mechatron.* **2008**, *13*, 158–168.

8. Ventrelli, L.; Beccai, L.; Mattoli, V.; Menciassi, A.; Dario, P. Development of A Stretchable Skin-Like Tactile Sensor Based on Polymeric Composites. In Proceedings of 2009 IEEE International Conference on Robotics and Biomimetics (ROBIO), Guilin, China, 19–23 December 2009; pp. 123–128.

9. Laschi, C.; Cianchetti, M.; Mazzolai, B.; Margheri, L.; Follador, M.; Dario, P. Soft robot arm inspired by the octopus. *Adv. Robot.* **2012**, *26*, 709–727.

10. Shepherd, R.F.; Ilievski, F.; Choi, W.; Morin, S.A.; Stokes, A.A.; Mazzeo, A.D.; Chen, X.; Wang, M.; Whitesides, G.M. Multigait soft robot. *Proc. Nat. Acad. Sci. USA* **2011**, *108*, 20400–20403.

11. Han, J.Y. Low-Cost Multi-Touch Sensing Through Frustrated Total Internal Reflection. In Proceedings of the 18th Annual ACM Symposium on User Interface Software and Technology, Seattle, WA, USA, 23–26 October 2005.

12. Dai, X.; Mihailov, S.J.; Blanchetière, C. Optical evanescent field waveguide Bragg grating pressure sensor. *Opt. Eng.* **2010**, *49*, doi:10.1117/1.3319819.

13. Rothmaier, M.; Luong, M.; Clemens, F. Textile pressure sensor made of flexible plastic optical fibers. *Sensors* **2008**, *8*, 4318–4329.

14. Heo, J.S.; Kim, J.Y.; Lee, J.J. Tactile Sensors Using the Distributed Optical Fiber Sensors. In Proceedings of 3rd International Conference on Sensing Technology (ICST 2008), Tainan, Taiwan, 30 November–3 December 2008; pp. 486–490.

15. Missinne, J.; van Steenberge, G.; van Hoe, B.; Bosman, E.; Debaes, C.; van Erps, J.; Yan, C.; Ferraris, E.; van Daele, P.; Vanfleteren, J.; *et al.* High density optical pressure sensor foil based on arrays of crossing flexible waveguides. *Proc. SPIE Micro. Opt.* **2010**, *7716*, doi:10.1117/12.854578.

16. Kulkarni, A.; Kim, H.; Choi, J.; Kim, T. A novel approach to use of elastomer for monitoring of pressure using plastic optical fiber. *Rev. Sci. Instrum.* **2010**, *81*, doi:10.1063/1.3386588.

17. Massaro, A.; Spano, F.; Lay-Ekuakille, A.; Cazzato, P.; Cingolani, R.; Athanassiou, A. Design and Characterization of a Nanocomposite Pressure Sensor Implemented in a Tactile Robotic System. *IEEE Trans. Instrum. Meas.* **2011**, *60*, 2967–2975.

18. Ramuz, M.; Tee, B.C.K.; Tok, J.B.H.; Bao, Z. Transparent, optical, pressure-sensitive artificial skin for large-area stretchable electronics. *Adv. Mater.* **2012**, *24*, 3223–3227.

19. Schneider, F.; Draheim, J.; Kamberger, R.; Wallrabe, U. Process and material properties of polydimethylsiloxane (PDMS) for optical MEMS. *Sens. Actuators A Phys.* **2009**, *151*, 95–99.

20. Chang-Yen, D.A.; Eich, R.K.; Gale, B.K. A monolithic PDMS waveguide system fabricated using soft-lithography techniques. *J. Lightw. Technol.* **2007**, *23*, 2088–2093.

21. Takei, K.; Takahashi, T.; Ho, J.C.; Ko, H.; Gillies, A.G.; Leu, P.W.; Fearing, R.S.; Javey, A. Nanowire active-matrix circuitry for low-voltage macroscale artificial skin. *Nat. Mater.* **2010**, *9*, 821–826.

22. Someya, T.; Sekitani, T.; Iba, S.; Kato, Y.; Kawaguchi, H.; Sakurai, T. A large-area, flexible pressure sensor matrix with organic field-effect transistors for artificial skin applications. *Proc. Natl. Acad. Sci. USA* **2004**, *101*, 9966–9970.

23. Kak, A.C.; Slaney, M. Principles of Computerized Tomographic Imaging. In *Classics in Applied Mathematics*; Society for Industrial and Applied Mathematics: New York, NY, USA, 2001; p. 335.

24. Ghatak, A.; Thyagarajan, K. *An Introduction to Fiber Optics*; Cambridge University Press: Cambridge, UK, 1998.

25. Court, I.N.; Willisen, F.K.V. Frustrated total internal reflection and application of its principle to laser cavity design. *Appl. Opt.* **1964**, *3*, 719–726.

26. Lipson, D.A.; Lipson, S.G.; Lipson, H. *Optical Physics*; University Cambridge Press: Cambridge, UK, 2011.

27. Novotny, D.L.; Hecht, B. *Principles of Nano-Optics*; Cambridge University Press: Cambridge, UK, 2006.

28. Kaufman, K.S.; Terras, R.; Mathis, R.F. Curvature loss in multimode optical fibers. *J. Opt. Soc. Am.* **1981**, *71*, 1513–1518.

29. Takahashi, T.; Takei, K.; Gillies, A.G.; Fearing, R.S.; Javey, A. Carbon nanotube active-matrix backplanes for conformal electronics and sensors. *Nano Lett.* **2011**, *11*, 5408–5413.

30. Mannsfeld, S.C.B.; Tee, B.C.K.; Stoltenberg, R.M.; Chen, C.V.H.H.; Barman, S.; Muir, B.V.O.; Sokolov, A.N.; Reese, C.; Bao, Z. Highly sensitive flexible pressure sensors with microstructured rubber dielectric layers. *Nat. Mater.* **2010**, *9*, 859–864.

31. Lipomi, D.J.; Vosgueritchian, M.; Tee, B.C.-K.; Hellstrom, S.L.; Lee, J.A.; Fox, C.H.; Bao, Z. Skin-like pressure and strain sensors based on transparent elastic films of carbon nanotubes. *Nat. Nano* **2011**, doi:10.1038/nnano.2011.184.

32. Pawluk, D.T.V.; Howe, R.D. Dynamic lumped element response of the human fingerpad. *J. Biomech. Eng.* **1999**, *121*, 178–183.

33. Jones, L.A.; Lederman, S.J. Tactile Sensing. In *Human Hand Function*; Oxford University Press: New York, NY, USA, 2006; pp. 44–74.

Reprinted from *Sensors*. Cite as: Kwon, Y.S.; Ko, M.O.; Jung, M.S.; Park, I.G.; Kim, N.; Han, S.-P.; Ryu, H.-C.; Park, K.H.; Jeon, M.Y. Dynamic Sensor Interrogation Using Wavelength-Swept Laser with a Polygon-Scanner-Based Wavelength Filter. *Sensors* **2013**, *13*, 9669–9678.

Article

Dynamic Sensor Interrogation Using Wavelength-Swept Laser with a Polygon-Scanner-Based Wavelength Filter

Yong Seok Kwon [1], **Myeong Ock Ko** [1], **Mi Sun Jung** [1], **Ik Gon Park** [1], **Namje Kim** [2], **Sang-Pil Han** [2], **Han-Cheol Ryu** [3], **Kyung Hyun Park** [2] and **Min Yong Jeon** [1,*]

[1] Department of Physics, Chungnam National University, Daejeon 305-764, Korea;
E-Mails: kyss4133@gmail.com (Y.S.K.); tjdwjdwnd@naver.com (M.O.K.);
misun6857@gmail.com (M.S.J.); ikgonss@naver.com (I.G.P.)

[2] THz Photonics Creative Research Center, ETRI, Daejeon 305-700, Korea;
E-Mails: namjekim@etri.re.kr (N.K.); sphan@etri.re.kr (S.-P.H.); khp@etri.re.kr (K.H.P.)

[3] Department of Car-Mechatronics, Sahmyook University, Seoul 139-742, Korea;
E-Mail: hcryu@syu.ac.kr

* Author to whom correspondence should be addressed; E-Mail: myjeon@cnu.ac.kr;
Tel.: +82-42-821-5459; Fax: +82-42-822-8011.

Received: 15 April 2013; in revised form: 30 June 2013 / Accepted: 15 July 2013 / Published: 29 July 2013

Abstract: We report a high-speed (~2 kHz) dynamic multiplexed fiber Bragg grating (FBG) sensor interrogation using a wavelength-swept laser (WSL) with a polygon-scanner-based wavelength filter. The scanning frequency of the WSL is 18 kHz, and the 10 dB scanning bandwidth is more than 90 nm around a center wavelength of 1,540 nm. The output from the WSL is coupled into the multiplexed FBG array, which consists of five FBGs. The reflected Bragg wavelengths of the FBGs are 1,532.02 nm, 1,537.84 nm, 1,543.48 nm, 1,547.98 nm, and 1,553.06 nm, respectively. A dynamic periodic strain ranging from 500 Hz to 2 kHz is applied to one of the multiplexed FBGs, which is fixed on the stage of the piezoelectric transducer stack. Good dynamic performance of the FBGs and recording of their fast Fourier transform spectra have been successfully achieved with a measuring speed of 18 kHz. The signal-to-noise ratio and the bandwidth over the whole frequency span are determined to be more than 30 dB and around 10 Hz, respectively. We successfully obtained a real-time measurement of the abrupt change of the periodic strain. The dynamic FBG sensor interrogation system can be read out with a WSL for high-speed and high-sensitivity real-time measurement.

Keywords: wavelength-swept laser; fiber Bragg grating; sensor interrogation; strain measurement; semiconductor optical amplifier

---

1. Introduction

Fiber optic sensors have been of considerable interest in many fields for structural health monitoring of civil infrastructures, buildings, aerospace, and the maritime area. Such sensors mainly use a fiber Bragg grating (FBG) for sensing physical quantities such as strain, temperature, pressure, and vibration in multipoint sensor interrogation systems [1–11]. In optical fiber sensing systems, FBGs have many advantages such as electromagnetic immunity, compactness, remote sensing capability, wavelength selectivity, and easy fabrication. FBGs have been employed as wavelength-selective components capable of selecting wavelengths on an absolute scale [12–16]. The fundamental basis for FBG sensors is interrogation of the shift in the Bragg wavelength of the FBG by using a broadband optical light source. The interrogation of the FBG sensor with a broadband optical source has been implemented with optical filtering techniques that are based on using either an interferometer or a passive optical filter [1–5]. The passive interrogation system using a broadband optical source has a low signal-to-noise ratio (SNR). It is difficult to achieve high-speed dynamic sensing with a broadband optical source. In order to obtain high-speed and high-sensitivity interrogation of a multiple FBG sensor system, the wavelength-swept laser (WSL) has been proposed as a suitable optical source [17]. The WSL has been developed as a promising optical source in optical coherence tomography, optical fiber sensors, and optical beat source generation [7–11,17–27]. The WSL approach has been demonstrated with various methods using a narrowband wavelength scanning filter inside a laser cavity, such as a rapidly rotating polygonal mirror, a diffraction grating on a galvo-scanner, and a scanning fiber Fabry-Perot tunable filter (FFP-TF) [19–24]. The main advantage of FBG sensor interrogation with a WSL is that it allows high-speed measurement in the temporal domain. When using a WSL in the FBG sensor interrogation system, there is a linear relationship between the wavelength measurement and the time measurement. The series of reflected wavelengths in the spectral domain exactly correspond to the series of pulse positions of the reflected signals in the temporal domain. Recently, dynamic strain FBG sensor interrogation using a Fourier-domain mode-locked WSL has been reported [8,10]. In these results, the measurement of the dynamic strain was limited to a few hundred Hz. Also, the WSL with a FFP-TF had a nonlinear response in the wave-number domain, since the response of the piezoelectric transducer in the FFP-TF has a nonlinear response to a sinusoidal modulation signal. Therefore, it requires a recalibration process in the wave-number domain [27–30].

In this paper, we propose a high-speed (~2 kHz) dynamic multiplexed FBG sensor interrogation using a WSL with a polygon-scanner-based wavelength filter around the 1,550-nm band. The output from the WSL is coupled into the multiplexed FBG array. The multiplexed FBG array consists of five FBGs that have different Bragg wavelengths. One of the multiplexed FBGs in the array is fixed on the stage of the piezoelectric transducer (PZT) stack to allow application of the dynamic periodic strain. The periodic reflected signals collected by the photo-detector are digitized using a data acquisition (DAQ) board. The pulse signal from each FBG is acquired using the peak search VI program that is built into LabVIEW® [31]. We successfully obtain a real-time measurement of the abrupt change of the periodic strain. A sinusoidal voltage waveform with an amplitude of 50 V and with a frequency that is varied from 500 Hz to 2 kHz is applied to the PZT stack to assess the dynamic performance. We obtain the fast Fourier transform (FFT) spectra from the sinusoidal waveforms ranging from 500 Hz to 2 kHz.

2. Experiments

Figure 1 shows the schematic diagram of the experimental setup for a high-speed dynamic sensor interrogation system using a WSL with a polygon-scanner-based wavelength filter. Basically, the WSL consisted of a semiconductor optical amplifier (SOA) as an optical gain medium, two polarization controllers, a 10% output coupler, an optical circulator (labeled as Circulator 1), and a polygon-scanner-based wavelength filter. The center wavelength of the SOA was 1540 nm with a full width at half maximum of 60 nm. The polygon-scanner-based wavelength filter was comprised of a fiber collimator, a blazed diffraction grating with 600 lines/mm at 1,500 nm, two achromatic doublet lenses, and a polygon scanner mirror with 36 facets. The blazed diffraction grating dispersed the collimated beam from the SOA and then recombined the reflected light from the polygon mirror facet [19,20,25]. The output from the WSL was coupled into the multiplexed FBG array through another optical circulator (labeled Circulator 2). The multiplexed FBG array consisted of five FBGs, which had different Bragg wavelengths. The reflected Bragg wavelengths of the multiplexed FBG array were 1,532.02 nm, 1,537.84 nm, 1,543.48 nm, 1,547.98 nm, and 1,553.06 nm. The reflected output from the FBG array was monitored with an optical spectrum analyzer (OSA) via Circulator 2 and with an oscilloscope via a photodiode.

Figure 1. Schematic diagram of the experimental setup.

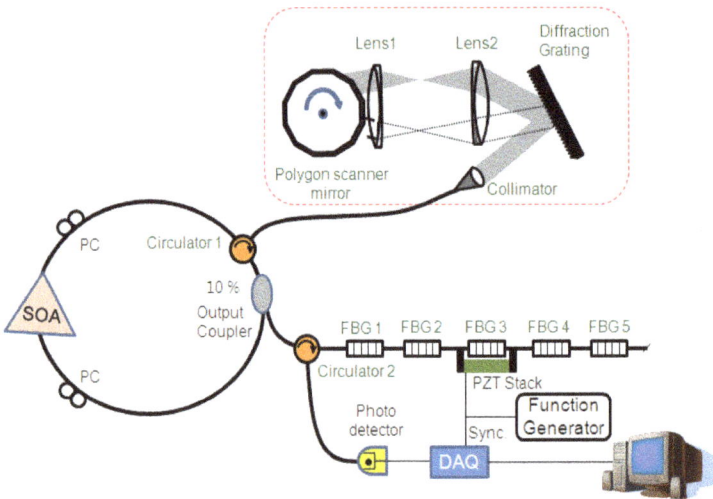

One of the FBGs was fixed on the stage of the PZT stack in order to allow dynamic strain to be applied to it. The reflected Bragg wavelength was shifted when a force is applied to the FBG, changing one of its physical parameter. The reflected signals from the FBGs were acquired via a high-speed photo-detector and a DAQ board (NI5122, National Instruments) that was operated at 100 Msample/s with 14-bit resolution. The trigger signal from a function generator was used to synchronize the DAQ board. The reflected outputs from the five FBGs were simultaneously detected as a series of reflected wavelengths in the spectral domain and as a series of pulses in the temporal domain by scanning over the spectral range. Since there is a correspondence between the time intervals and spectral intervals between the reflected signals from the multiplexed FBGs, the variation of the wavelength of each FBG can be easily converted to account for the sweeping speed of the WSL [7–11,17]. In order to find the period of the variation for the peak points of the reflected signals, the DAQ assistance tool of the LabVIEW® program was used. The pulse signal of each FBG was acquired using the peak search VI of the LabVIEW® program. For tracking of multiple pulse peaks simultaneously, the boundary of the wavelength region should be defined based on the optical bandwidth of the multiplexed FBG array.

Figure 2a shows the typical optical spectrum of the output of the WSL. The scanning frequency of the WSL was 18 kHz, and the 10-dB scanning bandwidth was more than 90 nm from 1,475 nm to 1,565 nm at that scanning rate. This covers the full optical bandwidth of the multiplexed FBG array. The instantaneous linewidth of the WSL was about 0.15 nm. This is almost the same as the 3 dB linewidth of the FBGs used. The output power of WSL is more than 0 dBm. Figure 2b shows the optical spectrum of the reflected wavelengths from the multiplexed FBG array. The reflected center wavelengths for FBGs 1–5 were 1,532.02 nm, 1,537.84 nm, 1,543.48 nm, 1,547.98 nm, and 1,553.06 nm, respectively. All of the FBGs had a reflectivity of more than 90%, and the narrow 3-dB bandwidth was measured to be 0.15 nm with the OSA (resolution: 0.1 nm). The reflected outputs from the multiplexed FBG array were converted to a single electrical signal using a high-speed photo-detector. Figure 2c shows the time-domain signal from the reflected pulses from the array of multiplexed FBGs, consisting of five FBGs. Five peaks were observed in the photo-detector output for a single period of the sinusoidal voltage driving the WSL. The positions of the series of pulses from the reflected signals in the temporal domain as shown in Figure 2b exactly correspond to the series of reflected wavelengths in the spectral domain as shown in Figure 2c.

Figure 2. (a) Optical spectrum of the WSL; (b) optical spectrum of the reflected wavelengths from the multiplexed FBG array; and (c) signal of the pulses reflected from the array of multiplexed FBGs.

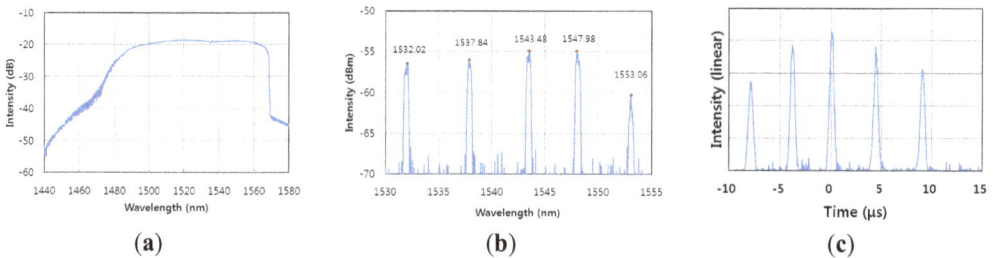

(a)   (b)   (c)

To measure the dynamic strain response, a periodic strain was applied to one of the multiplexed FBGs in the array via the PZT stack. Figure 3a shows a photograph of the oscilloscope trace for the output of the interrogation of the multiplexed FBG array without any dynamic strain. There are three pulses from the reflected signals from the FBGs on the screen of the oscilloscope. When the sinusoidal voltage is applied to the PZT stack, FBG 3 will experience a periodic strain from the PZT stack. A photograph of the oscilloscope trace for the dynamic strain of 1 kHz is shown in Figure 3b. The bandwidth of the center pulse in the photograph of Figure 3b is wider than that of Figure 3a. The pulses of FBG 2 and FBG 4 in the photograph, however, do not show any change for either of the cases of Figure 3a,b. Figure 4 shows the optical spectrum with and without the 1 kHz dynamic strain. The spectral bandwidth of the case with the dynamic strain is wider than that of the case without the dynamic strain. This is due to the modulation from the reflected wavelength of the dynamically strained FBG.

Figure 3. Photograph of the oscilloscope trace **(a)** without dynamic strain and **(b)** with dynamic strain.

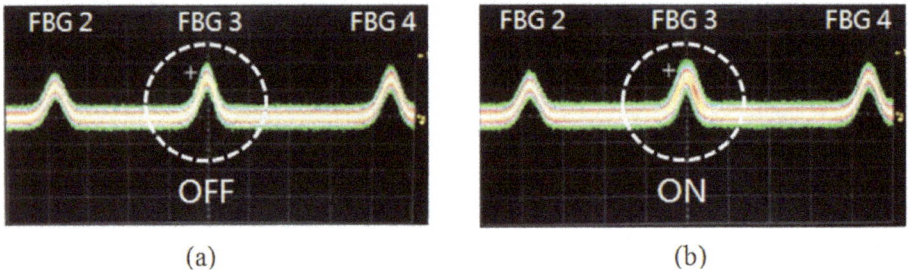

(a)   (b)

Figure 4. Optical spectrum when dynamic strain (1 kHz) is applied to FBG 3 and in the absence of dynamic strain (0 Hz).

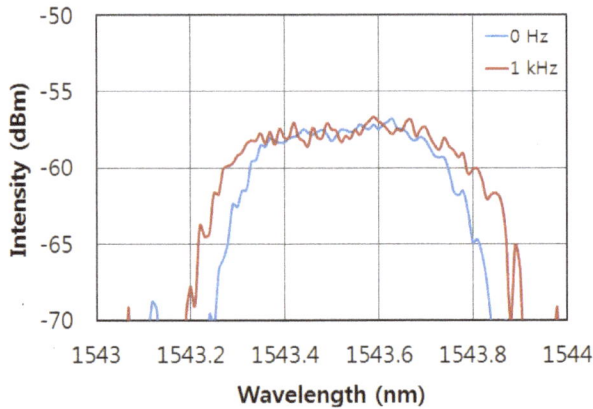

In order to confirm the possibility of real-time measurement, the frequency of the sinusoidal waveform applied to the PZT stack was changed abruptly. Figure 5a shows the results of this dynamic measurement when the frequency of the periodic strain in FBG 3 was changed. The periodic reflected signals collected by the photo-detector were digitized using a DAQ board with a high rate of 100 Msamples/s because the sampling rate of the DAQ board determines the accuracy of the acquired data points of the dynamic response. When the frequency of the sinusoidal waveform was changed abruptly from 500 Hz to 1 kHz, the abrupt variation of the periodic strain was captured from the screen of the LabVIEW program, as shown in Figure 5a. It was confirmed that this dynamic sensor system could read out the abrupt change of the periodic strain. The peak-to-peak amplitude of the dynamic applied strain was approximately 285 µε. Figure 5b shows the corresponding power spectral density of the FFT spectrum when the frequency of the sinusoidal waveform changed abruptly from 500 Hz to 1 kHz. The FFT spectrum was calculated by using Origin analysis software. The dynamic responses of the peak points at 500 Hz and 1 kHz modulation frequencies are displayed in Figure 5b. The bandwidths for both of them were determined to be around 10 Hz.

Figure 5. (a) Abrupt variation of the periodic strain from 500 Hz to 1 kHz; and **(b)** power spectral density of the FFT spectrum of (a).

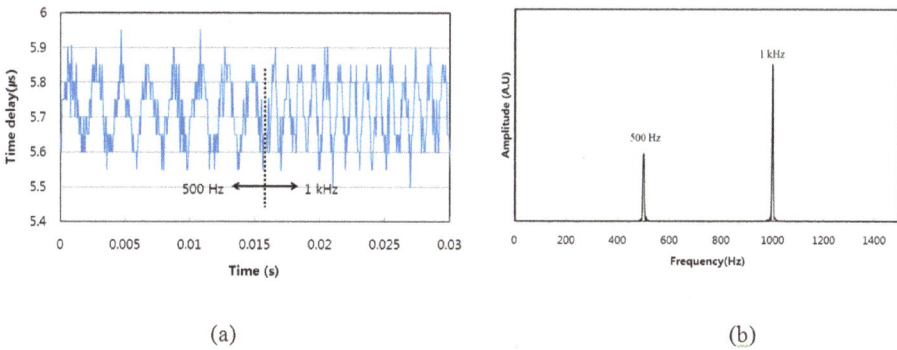

(a)   (b)

Figure 6. The effect of increasing the number of samples averaged over in measuring the periodic dynamic strain signals.

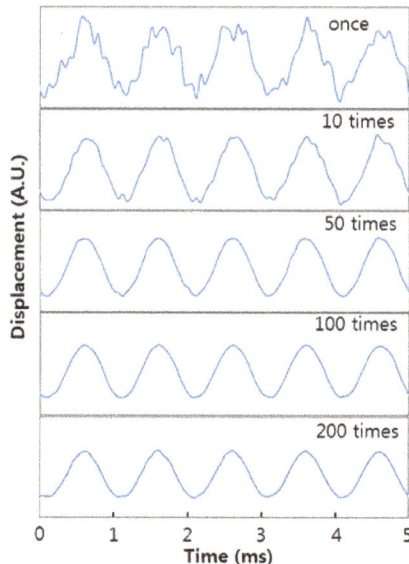

There is some spectral noise in Figure 5a. This noise is due to the slow response of the dynamic wavelength variation of the FBG. In order to remove the spectral noise, we repeatedly performed the dynamic sensing measurement and then averaged over several tens of samples. The periodic dynamic strain signals were measured for several cases by repeated sampling, as shown in Figure 6. The real-time measurement could be carried out when only one measurement was made. However, successively clearer sinusoidal signals were achieved as the number of measurement samples was increased.

As an example, a sinusoidal waveform with a frequency of 2 kHz and a voltage of 50 V was applied to the PZT stack. By repeatedly measuring the multiple peak positions, the temporal variation of the difference between multiple peaks could be obtained using a LabVIEW peak searching program. The temporal variation of the peak points was repeatedly measured over a total of 100 iterations. The periodic sensor output signal of 2 kHz from FBG 3 in the multiplexed FBG array was achieved using the WSL at a sweeping rate of 18 kHz, as shown in Figure 7a. The DAQ board with the high sampling rate of 100 Msample/s was used to record the temporal variation of the difference between multiple peaks via the photo-detector. In order to improve the resolution, a large number of data points (4,096) are collected for every single sweep of the 18 kHz period. The corresponding FFT spectrum from the periodic output of Figure 7a is shown in Figure 7b. There is a peak for the 2-kHz frequency component in the FFT spectrum. The SNR and frequency bandwidth were determined to be more than 40 dB and around 10 Hz, respectively. The RMS value of the applied strain was calculated as 70.54 $\mu\varepsilon_{rms}$ at 2 kHz. From the 40-dB SNR at the 2-kHz frequency component, the minimum detectable dynamic strain was calculated as 0.22 $\mu\varepsilon/Hz^{1/2}$.

Figure 7. (**a**) Periodic sensor output signal at 2 kHz; and (**b**) power spectral density of the FFT spectrum of (a).

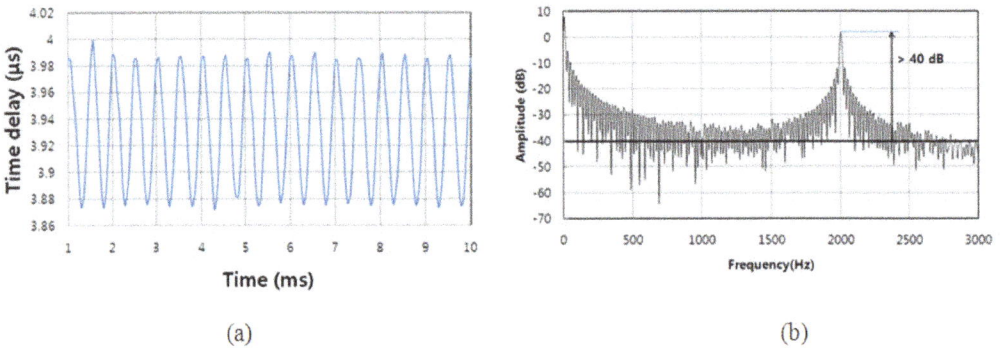

(a)   (b)

The sinusoidal waveform with an amplitude of 50 V was applied to the PZT stack with a frequency that was varied from 500 Hz to 2 kHz to assess the dynamic performance of the multiplexed FBG array. The measurement was carried out at intervals of 100 Hz from 500 Hz to 2 kHz. Figure 8 shows the FFT spectra for each of the various applied sinusoidal waveforms, with frequencies from 500 Hz to 2 kHz. The intensity variation of the FFT spectra is less than 2 dB over the whole frequency span. The SNR over the whole frequency span was determined to be more than 30 dB.

Figure 8. Power spectral density of the FFT spectrum based on varying the frequency of the applied sinusoidal waveform from 500 Hz to 2 kHz.

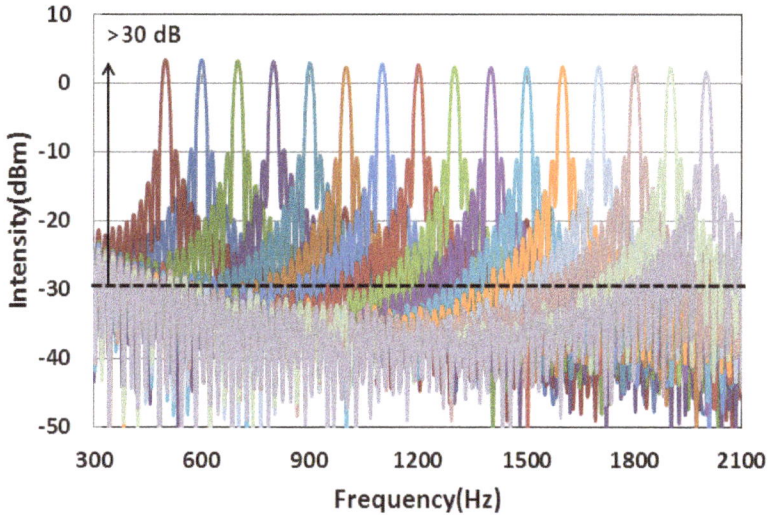

3. Conclusions

A high-speed (~2 kHz) dynamic multiplexed FBG sensor interrogation using a WSL with a polygon-scanner-based wavelength filter in the 1,550-nm band has been demonstrated. The scanning frequency of the WSL was 18 kHz, and the 10-dB scanning bandwidth was more than 90 nm at that scanning rate. The output from the WSL was coupled into the multiplexed FBG array, which consisted of five FBGs. A periodic strain was applied to one of the FBGs in the multiplexed array that was fixed on the stage of the PZT stack. A sinusoidal waveform with a frequency that was varied from 500 Hz to 2 kHz was applied to the PZT stack, and the dynamic performance was successfully characterized with a measuring speed of 18 kHz. The SNR and bandwidths over the whole frequency span were determined to be more than 30 dB and around 10 Hz, respectively. We achieved real-time measurement of the abrupt change of the periodic strain without any signal processing delay. Our results confirm that this dynamic FBG sensor interrogation system using WSL can be read out in real time.

Acknowledgments

This research was supported by the Korea Foundation for the Advancement of Science & Creativity (KOFAC) grant, by Nano-Material Technology Development Program through the NRF of Korea grant(2012M3A7B4035095), and by Basic Science Research Program through the National Research Foundation of Korea (NRF) (2010-0022645) funded by the Korean Government (MEST).

Conflicts of Interest

The authors declare no conflict of interest.

References

1. Kersey, A.D.; Berkoff, T.A.; Morey, W.W. High-resolution fiber Grating based strain sensor with interferometric wavelength-shift detection. *Electron. Lett.* **1992**, *28*, 236–238.
2. Melle, S.M.; Liu, K.; Measures, R.M. A passive wavelength demodulation system for guided-wave Bragg grating sensors. *IEEE Photon. Technol. Lett.* **1992**, *4*, 1539–1541.
3. Kersey, A.D.; Berkoff, T.A.; Morey, W.W. Multiplexed fiber Bragg grating strain-sensor system with a fiber Fabry-Perot wavelength filter. *Opt. Lett.* **1993**, *8*, 33–39.
4. Bang, H.-J.; Jun, S.-M.; Kim, C.-G. Stabilized interrogation and multiplexing techniques for fibre Bragg grating vibration sensors. *Meas. Sci. Technol.* **2005**, *16*, 813–820.
5. Kim, C.S.; Lee, T.H.; Yu, Y.S.; Han, Y.G.; Lee, S.B.; Jeong, M.Y. Multi-point interrogation of FBG sensors using cascaded flexible wavelength-division Sagnac loop filters. *Opt. Express* **2006**, *14*, 8546–8551.
6. Hongo, A.; Kojima, S.; Komatsuzaki, S. Applications of fiber Bragg grating sensors, and high-speed interrogation techniques. *Struct. Control Health Monit.* **2005**, *12*, 269–282.
7. Yun, S.H.; Richardson, D.J.; Kim, B.Y. Interrogation of fiber grating sensor arrays with a wavelength-swept fiber laser. *Opt. Lett.* **1998**, *23*, 843–845.
8. Jung, E.J.; Kim, C.-S.; Jeong, M.Y.; Kim, M.K.; Jeon, M.Y.; Jung, W.; Chen, Z. Characterization of FBG sensor interrogation based on a FDML wavelength swept laser. *Opt. Express* **2008**, *16*, 16552–16560.
9. Isago, R.; Nakamura, K. A high reading rate fiber Bragg grating sensor system using a high-speed swept light source based on fiber vibrations. *Meas. Sci. Technol.* **2009**, *20*, 034021.
10. Nakazaki, Y.; Yamashita, S. Fast and wide tuning range wavelength-swept fiber laser based on dispersion tuning and its application to dynamic FBG sensing. *Opt. Express* **2009**, *17*, 8310–8318.
11. Lee, B.C.; Jung, E.-J.; Kim, C.-S.; and Jeon, M.Y. Dynamic and static strain fiber Bragg grating sensor interrogation with a 1.3 μm Fourier domain mode-locked wavelength-swept laser. *Meas. Sci. Technol.* **2010**, *21*, 094008.
12. Ahmad, H.; Saat, N.K.; Harun, S.W. S-band erbium-doped fiber ring laser using a fiber Bragg grating. *Laser. Phys. Lett.* **2005**, *2*, 369–371.
13. Fu, H.Y.; Liu, H.L.; Dong, X.; Tam, H.Y.; Wai, P.K.A.; Lu, C. High-speed fibre Bragg grating sensor interrogation using dispersioncompensation fibre. *Electron. Lett.* **2008**, *44*, 618–619.
14. Fu, Z.H.; Wang, Y.X.; Yang, D.Z.; Shen, Y.H. Single-frequency linear cavity erbium-doped fiber laser for fiber-optic sensing applications. *Laser. Phys. Lett.* **2009**, *6*, 594–597.
15. Mohd Nasir, M.N.; Yusoff, Z.; Al-Mansoori, M.H.; Abdul Rashid, H.A.; Choudhury, P.K. Low threshold and efficient multi-wavelength Brillouinerbium fiber laser incorporating a fiber Bragg grating filter with intra-cavity pre-amplified Brillouin pump. *Laser. Phys. Lett.* **2009**, *6*, 54–58.
16. Schultz, S.; Kunzler, W.; Zhu, Z.; Wirthlin, M.; Selfridge, R.; Propst, A.; Zikry, M.; Peters, K. Full-spectrum interrogation of fiber Bragg grating sensors for dynamic measurements in composite laminates. *Smart Mater. Struct.* **2009**, *18*, 115015.

17. Yamashita, S.; Nakazaki, Y.; Konishi, R.; Kusakari, O. Wide and fast wavelength-swept fiber laser based on dispersion tuning for dynamic sensing. *J. Sens.* **2009**, *2009*, 572835.
18. Jeon, M.Y.; Kim, N.; Han, S.-P.; Ko, H.; Ryu, H.-C.; Yee, D.-S.; Park, K.H. Rapidly frequency-swept optical beat source for continuous wave terahertz generation. *Opt. Express* **2011**, *19*, 18364–18371.
19. Yun, S.H.; Boudoux, C.; Tearney, G.J.; Bouma, B.E. High-speed wavelength-swept semiconductor laser with a polygon-scanner-based wavelength swept filter. *Opt. Lett.* **2003**, *28*, 1981–1983.
20. Oh, W.Y.; Yun, S.H.; Tearney, G.J.; Bouma, B.E. 115 kHz tuning repetition rate ultrahigh-speed wavelength-swept semiconductor laser. *Opt. Lett.* **2005**, *30*, 3159–3161.
21. Huber, R.; Wojtkowski, M.; Fujimoto, J.G.; Jiang, J.Y.; Cable, A.E. Three-dimensional and C-mode OCT imaging with a compact, frequency swept laser source at 1300 nm. *Opt. Express* **2005**, *13*, 10523–10538.
22. Lee, S.-W.; Kim, C.-S.; Kim, B.-M. External-line cavity wavelength-swept source at 850 nm for optical coherence tomography. *IEEE Photon. Technol. Lett.* **2007**, *19*, 176–178.
23. Huber, R.; Wojtkowski, M.; Fujimoto, J.G. Fourier domain mode locking (FDML): A new laser operating regime and applications for optical coherence tomography. *Opt. Express* **2006**, *14*, 3225–3237.
24. Jeon, M.Y.; Zhang, J.; Wang, Q.; Chen, Z. High-speed and wide bandwidth Fourier domain mode-locked wavelength swept laser with multiple. *Opt. Express* **2008**, *16*, 2547–2554.
25. Lee, S.-W.; Song, H.-W.; Jung, M.-Y.; Kim, S.-H. Wide tuning range wavelength-swept laser with a single SOA at 1020 nm for ultrahigh resolution Fourier-domain optical coherence tomography. *Opt. Express* **2011**, *19*, 21227–21237.
26. Tsai, M.-T.; Chang, F.-Y. Visualization of hair follicles using high-speed optical coherence tomography based on a Fourier domain mode locking laser. *Laser Phys.* **2012**, *22*, 791–796.
27. Lee, B.-C.; Eom, T.-J.; Jeon, M.Y. k-domain linearization using fiber Bragg grating array based on Fourier domain optical coherence tomography. *Korean J. Opt. Photon.* **2011**, *22*, 72–76.
28. Eigenwillig, C.M.; Biedermann, B.R.; Palte, G.; Huber, R. K-space linear Fourier domain mode locked laser and applications for optical coherence tomography. *Opt. Express* **2008**, *16*, 8916–8937.
29. Park, I.G.; Choi, B.K.; Kwon, Y.S.; Jeon, M.Y. Performance comparison of fiber Bragg gratings sensor interrogation using two kinds of wavelength-swept lasers. *Proc. SPIE* **2012**, *8421*, 411–414.
30. Lee, S.-W.; Song, H.-W.; Kim, B.-K.; Jung, M.-Y.; Kim, S.-H.; Cho, J.D.; Kim, C.-S. Fourier Domain optical coherence tomography for retinal imaging with 800-nm swept source: Real-time resampling in *k*-domain. *J. Opt. Soc. Korea* **2011**, *15*, 293–299.
31. National Instruments. Available online: http://www.ni.com/white-paper/3770/en (accessed on 18 August 2012).

38

Reprinted from *Sensors*. Cite as: Hlubina, P.; Martynkien, T.; Olszewski, J.; Mergo, P.; Makara, M.; Poturaj, K.; Urbańczyk, W. Spectral-Domain Measurements of Birefringence and Sensing Characteristics of a Side-Hole Microstructured Fiber. *Sensors* **2013**, *13*, 11424–11438.

Article

Spectral-Domain Measurements of Birefringence and Sensing Characteristics of a Side-Hole Microstructured Fiber

Petr Hlubina [1,*], **Tadeusz Martynkien** [2], **Jacek Olszewski** [2], **Pawel Mergo** [3], **Mariusz Makara** [3], **Krzysztof Poturaj** [3] **and Waclaw Urbańczyk** [2]

[1] Department of Physics, Technical University Ostrava, 17. listopadu 15, Ostrava-Poruba 708 33, Czech Republic
[2] Institute of Physics, Wroclaw University of Technology, Wybrzeże Wyspiańskiego 27, Wroclaw 50-370, Poland; E-Mails: tadeusz.martynkien@pwr.wroc.pl (T.M.); jacek.olszewski@pwr.wroc.pl (J.O.); waclaw.urbanczyk@pwr.wroc.pl (W.U.)
[3] Laboratory of Optical Fibre Technology, Maria Curie-Sklodowska University, Pl. M. Curie-Sklodowskiej 3, Lublin 20-031, Poland; E-Mails: pawel.mergo@poczta.umcs.lublin.pl (P.M.); marmak@hermes.umcs.lublin.pl (M.M.); potkris@hermes.umcs.lublin.pl (K.P.)

* Author to whom correspondence should be addressed; E-Mail: petr.hlubina@vsb.cz; Tel.: +420-597-323-134; Fax: +420-597-323-403.

Received: 10 July 2013; in revised form: 12 August 2013 / Accepted: 27 August 2013 / Published: 28 August 2013

Abstract: We experimentally characterized a birefringent side-hole microstructured fiber in the visible wavelength region. The spectral dependence of the group and phase modal birefringence was measured using the methods of spectral interferometry. The phase modal birefringence of the investigated fiber increases with wavelength, but its positive sign is opposite to the sign of the group modal birefringence. We also measured the sensing characteristics of the fiber using a method of tandem spectral interferometry. Spectral interferograms corresponding to different values of a physical parameter were processed to retrieve the spectral phase functions and to determine the spectral dependence of polarimetric sensitivity to strain, temperature and hydrostatic pressure. A negative sign of the polarimetric sensitivity was deduced from the simulation results utilizing the known modal birefringence dispersion of the fiber. Our experimental results show that the investigated fiber has a very high polarimetric sensitivity to hydrostatic pressure, reaching -200 rad $\times$ MPa$^{-1}\times$ m^{-1} at 750 nm.

Keywords: microstructured fibers; birefringent fibers; fiber characterization; spectral interferometry; birefringence dispersion; fiber-optic sensors; polarimetric sensitivity

---

1. Introduction

Highly birefringent (HB) fibers, such as conventional elliptical-core, bow-tie or side-hole fibers, can be used as active elements of fiber-optic sensor configurations utilizing the interference of polarization modes. A physical quantity acting on HB fiber causes a change in the difference of propagation constants of the polarization modes or, equivalently, a change in the phase modal birefringence and, consequently, a change in the phase shift between the polarization modes at the fiber output. The phase shift can be measured by interferometric or polarimetric methods that can also be used in interferometric and polarimetric sensors for measuring different physical quantities, such as temperature, hydrostatic pressure and elongation [1,2]. As an example, conventional side-hole fibers are of interest for pressure sensing [3–5], because of a very high polarimetric sensitivity, which is due to two large air holes in the cladding adjacent to the core region. The holes break the mechanical symmetry of the fiber and are responsible for a high change of the phase modal birefringence when a symmetrical load induced by hydrostatic pressure applied to the fiber cladding is transferred into nonsymmetrical stress distribution in the core region.

Conventional HB fibers exhibit temperature-sensitive birefringence, so that when they are used for sensing other parameters than temperature, such as hydrostatic pressure, the temperature cross-sensitivity affects the measurement accuracy significantly. To overcome this limitation, HB microstructured (MS) fibers with much higher flexibility in shaping the phase modal birefringence and significantly less temperature dependence than conventional HB fibers have emerged as active elements of fiber-optic sensors. The birefringence in MS fibers originates either from breaking the hexagonal symmetry of the fiber structure [6,7] or from internal stress, causing an anisotropy of the refractive index in the fiber core [8]. HB MS fibers made of glass with a uniform composition in the entire cross-section have no thermal stress induced by the difference in thermal expansion coefficients between the doped fiber core and the fiber cladding. Consequently, the polarimetric sensitivity to temperature, which strongly depends on the geometry of the HB MS fiber, can be up to three orders of magnitude lower than in conventional elliptical-core fibers [9,10].

HB MS fibers with very low temperature cross-sensitivity are suitable for hydrostatic pressure sensing [11–14]. As an example, the effect of hydrostatic pressure on the phase modal birefringence has been analyzed for a commercially available HB MS fiber with two large air holes adjacent to the fiber core [15]. It has been shown that the polarimetric sensitivity to hydrostatic pressure is mostly related to pressure-induced stress birefringence in the core region or, equivalently, to the mechanical asymmetry of the microstructured region. In a commercially available HB MS fiber with two large air holes adjacent to the core region, the polarimetric sensitivity to pressure reaches a value of -14 rad $\times$ MPa$^{-1}\times$ m^{-1} at 780 nm [16].

In this paper, we present the results of the measurement of birefringence and the sensing characteristics of a side-hole MS fiber in the visible wavelength region (500–770 nm). First, the spectral dependence of the group and phase modal birefringence is measured using methods of spectral interferometry. Second, the sensing characteristics of the fiber are measured using a method of tandem spectral interferometry. We extend the use of a method [17] based on processing the spectral interferograms, including the equalization wavelength at which spectral interference fringes have the largest period, due to the zero overall group birefringence. From the spectral interferograms, the phase functions corresponding to different values of a physical parameter are retrieved, and the spectral dependence of polarimetric sensitivity to strain, temperature and hydrostatic pressure is determined. A negative sign of the polarimetric sensitivity is deduced from the simulation results, and the polarimetric sensitivity to strain reaches a value of -16 rad $\times$ m$\epsilon^{-1}\times$ m^{-1} at 750 nm. Similarly, the polarimetric sensitivity to temperature reaches a value of -0.25 rad $\times$ K$^{-1}\times$ m^{-1} at 750 nm. Moreover, the investigated fiber has a very high polarimetric sensitivity to hydrostatic pressure, reaching -200 rad $\times$ MPa$^{-1}\times$ m^{-1} at 750 nm.

2. Experimental Methods

2.1. Measurement of the Modal Birefringence Dispersion

HB fibers are characterized by the phase and group modal birefringence. The phase modal birefringence, $B(\lambda)$, is defined as:

$$B(\lambda) = n_x(\lambda) - n_y(\lambda) \tag{1}$$

where $n_x(\lambda)$ and $n_y(\lambda)$ are the wavelength-dependent phase effective indices of x and y polarization modes supported by the HB fiber. Similarly, the group modal birefringence, $G(\lambda)$, is defined as:

$$G(\lambda) = N_x(\lambda) - N_y(\lambda) = -\lambda^2 \frac{\mathrm{d}[B(\lambda)/\lambda]}{\mathrm{d}\lambda} \tag{2}$$

where $N_x(\lambda)$ and $N_y(\lambda)$ are the group effective indices of the polarization modes.

The group modal birefringence dispersion can be measured by a method of spectral-domain tandem interferometry [18,19] utilizing the experimental setup shown in Figure 1. In this setup, the path length difference, Δ_M, adjusted in a Michelson interferometer (MI) compensates for the group optical path difference, $G(\lambda)z$, between the polarization modes in a fiber under test (FUT) of length z. The transmission azimuth of both a polarizer and an analyzer is adjusted at $45°$ with respect to the polarization axes of the FUT, so that the interference of the polarization modes is observed. If we consider $G(\lambda) < 0$ for the FUT, the spectral intensity at the output of the tandem configuration can be expressed as [18,19]:

$$I(z;\lambda) = I_0(\lambda)\left\{1 + V(z;\lambda)\cos\left\{(2\pi/\lambda)[\Delta_\mathrm{M} + B(\lambda)z]\right\}\right\} \tag{3}$$

where $I_0(\lambda)$ is the reference spectral intensity and $V(z;\lambda)$ is the visibility term given by:

$$V(z;\lambda) = \exp\{-(\pi^2/2)\left\{[\Delta_\mathrm{M} + G(\lambda)z]\Delta\lambda_\mathrm{R}/\lambda^2\}^2\right\} \tag{4}$$

It results from Equation (3) that the spectral interference fringes can be resolved, due to the highest visibility in the vicinity of the equalization wavelength, λ_0 [18], given by:

$$\Delta_M = -G(\lambda_0)z \tag{5}$$

Thus, Δ_M adjusted in the MI and measured as a function of the equalization wavelength, λ_0, gives the wavelength dependence of the group modal birefringence, $G(\lambda_0) = \Delta_M/z$, in the FUT directly. Because the group modal birefringence, $G(\lambda)$, is related to the phase modal birefringence, $B(\lambda)$, via Equation (2), we can obtain the relative wavelength dependence of the phase modal birefringence. It can be combined with the known value at one specific wavelength to obtain the wavelength dependence of the phase modal birefringence, $B(\lambda)$ [18]. The benefits of our spectral methods, compared to the standard frequency-domain interferometer [20], which measures the periodicity of the interference fringes, include no need to use a source of short broadband optical pulses.

Figure 1. Experimental setup with a Michelson interferometer (MI) for measuring the dispersion of the group modal birefringence in a fiber under test (FUT). The remaining components: a white-light source (WLS), a collimating lens (CL), mirrors 1–2 (M1–2), a beam splitter (BS), a polarizer (P), microscope objectives 1–3 (MO1–3), a delay line (DL) and an analyzer (A).

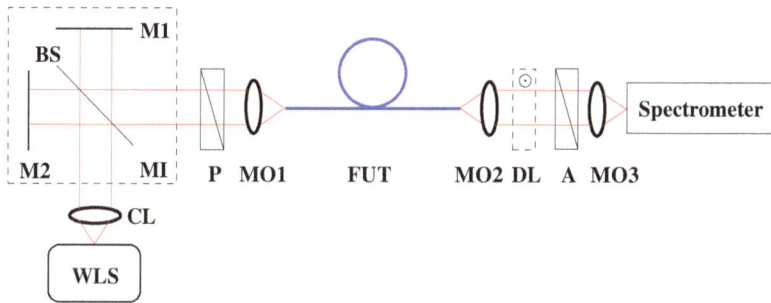

2.2. Measurement of Polarimetric Sensitivity to Strain, Temperature and Hydrostatic Pressure

To sense a physical parameter (strain, temperature or hydrostatic pressure) acting on the FUT, the configuration of the FUT in tandem with a birefringent crystal of the group birefringence, $G_c(\lambda)$, can be used [21], as shown in Figure 2. The spectral intensity at the output of the tandem configuration with a polarizer and an analyzer oriented $45°$ with respect to the eigenaxes of the FUT is given for $G(\lambda) < 0$ and $G_c(\lambda) > 0$ by [21]:

$$I(z;\lambda) = I_0(\lambda)\left\{1 + V(z;\lambda)\cos\left\{(2\pi/\lambda)[B_c(\lambda)d + B(\lambda)z]\right\}\right\} \tag{6}$$

where $I_0(\lambda)$ is the reference spectral intensity, $B_c(\lambda)$ is the phase birefringence of the crystal and $V(z;\lambda)$ is the visibility term given by:

$$V(z;\lambda) = \exp\{-(\pi^2/2)\left\{[G_c(\lambda)d + G(\lambda)z]\Delta\lambda_R/\lambda^2\right\}^2\} \tag{7}$$

It results from Equation (7) that the spectral interference fringes can be resolved, due to the highest visibility in the vicinity of the equalization wavelength, λ_0, given by:

$$G_c(\lambda_0)d = -G(\lambda_0)z \qquad (8)$$

Figure 2. Experimental setup with a fiber under test (FUT) to measure the polarimetric sensitivity to strain, temperature and hydrostatic pressure. The remaining components: a white-light source (WLS), microscope objectives 1–3 (MO1–3), a polarizer (P), a delay line (DL), an analyzer (A) and a slit (S).

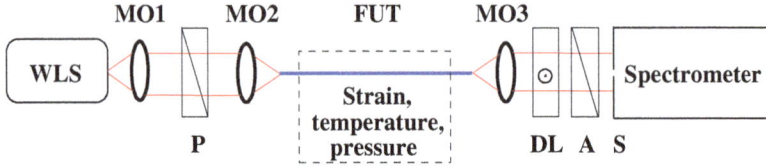

From the spectra recorded to sense a specific physical parameter, X, the phase shift, $\Delta[\phi_x(\lambda) - \phi_y(\lambda)]$, between the polarization modes can be retrieved. The polarimetric sensitivity, $K_X(\lambda)$, of the FUT to the physical quantity, X, is defined by the following relation:

$$K_X(\lambda) = \frac{1}{L}\frac{d[\phi_x(\lambda) - \phi_y(\lambda)]}{dX} \qquad (9)$$

and represents an increase in the phase shift between the polarization modes induced by the unit change of the physical quantity, X, acting on unit fiber length [16].

3. Experimental Configurations

The experimental setup used to measure the group modal birefringence dispersion is shown in Figure 1. It consists of a white-light source (WLS): a halogen lamp, collimating lens (C), a bulk-optic Michelson interferometer with a beam splitter and mirrors 1 and 2, polarizer (P) (LPVIS050, Thorlabs), microscope objectives MO1–3 (10×, 0.30 NA, Meopta), fiber under test (FUT), delay line (DL) represented by a birefringent quartz crystal of a suitable thickness, analyzer (A) (LPVIS050, Thorlabs), a spectrometer (S2000, Ocean Optics) and a personal computer. The polarizer and analyzer are oriented 45° with respect to the fiber eigenaxes.

The experimental setup used to measure the spectral dependence of the polarimetric sensitivity to strain, temperature and hydrostatic pressure is shown in Figure 2. It consists of a white-light source (WLS): a supercontinuum source (NKT Photonics), polarizer (P) (LPVIS050, Thorlabs), fiber under test (FUT), microscope objectives MO1–3 (10×, 0.30 NA), delay line (DL), analyzer (A) (LPVIS050, Thorlabs), a spectrometer (USB4000, Ocean Optics) with a 25 µm-wide slit (S) and a personal computer. To record the spectral intensity, $I(z;\lambda)$, the polarizer and analyzer are oriented 45° with respect to the fiber eigenaxes, and the optical axis of the birefringent quartz crystal is oriented 0° with respect to the major fiber eigenaxis. Even if many components employed in the experimental setup are bulky and good alignment is necessary, the stability of the measurements is satisfactory.

The FUT is a birefringent side-hole fiber with additional microstructure near the core region composed of six small holes of a diameter of about 0.7 μm, which prevent the guided mode from leaking out from the core. The fiber was drawn at the Department of Optical Fibers Technology, University of Marie Curie-Sklodowska in Lublin, Poland. As is shown in Figure 3, the fiber core has an elliptical shape of the dimensions 6.5 μm × 2.8 μm, while the cladding diameter is 129 μm. The longer axis of the core ellipse is perpendicular to the symmetry axis connecting the centers of the large holes. The initial GeO_2 concentration in the rod used to preform fabrication was 12 mol%, which corresponds to the refractive index contrast of 0.017 at 633 nm. However, due to material flow and GeO_2 diffusion during the fiber drawing process, it is expected that final GeO_2 concentration in the doped region of the fiber core is much lower. As the glass bridges between the large holes and the doped region of the core are very narrow (about 0.4–0.5 μm), this results in the cutoff of the fundamental mode at about 800 nm.

Figure 3. Structure of the FUT: general view of the fiber cross-section (**a**); Enlarged image of the central region with the green line showing the core boundary (**b**).

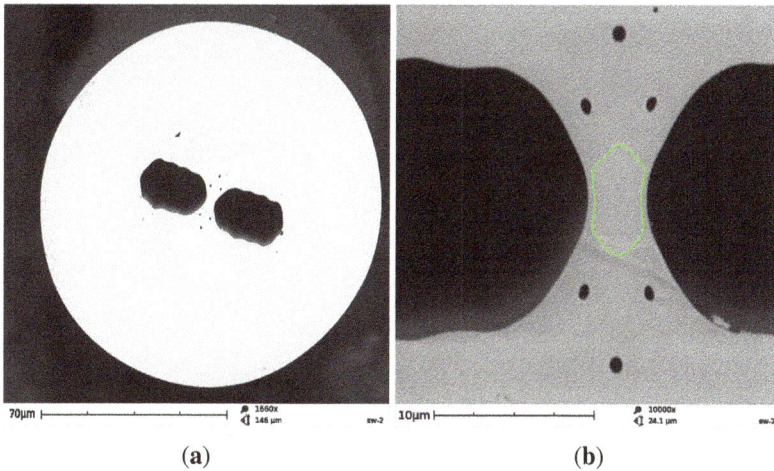

(**a**)   (**b**)

4. Experimental Results and Discussion

4.1. Phase and Group Modal Birefringence Dispersion

The results of the measurement of the group modal birefringence dispersion, $G(\lambda)$, in the experimental setup shown in Figure 1 are presented in Figure 4a by markers. The negative sign of the group modal birefringence is specified using a simple procedure with a delay line (DL) [19]. The precision in obtaining the group modal birefringence is better than 0.1% [19]. In the same figure is also shown a polynomial fit used to obtain the relative spectral dependence of the phase modal birefringence. It was combined with the known value, $B = 1.16 \times 10^{-4}$, measured at $\lambda = 665$ nm by a lateral force method applied in the spectral domain [22] to obtain the spectral dependence of the phase modal birefringence, $B(\lambda)$, which is shown in Figure 4b. The phase modal birefringence increases with wavelength, but its sign is opposite to the sign of the group modal birefringence, as is standard for HB MS fibers [19].

Figure 4. Measured spectral dependence of the group (**a**) and phase (**b**) modal birefringence in the FUT. The line crossing the markers is a polynomial fit.

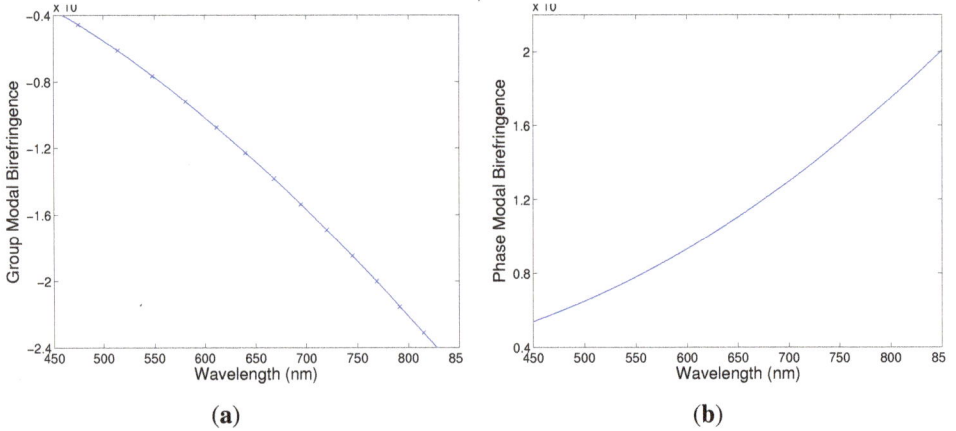

(**a**)   (**b**)

4.2. Polarimetric Sensitivity to Strain, Temperature and Hydrostatic Pressure

To determine the polarimetric sensitivity to strain, $K_\epsilon(\lambda)$, we recorded a sequence of spectral interferograms for increasing strain ϵ with a step small enough to assure unambiguity in retrieving the strain-induced phase changes, $\Delta[\phi_x(\lambda) - \phi_y(\lambda)]$. To measure the parameter, a fiber of length $L = 0.895$ m was attached with epoxy glue to two translation stages and elongated up to 800 μm (stretched up to 894 με). The measurements were repeated several times for increasing and decreasing strain, with no hysteresis observed. Figure 5a shows two recorded spectra obtained for a suitably chosen delay line and corresponding to two elongation,s $\Delta L_1 = 100$ μm and $\Delta L_2 = 150$ μm, of the fiber when the overall length of the fiber, including also the sensing part, with length $L = 0.895$ m, was $z = 2.295$ m. It is clearly seen from the figure that the interference of polarization modes in tandem with the delay line shows up as the spectral modulation with the wavelength-dependent period and the equalization wavelength, $\lambda_0 = 581.61$ nm. The spectral interference fringes for the two elongations are with the same equalization wavelength, but with different phases.

Using a new procedure [23], we retrieved from the two spectral interferograms the phase functions that are wavelength-dependent, with a minimum at the equalization wavelength, λ_0. From the phase functions retrieved from several recorded spectral interferograms, the spectral dependence of the polarimetric sensitivity to strain K_ϵ was obtained, as shown in Figure 5b by the red curve. The polynomial approximation of the absolute mean value of the polarimetric sensitivity as a function of wavelength is shown in the same figure by the blue line. The relative error of the measured polarimetric sensitivity to strain is about 5%.

The sign of the polarimetric sensitivity can be deduced from the results of the subsequent simulation. First, consider a positive sign of the polarimetric sensitivity to strain, as shown in

Figure 6a and simulate the spectral interferogram, $I(z; \lambda)$, according to Equation (6). First, we consider the fiber elongation, $\Delta L = 100$ µm, so that the strain-induced phase change is:

$$\Delta[\phi_x(\lambda) - \phi_y(\lambda)] = K_\epsilon(\lambda)\Delta L \qquad (10)$$

Figure 5. Examples of the recorded spectra corresponding to two elongations, $\Delta L_1 = 100$ µm (blue) and $\Delta L_2 = 150$ µm (red), of the FUT (**a**); The spectral dependence of the absolute value of the polarimetric sensitivity to strain (**b**). The line is the approximate quadratic dependence.

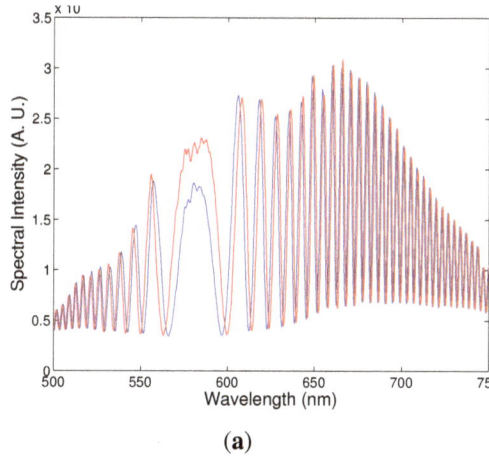

(**a**)

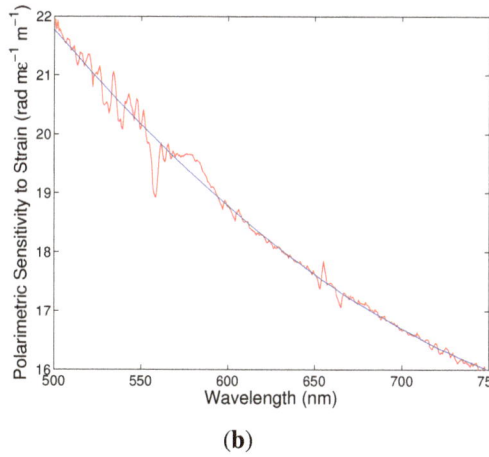

(**b**)

46

Figure 6. The spectral dependence of the polarimetric sensitivity to strain: a positive sign (**a**); Two theoretical spectra corresponding to the fiber elongations, $\Delta L = 100$ μm (blue) and $\Delta L = 150$ μm (red) (**b**).

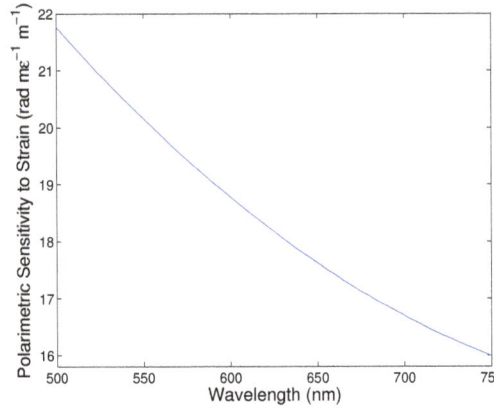

(a)

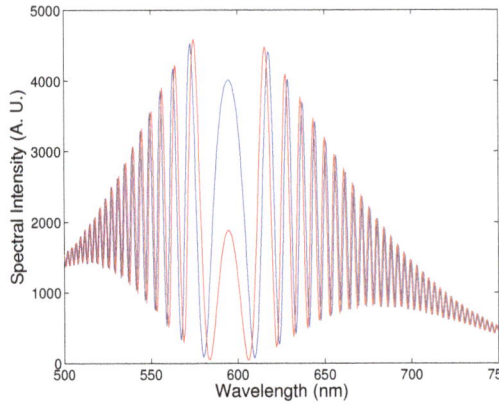

(b)

Using the well known birefringence dispersion for the quartz crystal [21] and the measured birefringence dispersion for the FUT (see Figures 4), we obtained for the fiber length, $z = 2.35$ m, and for the crystal thickness, $d = 23.5$ mm, the spectral interferogram, $I(z; \lambda)$, shown in Figure 6b by the blue curve. Next, we consider the fiber elongation, $\Delta L = 150$ μm, and the corresponding spectral interferogram, $I(z; \lambda)$, is shown in Figure 6b by the red curve. It is clearly seen that the spectral interference fringes are shifting to the equalization wavelength with the increasing elongation.

Second, consider a negative sign of the polarimetric sensitivity to strain, as shown in Figure 7a, and simulate the change of the spectral interferogram, $I(z; \lambda)$, with elongation. Using the same parameters—the fiber length, $z = 2.35$ m, and the crystal thickness, $d = 23.5$ mm—we obtained for the fiber elongation, $\Delta L = 100$ μm, the spectral interferogram, $I(z; \lambda)$, shown in Figure 7b by the blue curve. Next, we consider the fiber elongation, $\Delta L = 150$ μm, and Figure 7b shows the corresponding spectral interferogram, $I(z; \lambda)$, by the red curve. In this case, the spectral interference fringes are shifting from the equalization wavelength with the increasing elongation.

Figure 7. The spectral dependence of the polarimetric sensitivity to strain: a negative sign (**a**); Two theoretical spectra corresponding to the fiber elongations, $\Delta L = 100 \ \mu\text{m}$ (blue) and $\Delta L = 150 \ \mu\text{m}$ (red) (**b**).

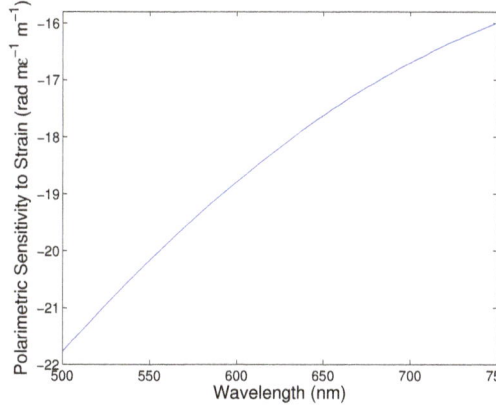

(**a**)

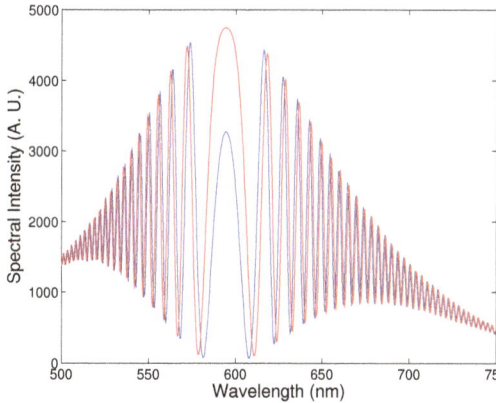

(**b**)

Comparing the spectral interferograms recorded for the increasing elongation (see Figure 5a) to the simulated spectral interferograms corresponding to different signs of the polarimetric sensitivity to strain (see Figures 6b and 7b), we conclude that the polarimetric sensitivity to strain is negative, and its absolute value decreases with wavelength, from a value of 21.8 rad $\times$ m$\epsilon^{-1} \times$ m^{-1} to a value of 16.0 rad $\times$ m$\epsilon^{-1} \times$ m^{-1} (in a range from 500 to 750 nm). A negative sign of the polarimetric sensitivity to strain indicates a decrease of the phase modal birefringence with strain. The polarimetric sensitivity of the investigated fiber to strain is comparable to that of conventional elliptical-core HBs [24].

To determine the polarimetric sensitivity to temperature, $K_T(\lambda)$, we recorded a sequence of spectral interferograms for increasing temperature, T, with a step small enough to assure unambiguity in retrieving the temperature-induced phase changes, $\Delta[\phi_x(\lambda) - \phi_y(\lambda)]$. To measure $K_T(\lambda)$, a bare fiber of length $L = 0.225$ m was immersed in water heated in a chamber with increasing temperature up to 373 K. Figure 8a shows two recorded spectra obtained for a suitably chosen thickness of the birefringent crystal and corresponding to temperatures $T_1 = 297$ K and $T_2 = 333$ K, when the overall length of the fiber was $z = 1.620$ m. It is clearly seen from the figure that the interference of polarization modes in tandem with the birefringent crystal shows up as the spectral modulation with the wavelength-dependent period and the equalization wavelength, $\lambda_0 = 583.02$ nm. The spectral interference fringes for the two temperatures are with the same equalization wavelength, but with different phases.

Figure 8. Two recorded spectra corresponding to temperatures $T_1 = 297$ K (blue) and $T_2 = 333$ K (red) (**a**); The spectral dependence of the polarimetric sensitivity to temperature (**b**). The line is the approximate quadratic dependence.

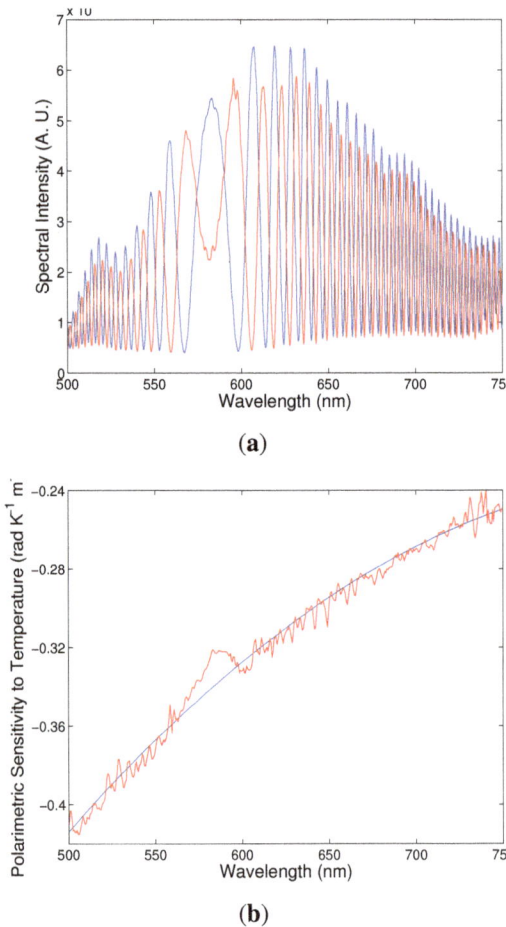

(**a**)

(**b**)

We retrieved from the two spectral interferograms the phase functions that are wavelength-dependent with a minimum at the equalization wavelength, λ_0. From the phase functions retrieved from several recorded spectral interferograms, the spectral dependence of the polarimetric sensitivity to temperature, $K_T(\lambda)$, was obtained, as shown in Figure 8b by the red curve. The polynomial approximation of the mean value of the polarimetric sensitivity as a function of wavelength is shown in the same figure by the blue line. The relative error of the measured polarimetric sensitivity to temperature is about 5%. The negative sign of the polarimetric sensitivity can be once again deduced from the results of the simulation. Similarly, as for the strain, the spectral interference fringes are shifting from the equalization wavelength with increasing temperature. A negative sign of the polarimetric sensitivity to temperature indicates a decrease of the phase modal birefringence with temperature, due to the release of thermal stress in the core region. The polarimetric sensitivity of the investigated fiber to temperature is comparable to that of conventional elliptical-core HBs [25].

To determine the polarimetric sensitivity to hydrostatic pressure, $K_p(\lambda)$, we used a rather different procedure. We recorded a sequence of spectral interferograms for increasing pressure, p, with the known pressure-induced phase changes, $\Delta[\phi_x(\lambda_0)-\phi_y(\lambda_0)]$, e.g., 10π at the equalization wavelength, λ_0. The sensing length of the fiber in a pressure chamber was $L = 0.648$ m, and measurement of the polarimetric sensitivity to hydrostatic pressure was performed up to 2.3 MPa. Figure 9a shows two recorded spectra obtained for a suitably chosen thickness of the birefringent crystal and corresponding to hydrostatic pressures $p_1 = 0.72$ MPa and $p_2 = 1.05$ MPa when the overall length of the fiber was $z = 1.129$ m. It is clearly seen from the figure that for the hydrostatic pressure, $p_1 = 0.72$ MPa, the interference of polarization modes in tandem with the birefringent crystal shows up as the spectral modulation with the wavelength-dependent period and the equalization wavelength, $\lambda_0 = 597.05$ nm. For the hydrostatic pressure, $p_1 = 1.05$ MPa, the equalization wavelength is changed, due to the group modal birefringence change [26].

We retrieved from the spectral interferogram the phase function, and the same procedure was used for the spectral interferogram corresponding to the hydrostatic pressure, $p_2 = 1.05$ MPa, when the known phase change was 10π at the equalization wavelength, λ_0. From the phase functions retrieved from several spectral interferograms recorded for the known phase change at the equalization wavelength, λ_0, the spectral dependence of the polarimetric sensitivity to hydrostatic pressure, $K_p(\lambda)$, was obtained, as shown in Figure 9b by the red curve. The quadratic approximation (blue line) of the mean value of the spectral dependence of the polarimetric sensitivity to hydrostatic pressure is shown in Figure 9b, and the relative error is about 1%. A negative sign of the polarimetric sensitivity to hydrostatic pressure can be deduced from the simulation results. The sign indicates that pressure-induced material birefringence in the core region has an opposite sign to the geometrical birefringence, due to core ellipticity. The absolute value of the polarimetric sensitivity to hydrostatic pressure increases with wavelength from a value of 140.6 rad $\times$ MPa$^{-1}\times$ m^{-1} to a value of 203.7 rad $\times$ MPa$^{-1}\times$ m^{-1} (in a range from 530 to 770 nm). For comparison, a side-hole MS fiber with the polarimetric sensitivity of about -220 rad $\times$ MPa$^{-1}\times$ m^{-1} at $\lambda = 700$ nm and -100 rad $\times$ MPa$^{-1}\times$ m^{-1} at $\lambda = 1.55$ μm was reported [14].

Figure 9. Two recorded spectra corresponding to the first delay and hydrostatic pressures, $p_1 = 0.72$ MPa (blue) and $p_2 = 1.05$ MPa (red) **(a)**; The spectral dependence of the polarimetric sensitivity to hydrostatic pressure **(b)**. The red line is the approximate quadratic dependence.

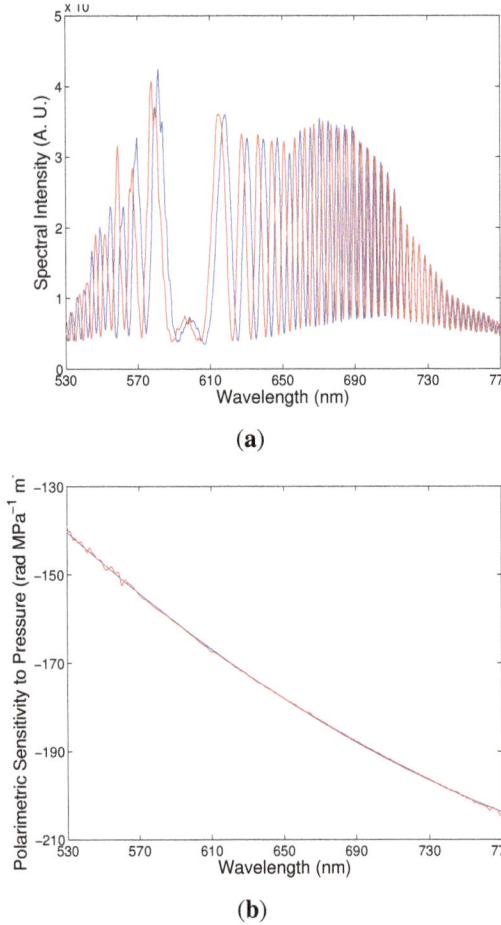

(a)

(b)

We also measured the polarimetric sensitivity to hydrostatic pressure, $K_p(\lambda)$, with a different thickness of the birefringent crystal when no equalization wavelength is resolvable in the recorded spectrum and when the pressure-induced phase change is known at a specific wavelength. In this case, the spectral fringes of slightly wavelength-dependent period are resolved in the recorded spectrum, and the spectral polarimetric sensitivity to hydrostatic pressure, $K_p(\lambda)$, is determined from the change of spectral phase with the hydrostatic pressure. As an example, Figure 10b shows two examples of the recorded spectra corresponding to hydrostatic pressures $p_1 = 0.3$ MPa and $p_2 = 0.99$ MPa, when the specific wavelength is 641.93 nm and the spectral fringes are shifted with the increasing hydrostatic pressure to longer wavelengths. It is clearly seen from the figure that the interference of polarization modes in tandem with the delay line shows up as the spectral modulation with no equalization wavelength in the considered spectral region. The spectral interference fringes for the two pressures are with different phases, which were retrieved from the two spectral interferograms using a procedure based on the application of a windowed Fourier transform [27]. Figure 10b shows by the red curve the spectral dependence of the polarimetric sensitivity to hydrostatic pressure, $K_p(\lambda)$, together with the approximate quadratic dependence (blue line). This approach gives the polarimetric sensitivity with errors smaller in comparison with those obtained by the first approach.

Figure 10. Two recorded spectra corresponding to the second delay and hydrostatic pressures, $p_1 = 0.3$ MPa (blue) and $p_2 = 0.99$ MPa (red) (**a**); The spectral dependence of the polarimetric sensitivity to hydrostatic pressure (**b**). The red line is the approximate quadratic dependence.

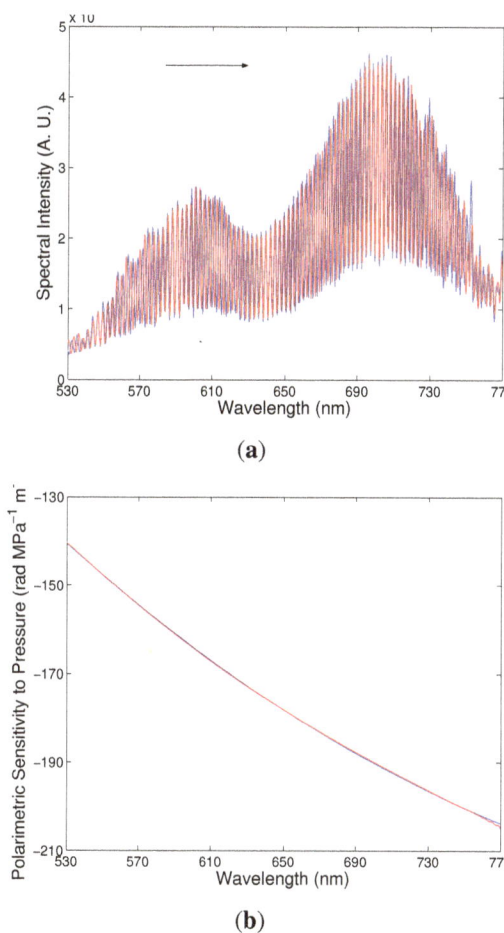

(**a**)

(**b**)

5. Conclusions

We measured the birefringence and sensing characteristics of a side-hole MS fiber in the visible wavelength region (500–770 nm). The spectral dependence of the group and phase modal birefringence is measured using the methods of spectral interferometry. The sensing characteristics of the fiber are measured using a method of tandem spectral interferometry. Spectral interferograms corresponding to different values of a physical parameter are processed to retrieve the spectral phase functions. These are used to determine the spectral dependence of polarimetric sensitivity to strain, temperature and hydrostatic pressure. A negative sign of the polarimetric sensitivity is deduced from the simulation results utilizing the known modal birefringence dispersion of the fiber.

The polarimetric sensitivity of the investigated fiber to strain reaches $K_\epsilon = -16$ rad $\times$ m$\epsilon^{-1} \times$ m^{-1} at $\lambda = 750$ nm, which is comparable to that of conventional HBs with an elliptical core. Similarly, the polarimetric sensitivity to temperature reaches $K_T = -0.25$ rad $\times$ K$^{-1} \times$ m^{-1} at $\lambda = 750$ nm, which is comparable to that of conventional HBs with an elliptical core. The investigated fiber has a very high polarimetric sensitivity to hydrostatic pressure, reaching $K_p = -200$ rad $\times$ MPa$^{-1} \times$ m^{-1} at $\lambda = 750$ nm. The ratio, K_p/K_T, which is an important figure of merit for the investigated fiber, reaches a value of 800 K$\times$ MPa^{-1} at $\lambda = 750$ nm, so that the fiber is suitable for hydrostatic pressure measurements with low cross-sensitivity to temperature.

Acknowledgments

The research was partially supported by the Grant Agency of the Czech Republic through grant P102/11/0675 and by the COST TD1001 action "OFSeSa" through project LD12003.

Conflicts of Interest

The authors declare no conflict of interest.

References

1. Fürstenau, N.; Schmidt, M.; Bock, W.J.; Urbańczyk, W. Dynamic pressure sensing with a fiber-optic polarimetric pressure transducer with two-wavelength passive quadrature readout. *Appl. Opt.* **1998**, *37*, 663–671.
2. Bock, W.J.; Urbańczyk, W. Temperature-desensitization of fiber-optic pressure sensor by simultaneous measurement of pressure and temperature. *Appl. Opt.* **1998**, *37*, 3897–3901.
3. Xie, H.M.; Dabkiewicz, P.; Ulrich, R.; Okamoto, K. Side-hole fiber for fiber-optic pressure sensing. *Opt. Lett.* **1986**, *11*, 333–335.
4. Wojcik, J.; Mergo, P.; Urbańczyk, W.; Bock, W. Possibilities of application of the side-hole circular core fibre in monitoring of high pressures. *IEEE Trans. Instrum. Meas.* **1998**, *47*, 805–808.
5. Tanaka, S.; Yoshida, K.; Kinugasa, S.; Ohtsuka, Y. Birefringent side-hole fiber for use in strain sensor. *Opt. Rev.* **1997**, *4*, A92–A95.

6. Ortigosa-Blanch, A.; Knight, J.C.; Wadsworth, W.J.; Arriaga, J.; Mangan, B.J.; Birks, T.A.; Russell, P.St.J. Highly birefringent photonic crystal fibers. *Opt. Lett.* **2000**, *25*, 1325–1327.

7. Suzuki, K.; Kubota, H.; Kawanishi, S.; Tanaka, M.; Fujita, M. Optical properties of a low-loss polarization maintaining photonic crystal fiber. *Opt. Express* **2001**, *9*, 676–680.

8. Folkenberg, J.R.; Nielsen, M.D.; Mortensen, N.A.; Jakobsen, C.; Simonsen, H.R. Polarization maintaining large mode area photonic crystal fiber. *Opt. Express* **2004**, *12*, 956–960.

9. Kim, D.H.; Kang, J.U. Sagnac loop interferometer based on polarization maintaining photonic crystal fiber with reduced temperature sensitivity. *Opt. Express* **2004**, *12*, 4490–4495.

10. Zhao, C.H.L.; Yang, X.; Lu, C.; Jin, W.; Demokan, M.S. Temperature-insensitive interferometer using a highly birefringent photonic crystal fiber loop mirror. *IEEE Photon. Technol. Lett.* **2004**, *16*, 2535–2537.

11. Bock, W.J.; Chen, J.; Eftimov, T.; Urbańczyk, W. A photonic crystal fiber sensor for pressure measurements. *IEEE Trans. Instrum. Meas.* **2006**, *55*, 1119–1123.

12. Fu, H.Y.; Tam, H.Y.; Shao; L.Y.; Dong, X.; Wai, P.K.; Lu, C.; Khijwania, S.K. Pressure sensor realized with polarization-maintaining photonic crystal fiber-based Sagnac interferometer. *Appl. Opt.* **2008**, *47*, 2835–2839.

13. Martynkien, T.; Statkiewicz-Barabach, G.; Olszewski, J.; Wojcik, J.; Mergo, P.; Geernaert, T.; Sonnenfeld, C.; Anuszkiewicz, A.; Szczurkowski, M. K.; Tarnowski, K.; *et al.* Highly birefringent microstructured fibers with enhanced sensitivity to hydrostatic pressure. *Opt. Express* **2010**, *18*, 15113–15121.

14. Wu, C.; Li, J.; Feng, X.H.; Guan, B.O.; Tam, H.Y. Side-hole photonic crystal fiber with ultrahigh polarimetric pressure sensitivity. *J. Lightwave Technol.* **2011**, *29*, 943–948.

15. Szpulak, M.; Martynkien, T.; Urbańczyk, W. Effects of hydrostatic pressure on phase and group modal birefringence in microstructured holey fibers. *Appl. Opt.* **2004**, *43*, 4739–4744.

16. Statkiewicz, G.; Martynkien, T.; Urbańczyk, W. Measurements of modal birefringence and polarimetric sensitivity of the birefringent holey fiber to hydrostatic pressure and strain. *Opt. Commun.* **2004**, *241*, 339–348.

17. Hlubina, P.; Olszewski, J.; Martynkien, T.; Mergo, P.; Makara, M.; Poturaj, K.; Urbańczyk, W. Spectral-domain measurement of strain sensitivity of a two-mode birefringent side-hole fiber. *Sensors* **2012**, *12*, 12070–12081.

18. Hlubina, P.; Martynkien, T.; Urbańczyk, W. Dispersion of group and phase modal birefringence in elliptical-core fiber measured by white-light spectral interferometry. *Opt. Express* **2003**, *11*, 2793–2798.

19. Hlubina, P.; Ciprian, D.; Kadulová, M. Wide spectral range measurement of modal birefringence in polarization-maintaining fibres. *Meas. Sci. Technol.* **2009**, *20*, 025301, doi:10.1088/0957-0233/20/2/025301.

20. Cao, X.D.; Meyerhofer, D.D. Frequency-domain interferometer for measurement of the polarization mode dispersion in single-mode optical fibers. *Opt. Lett.* **1994**, *19*, 1837–1839.

21. Hlubina, P.; Ciprian, D.; Knyblová, L. Interference of white light in tandem configuration of birefringent crystal and sensing birefringent fiber. *Opt. Commun.* **2006**, *260*, 535–541.

22. Hlubina, P.; Ciprian, D. Spectral-domain measurement of phase modal birefringence in polarization-maintaining fiber. *Opt. Express* **2007**, *15*, 17019–17024.

23. Hlubina, P.; Olszewski, J. Phase retrieval from spectral interferograms including a stationary-phase point. *Opt. Commun.* **2012**, *285*, 4733–4738.

24. Huang, S.Y.; Blake, J.N.; Kim, B.Y. Perturbation effects on mode propagation in highly elliptical core two mode fibers. *J. Lightwave Technol.* **2010**, *8*, 23–33.

25. Urbańczyk, W.; Martynkien, T.; Bock, W.J. Dispersion effects in elliptical-core highly birefringent fibers. *Appl. Opt.* **2001**, *40*, 1911–1920.

26. Anuszkiewicz, A.; Martynkien, T.; Mergo, P.; Makara, M.; Urbańczyk, W. Sensing and transmission characteristics of a rocking filter fabricated in a side-hole fiber with zero group birefringence. *Opt. Express* **2013**, *21*, 12657–12667.

27. Hlubina, P.; Luňáček, J.; Ciprian, D.; Chlebus, R. Windowed fourier transform applied in the wavelength domain to process the spectral interference signals. *Opt. Commun.* **2008**, *281*, 2349–2354.

Reprinted from *Sensors*. Cite as: Chen, C.-H.; Yeh, B.-K.; Tang, J.-L.; Wu, W.-T. Fabrication Quality Analysis of a Fiber Optic Refractive Index Sensor Created by CO2 Laser Machining. *Sensors* **2013**, *13*, 4067–4087.

Article

Fabrication Quality Analysis of a Fiber Optic Refractive Index Sensor Created by CO_2 Laser Machining

Chien-Hsing Chen [1], Bo-Kuan Yeh [2], Jaw-Luen Tang [1] and Wei-Te Wu [2],*

[1]   Department of Physics, National Chung Cheng University, Chiayi 621, Taiwan;
     E-Mails: saesozj@yahoo.com.tw (C.-H.C.); jawluen@gmail.com (J.-L.T.)
[2]   Department of Biomechatronics Engineering, National Pingtung University of Science and Technology, Pingtung 912, Taiwan; E-Mail: bo.kuang1@gmail.com

*   Author to whom correspondence should be addressed; E-Mail: weite@mail.npust.edu.tw; Tel.: +886-8-770-3202 (ext. 7599); Fax: +886-8-774-0420.

Received: 28 December 2012; in revised form: 14 March 2013 / Accepted: 22 March 2013 / Published: 26 March 2013

Abstract: This study investigates the CO_2 laser-stripped partial cladding of silica-based optic fibers with a core diameter of 400 μm, which enables them to sense the refractive index of the surrounding environment. However, inappropriate treatments during the machining process can generate a number of defects in the optic fiber sensors. Therefore, the quality of optic fiber sensors fabricated using CO_2 laser machining must be analyzed. The results show that analysis of the fiber core size after machining can provide preliminary defect detection, and qualitative analysis of the optical transmission defects can be used to identify imperfections that are difficult to observe through size analysis. To more precisely and quantitatively detect fabrication defects, we included a tensile test and numerical aperture measurements in this study. After a series of quality inspections, we proposed improvements to the existing CO_2 laser machining parameters, namely, a vertical scanning pathway, 4 W of power, and a feed rate of 9.45 cm/s. Using these improved parameters, we created optical fiber sensors with a core diameter of approximately 400 μm, no obvious optical transmission defects, a numerical aperture of 0.52 ± 0.019, a 0.886 Weibull modulus, and a 1.186 Weibull-shaped parameter. Finally, we used the optical fiber sensor fabricated using the improved parameters to measure the refractive indices of various solutions. The results show that a refractive-index resolution of 1.8×10^{-4} RIU (linear fitting $R^2 = 0.954$) was achieved for sucrose solutions with refractive indices ranging between 1.333 and 1.383. We also adopted the particle plasmon resonance sensing scheme using the fabricated optical fibers.

56

The results provided additional information, specifically, a superior sensor resolution of 5.73×10^{-5} RIU, and greater linearity at $R^2 = 0.999$.

Keywords: CO_2 laser machining; optical fiber sensor; refractive index sensing

---

1. Introduction

Various biosensors, such as the electrochemical sensors developed by Clark and Lyons [1], use a galvanometer to measure the glucose concentration of a solution and achieve the measurement objectives. Additionally, semiconductor ion-sensitive biosensors adopt a semiconducting structure that comprises metal-insulating field effect transistors (MISFET) [2]. Another example are the optical fiber sensors that exploit the optical fiber transmission characteristics to achieve sensing objectives, such as evanescent wave and surface plasmon resonance technologies [3].

Of the various biosensor types, optical fiber biosensors offer the unique characteristic of no electromagnetic interference. Small, lightweight, and with the potential for miniaturization, optic fibers can be used not only to transmit light signals, but also as the primary sensing element. Optical fibers are widely employed for engineering and environmental control and in mechanical and biological developments [4].

Optical fibers have a three-layer structure that comprises a silica-based fiber core, a polymer cladding, and a coating of harder polymer as the outermost layer that protects the fiber. Various methods and structures to provide optical fibers with sensing capabilities have been developed, including fiber Bragg grating [5], fiber-optic interferometers [6], and window-type optical fiber sensors [7]. Among them, window-type optical fiber sensors, as shown in Figure 1, have the simplest structure; only partial removal of the coating material is required to expose the fiber core beneath. Once exposed, the window-type optical fiber structure allows sensors in a test environment to conduct ambient refractive index sensing using the attenuated total reflection (ATR).

The current methods for stripping part of the optical fiber material can be broadly divided into mechanical and chemical methods. The most common of the many mechanical fiber optic stripping methods involve polishing the stripper or fiber [8]. However, the fact that the fiber optic stripper can potentially damage the fiber core presents a significant disadvantage. The fiber polishing method typically requires more expensive equipment, although it does offer high machining accuracy. The chemical method involves the use of various solutions such as sulfuric acid, which was employed by Matthewson [9]. The optical fiber was soaked in sulfuric acid before heating it to between 180 and 200 °C to soften and strip the outer coating material. Nonetheless, etching quality is also difficult to control because a slight error can generate unexpected processing phenomena that affect the sensing quality. Researchers have also employed the flame vaporization technique by exploiting the melting point characteristics of various layers of the optical fiber cable. This technique is used to vaporize the outer cladding material, exposing the glass fiber core. Although easy to process, the processing scope and duration of this method is difficult to control, rendering it unsuitable for extended research [10]. Other research teams have employed lasers and precision lenses on laser processing platforms with a moving mechanism. This method of removal is less time-consuming compared to the other two methods and the parameters are easier to control. Regarding laser processing, the precision lens on the laser processing platform tends to age, which may affect the accuracy of the moving platform and lead to cause experimental errors. This can also lead to inferior processing results, problems such as an inability to correctly remove materials, and/or changes in the material properties and costs of heating the area because of excessive laser energy [11,12]. However, regardless of which method is employed, they all provide the same disadvantage, that is, a lack of comprehensive post-processing quality control procedures.

Based on the above discussion, and to further understand the basic characteristics of window-type optical fiber sensors, we used CO_2 laser processing as the fiber optic sensor processing method in this study because the laser-processing parameters are convenient to configure and provide a wider range of basic characteristics. Studies of common optical fiber sensors typically investigate only fiber optic sensor fabrication methods or the resolution and sensitivity of back-end sensing applications; the processing quality of the sensing area is rarely examined [13–15]. Poor-quality processing, such as over-processing resulting in excessive removal or modification of material, can reduce the sensor resolution and sensitivity, cause light scattering in the sensing area, insufficient sensing power, or functional surface coverage during subsequent surface plasmon resonance (SPR) or particle plasmon resonance (PPR) detection [16,17]. Therefore, the purpose of this study was to eliminate defects or residue from the sensing area of window-type optical fiber sensors. We examined the CO_2 laser processing results for the sensing area and established a CO_2 laser processing quality inspection method. Finally, a window-type optical fiber sensor was developed according to the optimal processing conditions identified in this study, and the sensor performance was subsequently verified.

Figure 1. Schematic of the fiber sensor: (**a**) crude fiber; and (**b**) fiber sensors (window type).

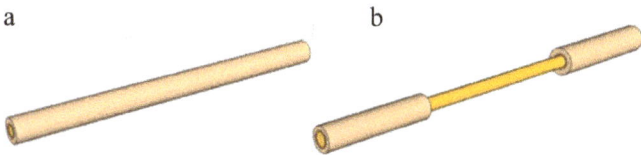

2. Experimental Section

The adjustable parameters of the CO_2 laser machine (Model Mercury-II M-12, LaserPro Inc., New Taipei, Taiwan) employed in this study included the processing power (1 W to 10 W), processing speed (0.63 cm/s to 63 cm/s), laser-sourced cooling nozzle pressure (0.1 MPa at less than 60 psi), and focusing position (adjusted by altering the Z-axis in the machine's three-axis displacement platform). The non-adjustable parameters were the laser pulse width (±0.2 μs) and pulse frequency (5 KHz). This study primarily analyzed the sensing area quality of laser-processed optical fiber sensors. To conduct various quality analyses more accurately, we set the fixed length of the sensing region to 1 cm, the focusing position on the fiber core to 0.1 mm, the laser pulse width to 1 ± 0.2 μs, the laser pulse frequency to 5 KHz, and the air nozzle gauge pressure to 0.3 MPa for air processing. In this study, we considered the laser processing power and speed, in addition to self-developed fixtures, to explore the laser processing path. Figure 2a shows the processing fixture used to attach the processed optical fibers; this fixture is capable of attaching five optical fibers simultaneously. Figure 2b shows a rotating fixture with a central hole packed tightly with optical fibers ready for processing in lockstep rotation. Using the preset structure highlighted at every 60°, the operator can process optical fibers every 60° a total of six times.

Figure 2. Schematic of the laser machining fixtures: (**a**) the processing fixture; and (**b**) the rotating and fixed optical fiber fixture.

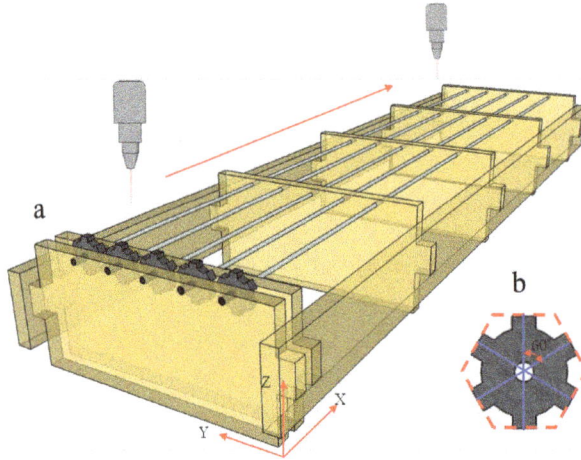

The yellow portion of the structure shown in Figure 3 is the optical fiber to be processed, the blue dotted line denotes the established processing direction of the optical fiber, and the red arrows and circular patterns represent the moving path and processing area of the laser source. When the laser source and the blue dotted reference line move horizontally during processing, the structure adopts the parallel machining condition; otherwise, the vertical machining condition is employed.

Figure 3. Schematic of the laser processing path: (**a**) parallel machining; and (**b**) vertical machining.

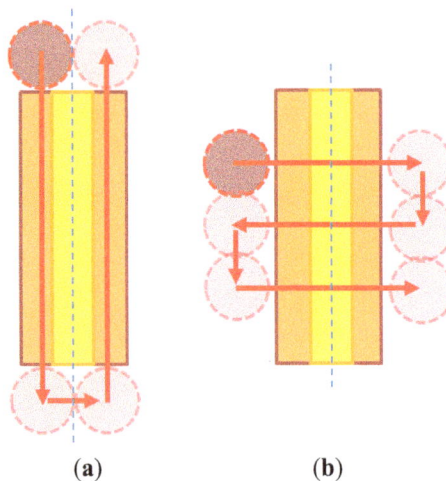

(**a**)   (**b**)

The research goals of this study were to propose a comprehensive method for assessing the quality of window-type optical fibers processed by lasers, the convenience of light coupling in subsequent sensing measurements, and the mechanical strength of manufactured window optical fibers. Therefore, we employed a multi-mode glass optical fiber with a 400-μm fiber core, manufactured by Newport® under the model number F-MBC, as the optical fiber. Regarding the size and structure, the optical fiber comprised a 400-μm fiber core, 430-μm cladding, and 730-μm coating, as shown in Figure 4. The fiber core was made of silica material, the cladding was made of hard and brittle polymer, and the coating was made of Tefzel material. These materials were used to provide the optical fiber with superior mechanical protection.

Figure 4. Schematic of the optical fiber.

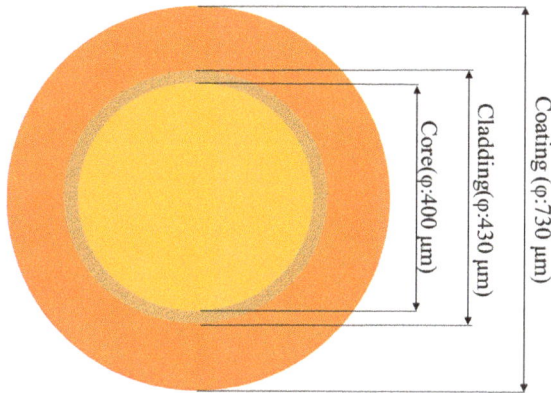

By adjusting various methods for deploying the laser power, processing speed, and processing path, we defined the following processing defects and circumstances:

1. The fiber core and coating material is either completely or partially removed, as shown in Figure 5.

    (a) The fiber core and coating material are partially removed.
    (b) The fiber core and coating material are insufficiently removed.
    (c) The fiber core and coating material are completely (maybe even excessively) removed.

2. The surface of the fiber core is altered, as shown in Figure 6.

    (a) The surface changes are not uniform.
    (b) The surface changes are uniform.

3. The fiber core shape changes, as shown in Figure 7.

    (a) Wavy pattern
    (b) Excessive removal

Figure 5. Schematic of an optical fiber cable: (**a**) partially removed; (**b**) insufficiently removed; and (**c**) completely removed.

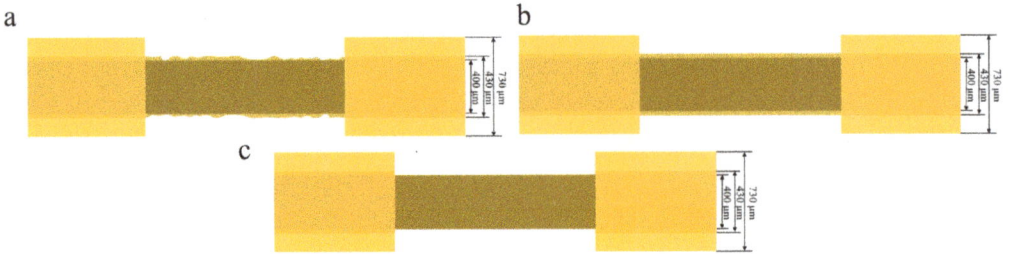

Figure 6. Schematic of optical fiber defects: (**a**) non-uniform changes in property; and (**b**) uniform changes in property.

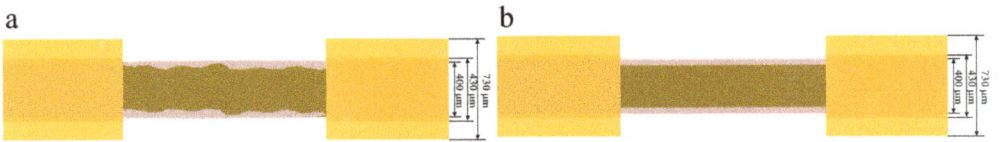

Figure 7. Schematic of an optical fiber: (**a**) wavy pattern; and (**b**) excessive removal.

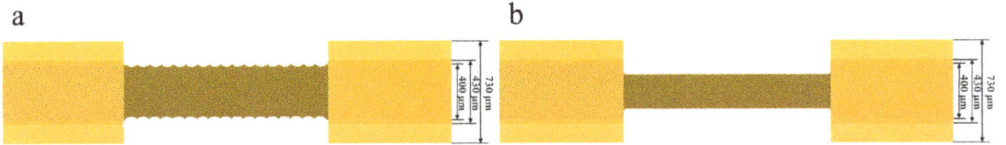

Therefore, to avoid the defects caused by laser processing, which can affect the sensing capability of optical fiber sensors, negatively influence the sensor performance, or modify other functionalized surfaces, we established a quality analysis method for identifying these defects. This quality analysis method includes the following four items:

(1)   Size measurement

This element can identify lacking, incomplete, or excessive removals and distinguish between the two defect types, namely, wavy pattern and excessive removal, during fiber core modifications. The measuring instrument is shown in Figure 8. The optical fiber sensor location point measurements are shown in Figure 9. First, data from Points 1 and 6 were removed because the cumulative thermal effect at these two points was relatively lower than the processing level in other areas. To prevent the values measured at these two points from affecting the values measured at normal processing areas, we use them to verify the total processing length at the sensing area. Then, the additional diameter of the processing core was used as the average value of the four remaining data groups (the remainder of the core, D_c^R).

Figure 8. Schematic of the measuring instrument: (**a**) fiber connecter; (**b**) measuring fixture; (**c**) processed fiber; and (**d**) micro-stage.

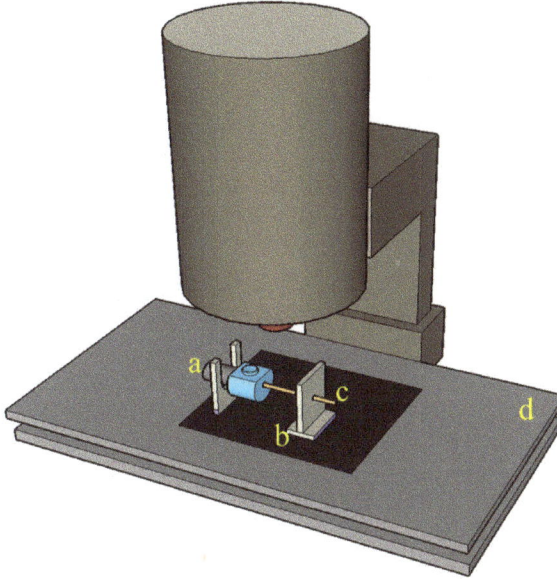

Figure 9. Schematic of the measurement position.

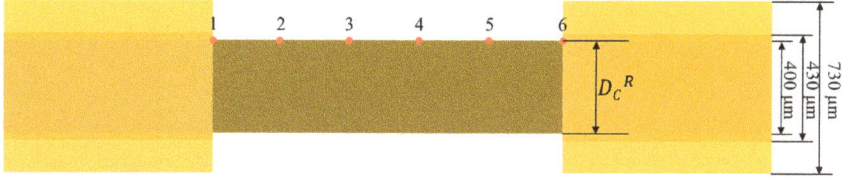

(2)   Light transmission defect detection method

The light transmission defect detection method was primarily based on internal optical fiber core propagation by optically coupling the laser to the process. For this study, we used the structure shown in Figure 10 to guide the laser light inside the optical fiber sensor as it propagates. First, an optical collimator was used to focus the laser source (wavelength: 532 nm; power: 10 mW) inside the optical fiber transmission line (NA = 0.27) during light propagation. At this time, light from the source was transmitted to the end with an NA value of 0.27 and then guided through the optical fiber adapter as it propagated within the optical fiber core (NA = 0.37). The Optical fiber core was then examined to confirm that the confined light source was guided to the core of the optical fiber when propagating. A diagram of the measurement is shown in Figure 11. When the surface is irregular or contains debris from processing, according to the optical transmission principle, irregular surfaces in the path of light transmissions cause the light to diffuse. Observation of the light diffusion phenomenon facilitates the achievement of quality analysis objectives.

Figure 10. Diagram of the components of an optical light transmission defect: (**a**) laser source; (**b**) collimator; (**c**) fiber cable (NA = 0.27); (**d**) fiber adapter; (**e**) fiber connecter; and (**f**) fiber sensor (NA = 0.37).

Figure 11. Schematic of the tool used for qualitative analysis of the optical transmission defects: (**a**) laser source; (**b**) collimator; (**c**) fiber cable; (**d**) fiber adapter; (**e**) fiber connecter; (**f**) processed fiber; and (**g**) micro-stage.

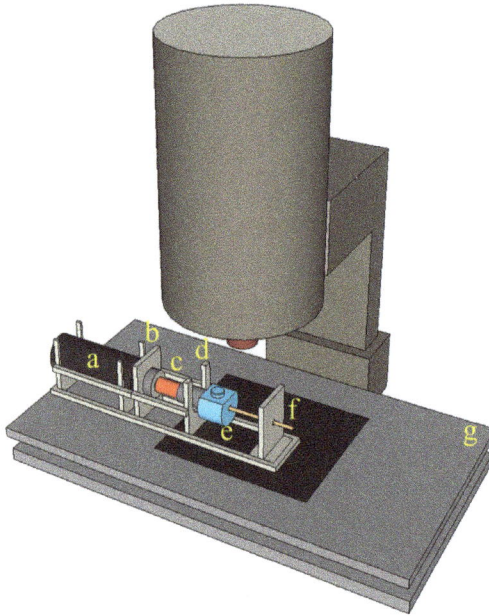

(3) Numerical aperture (NA)

The two quality analysis methods previous mentioned were used to perform preliminary quality analysis of the laser-processed optical fiber sensors. However, to ensure that no defects remain undetected, additional in-depth investigations of the internal sensing area may be necessary. Therefore, we established a numerical aperture measurement platform to measure the post-processing numerical aperture of optical fiber sensors. The architecture in Figure 12 shows that parallel optical modulation was first conducted. Subsequently, the focusing lens and optical fiber coupler were used to focus and couple the beam onto the core inside the processed optical fiber sensor for light propagation. To measure the NA value of the optical fiber sensor area, we partially cut the processed optical fiber sensor, as shown in Figure 12a,b, and employed the grinding method to smooth the sectioned surface for further measurement. After the light coupled to the core layer was propagated, it scattered from the back-end of the truncated plane to the rear, forming a circular light spot with a diameter of D. This was then attached to the optical sensor using the moving platform to measure the size of the circular light spot with a diameter of D at various lengths L. According to Equation (1) [18], the L and D values measured can be employed to determine the numerical aperture value. If the sensing area of the processed optical fiber sensor shows homogeneous surface modification, the NA value of the sensing area decreases. This is because the width of the possible transmission light path is reduced by surface modifications. In this study, we used this method as a basis for identifying defects.

$$NA = \sin\theta = \frac{D}{\sqrt{D^2 + 4L^2}} \tag{1}$$

Figure 12. Schematic of the numerical aperture measurement instrument.

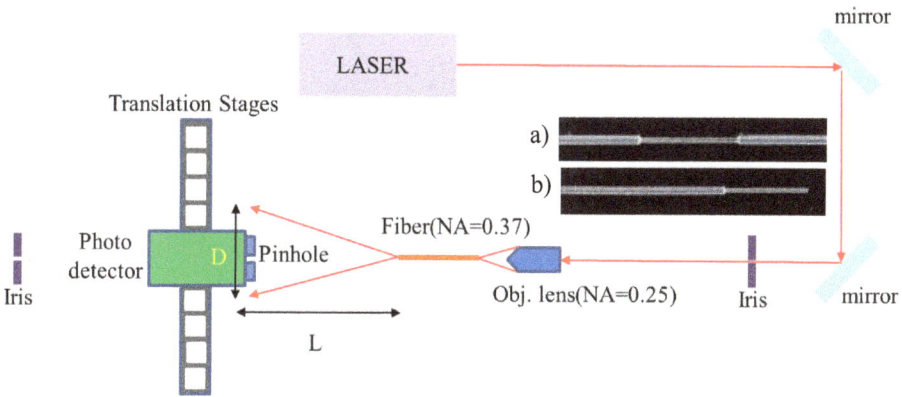

(4)   Weibull tensile test [19,20]

The Weibull tensile test is suitable for analyzing the tensile properties of hard and brittle materials. The partial exposure of the optical fiber's core silica material during the laser-stripping procedure either alters the material properties or results in internally generated processing defects. If window-type optical fiber sensors contain defects not detected by the previously described quality analysis method, a Weibull tensile test can be employed to exploit the changing trends in tensile failure characteristics. Figure 13 shows a tensile test diagram. Data obtained from the tensile test are analyzed using the Weibull distribution. The mathematical formula for the Weibull distribution is shown as Equation (2), where P is the cumulative probability of failure, σ is the applied tension force (kgf), η is the scale parameter, σ_0 is the minimum breaking tension force (kgf), and ω is the Weibull modulus:

$$P = 1 - \exp[-(\frac{\sigma - \sigma_0}{\eta})^\omega] \tag{2}$$

Figure 13. Schematic of the tensile test (Model FS-1002, Lutron Inc., New Taipei, Taiwan): (**a**) force gauge; (**b**) measuring fixture; (**c**) translation stage; (**d**) gauge value display; and (**e**) processed fiber.

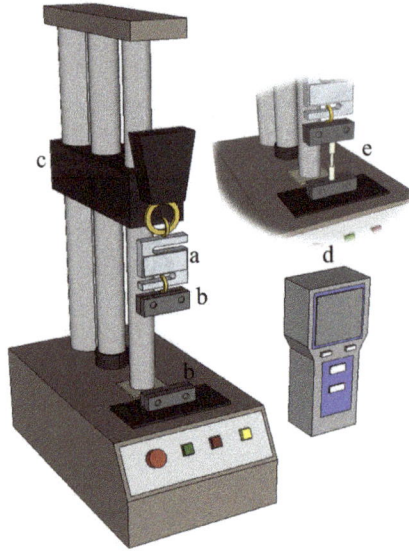

(5) Sensing experiment

This study investigated whether the processing parameters selected based on the quality analysis results offer sensing capabilities. The setup of the sensing experiment employed for this study is shown in Figure 14. First, a function generator (Model 33220A, Agilent Inc., Santa Clara, CA, USA) was used to generate a direct current pulse with square waves of a 1-KHz frequency and 1-V voltage to drive the green light-emitting diode (LED) light source (Model EHP-AX08LS-HA/SUG01-P01, Everlight Inc., New Taipei, Taiwan). Next, the total reflection characteristic of optical fiber was used to direct the light into the sensing area for ATR sensing. Subsequently, the optical signal was further directed through the optical fiber to the back-end signal capture device for photoelectric signal conversion and processing. The microfluidic chip was made of poly(methyl methacrylate): PMMA (chip size: $50 \times 20 \times 8$ mm; micro-channel size: $50 \times 1 \times 0.9$ mm) and fabricated using a computer numerical control engraving machine by our group [21]. The sensing fiber was packaged inside the microfluidic chip. The input and output ports of the microfluidic chip were used to infuse the solution flowing through the sensing fiber. The sensing environment involved the injection of various sucrose solution concentrations into a sensing microfluidic chip using a syringe for further sensing testing. For the experiment, we first prepared deionized (DI) water. The relationship between the sucrose concentration and refractive index of 1.333 to 1.383 RIU is shown in Table 1 [7]. After the sensing experiment, to compare our results with those reported in previous literature [13], we calculated the resolution of the sensor. Then, we modified the gold nanoparticles in the sensing area to conduct PPR-sensed environmental refractive index measurements [13]. The gold nanoparticles modified in the optical fiber sensing area were prepared by the study researchers [7].

Figure 14. Schematic of the experimental setup for creating sensing measurements: (**a**) function generator (Agilent Inc. Model: 33220A); (**b**) LED light source (Model: EHP-AX08LS-HA/SUG01-P01, Everlight Inc.); (**c**) sensing chip; (**d**) photo diode (Model PD-ET2040, EOT Inc., Traverse, MI, USA); (**e**) lock-in amplifier (Model 7225, Signal Recovery Inc., Oak Ridge, TN, USA); and (**f**) computer.

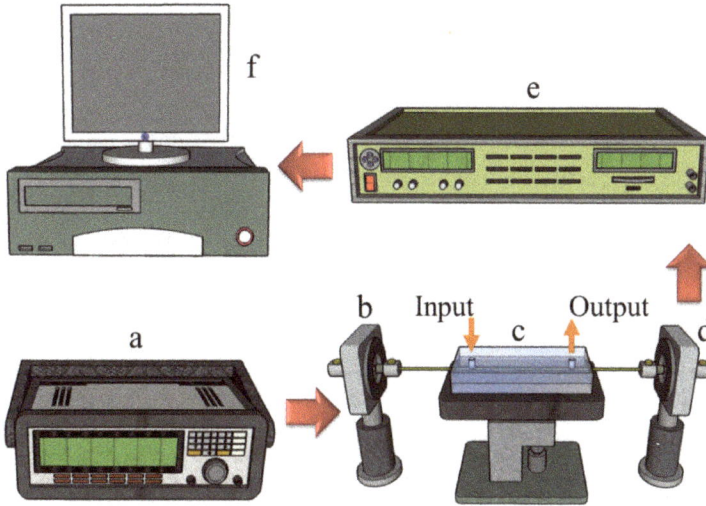

Table 1. Refractive index of various sucrose solution concentrations [7].

	RIU	Wt %
DI water	1.333	0
No.1	1.343	6.8
No.2	1.353	13.25
No.3	1.363	19.45
No.4	1.373	25.4
No.5	1.383	31.05

3. Results and Discussion

3.1. Size Analysis

This study used the parameter scanning method to identify superior parameters. First, the result of a large-range parameter measurement was employed to obtain a near residual diameter (D_c^R) of 400 μm at a processing speed of 9.45 to 25.2 cm/s and to arrange a parameter scanning experiment in this processing speed range. Figure 15 is a diagram showing the relationship between the power of various parallel spindle paths and D_c^R; the results indicate that as the processing power increases, the residual diameter gradually decreases under various processing speeds. From the same perspective of processing energy, as the processing speed increases, the total energy loss per unit area per unit of time causes D_c^R to gradually increase. Figure 16 is a diagram of the vertical spindle processing relationship, which exhibits the same conditions. The line graph results in Figures 15 and 16 show that the sloping trend of parallel spindle processing exceeds that of vertical spindle processing. This suggests that the material removal processing change rate under horizontal conditions is comparatively greater. In addition, the standard deviations of the data points from the two processing paths show that parallel spindle path processing is greater than vertical spindle path processing. The cause of this phenomenon may be the greater horizontal processing path displacement, because cumulative structure errors can lead to inconsistent processing quality.

Figure 15. Schematic of the size analysis results (Scanning path: parallel).

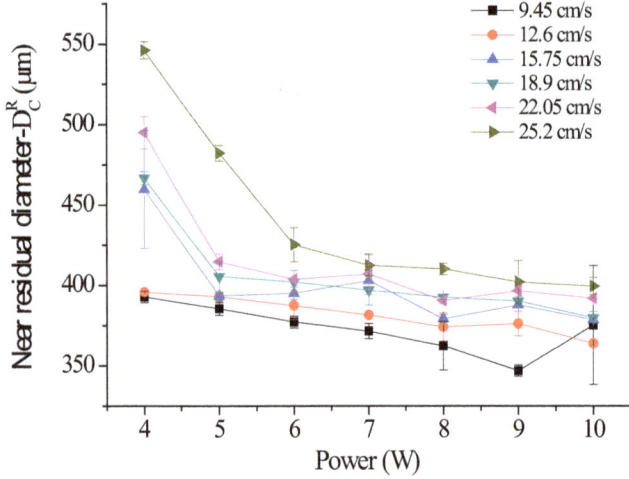

Figure 16. Schematic of the size analysis results (Scanning path: vertical).

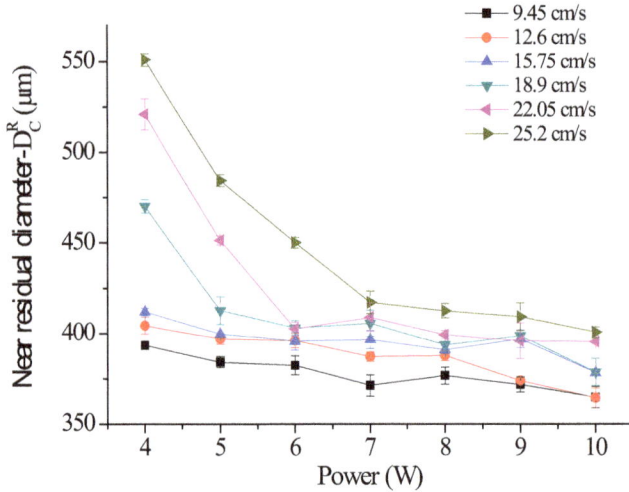

To ensure a residual diameter of nearly 400 μm using laser processing while avoiding the non-processed removal of material from the fiber core layer, we established an acceptable range for the residual diameter:

$$395 \ \mu m < (\text{average value} + |\text{standard deviation}|) < 400 \ \mu m$$

This range was used to select the following parameters for the second quality analysis process, as shown in Table 2.

Table 2. Size analysis access parameters.

Scanning Path	Power (W)	Velocity (cm/s)	Avg. D_C^R (µm)	Std. D_C^R (µm)
Parallel	4	9.45	392.9	3.48
	4	12.6	395.8	0.88
	5	12.6	393.1	2.18
	5	15.75	393.6	2.51
	6	15.75	395.3	3.75
Vertical	4	9.45	393.5	1.77
	5	12.6	397.1	2.77
	8	18.9	393.8	3.36
	10	22.05	395.6	1.12

3.2. Qualitative Analysis of Optical Transmission Defects

The nine sets of data obtained through size measurements satisfy the criteria for superior parameters (within the expected range). The optical defect transmission method was used to qualitatively analyze the quality of the processing area. The results in Table 3 show that under parallel processing machining conditions, with a laser processing power of 4 W and a processing speed of 9.45 cm/s, excessively low processing speeds result in exorbitant removal of material, generating wavy patterns on the surface of the processing area, as shown in Figure 17. Additionally, regarding the vertical processing machining condition, for optical fiber sensors processed with a power of 5 to 10 W and a speed of 12.6 to 22.05 cm/s, some of the material could not be removed correctly, resulting in the optical leakage phenomenon, as shown in Figure 18. Figure 19 is a diagram of the schematic without defects.

Table 3. Results of the qualitative analysis of optical transmission defects.

Machining parameter			Result	
Scanning Path	Power (W)	Velocity (cm/s)	Result	Reason
Parallel	4	9.45	×	Wavy pattern
	4	12.6	O	Without light leakage
	5	12.6	O	Without light leakage
	5	15.75	O	Without light leakage
	6	15.75	O	Without light leakage
Vertical	4	9.45	O	Without light leakage
	5	12.6	×	Removal incomplete
	8	18.9	×	Removal incomplete
	10	22.05	×	Removal incomplete

68

Figure 17. Schematic of the wavy pattern defect.

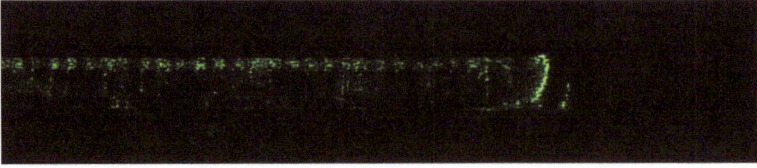

Figure 18. Schematic of the incomplete removal defect.

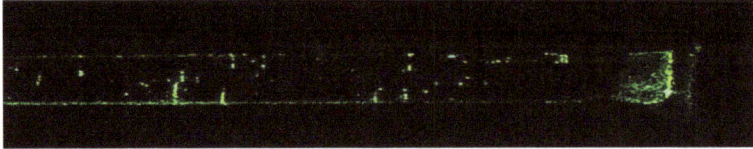

Figure 19. Schematic without any obvious optical transmission defect.

3.3. Numerical Aperture Measurement

Preliminary analysis of the quality of laser-processed optical fiber sensors was conducted using two quality analysis methods. We then established an NA measurement platform to measure the NA of the processed optical fiber sensors. During the experiment, we measured the NA of preprocessed optical fibers, obtaining an average value of 0.385 ± 0.01. Compared to the factory specification of 0.37 ± 0.02 with $\alpha = 0.01$, the average value obtained using the two quality analysis methods did not differ significantly. Therefore, the measurement platform developed in this study is feasible.

After conducting the above experiment, we used the quality analysis results to perform separate NA measurement experiments. The results are shown in Table 4 and are relatively small for several NA groups. Because the refractive index between the fiber coating and core layers was modified, according to Equation (3), where $n_{cladding}$ decreases and n_{core} remains the same, the NA value of the material increases [18]. The F-test in an analysis of variance (ANOVA) was used to analyze each group at a significance level of $\alpha = 0.05$. Using the right-tailed test method with an F significance value as the determination basis, we found that a significant relationship existed between Groups 1, 2, and 3. The value of Group 4 was relatively small, whereas that for Group 5 was large in comparison to the other groups. This result was included in the next quality analysis process to identify defects:

$$NA \equiv \sin \theta_c = \sqrt{n_{core}^2 - n_{cladding}^2} \tag{3}$$

Table 4. Results of numerical aperture measurements.

	Scanning Path	Power (W)	Velocity (cm/s)	Number of samples	NA Avg.	NA Std.
	Crude fiber				0.385	0.01
No. 1		4	12.6		0.482	0.023
No. 2	Parallel	5	12.6	10	0.497	0.021
No. 3		5	15.75		0.491	0.023
No. 4		6	15.75		0.449	0.032
No. 5	Vertical	4	9.45		0.520	0.019

3.4. Weibull Tensile Test

By converting the diagram of stress and cumulative damage probability created using the Weibull tensile test, we can plot a Weibull graph. Figure 20 shows that the information in Table 5 can be obtained after linearly fitting the five parameter groups. The larger the Weibull modulus equation ω in the Weibull distribution, the greater the reliability. Another important parameter is η, which belongs to the scale parameter. The larger this value, the greater the damage to the material. Additionally, the wider the fracture stress distribution, the less homogeneous the material. By contrast, the smaller the value, the more homogeneous the material. Therefore, the resulting table shows that Group 4 has a superior Weibull modulus. However, regarding the results of NA measurements, this group had the smallest value. Thus, to prevent defects, Group 4 was ignored. Statistical analysis of the NA measurement results indicated that Groups 1, 2, and 3 had no significant differences. However, the Weibull tensile analysis results showed that various parameters changed excessively. To avoid the occurrence of non-detected defects, these groups were also eliminated. Therefore, the parameters of Group 5 were selected as the optimum processing parameters for this study.

Figure 20. Weibull distribution.

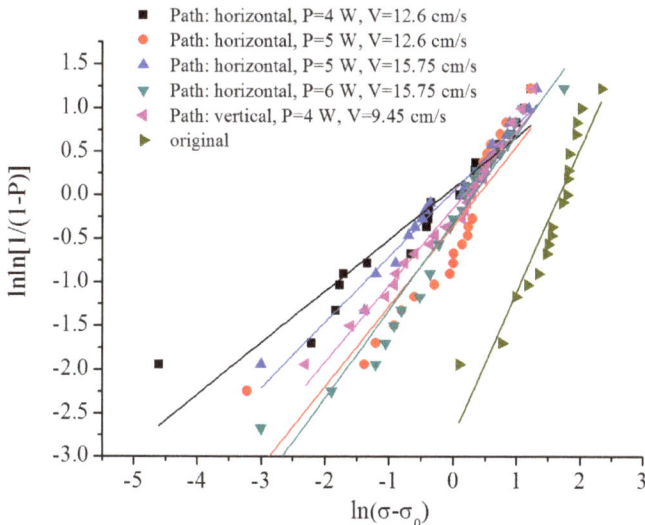

Table 5. Weibull tensile test results.

| | Machining parameter | | | Number of samples | ω | η | σ_{min} (kgf) | σ_{max} (kgf) | R^2 |
	Scanning Path	Power (W)	Velocity (cm/s)						
No. 1	Parallel	4	12.6	30	0.591	0.878	0.14	3.95	0.913
No. 2		5	12.6		0.922	1.476	0.25	4.05	0.833
No. 3		5	15.75		0.753	0.942	0.1	4.1	0.969
No. 4		6	15.75		1.008	1.368	0.1	5.9	0.963
No. 5	Vertical	4	9.45		0.886	1.186	0.25	4.15	0.974

3.5. Refractive Index Sensing Measurements

The quality inspection method previously described was used to select processing parameters for the vertical spindle processing path. A superior set of parameters, including a processing power of 4 W and a processing speed of 9.45 cm/s, was used to create optical fiber sensors for the refractive index sensing experiments. In Figure 21, the X-axis represents the time and the Y-axis represents the signal. Figure 21 shows that the experimental intensity decreased as the injection concentration increased (refractive index increment). When the signal measurement was complete, the sensor resolution was calculated. In Figure 22, the X-axis represents the refractive index value and the Y-axis represents the signal level of each concentration after the average signal was normalized. The relationship diagram in Table 6 shows the sensor slope (m) used to calculate the sensor resolution. After repeating the experiment three times, we obtained an average sensor resolution value of 1.8×10^{-4} RIU.

Figure 21. Plot of the ATR fiber sensors' temporal responses to injections of increasing sucrose solution concentrations and refractive indices.

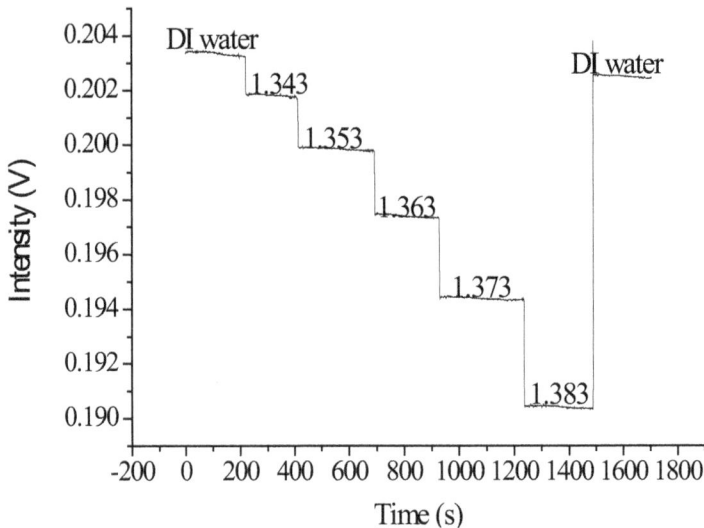

Figure 22. Plot of the ATR fiber sensor response *versus* the refractive index of the sucrose solution.

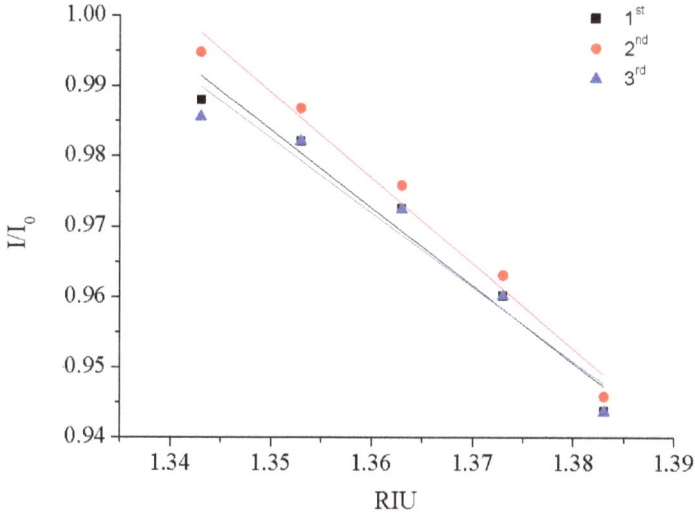

Table 6. The sensing experiment results for ATR fiber sensor.

No.	Sensing slope (m)	Coefficient of variation (σ = *Standard* of deviation/m)	RI resolution 3 σ/m (RIU)
1	1.106	7.65×10^{-5}	2.08×10^{-4}
2	1.217	4.23×10^{-5}	2.23×10^{-4}
3	1.058	1.07×10^{-5}	1.09×10^{-4}
		Average sensor resolution values	1.8×10^{-4}

This study also used the superior processing parameters obtained through quality analysis to create window-type optical fiber sensors by modifying gold nanoparticles [13]. The PPR sensing method was used to conduct three iterations of the refractive index sensing experiment. A diagram of the relationship between the experiment time and signal is shown in Figure 23. The relationship between the average intensity of the refractive index of a measured signal and the refractive index is shown in Figure 24. The results of the sensing resolution after linear regression are shown in Table 7. The average sensor resolution was approximately 5.73×10^{-5} RIU. Compared to unmodified ATR sensors, the resolution of the sensor was an order of magnitude smaller as that reported in previous literature, although the sensor length was similar [13]. The results of the linearity comparison modified sensing method are shown in Figure 25. The quality analyzed optical fiber sensors have a superior degree of linearity compared to the non-quality analyzed optical fiber sensors reported in previous studies. This may be because the removal of optical fiber material could not be confirmed in related literature. Thus, quality-analyzed window-type optical fiber sensors possess basic ATR sensing abilities, with the sensor resolution reaching a 10^{-4} RIU level. Once the modified gold nanoparticles sensed using the PPR method were excited, they created a regional plasma resonance response that effectively increased the sensor resolution and linearity.

Figure 23. Plot of the PPR fiber sensors' temporal response to injections of increasing sucrose solution concentrations and refractive indices.

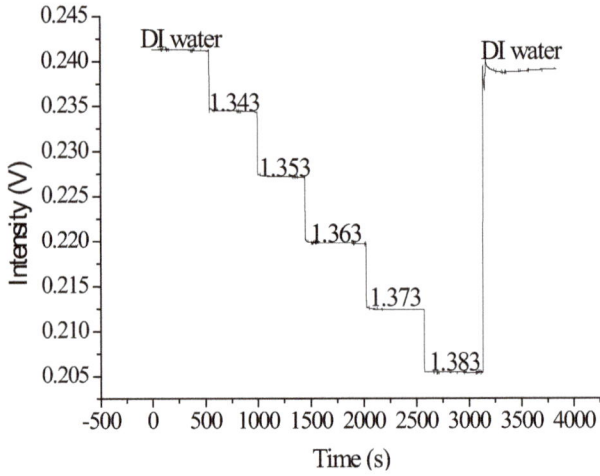

Figure 24. Plot of the PPR fiber sensor response *versus* the refractive index of the sucrose solution.

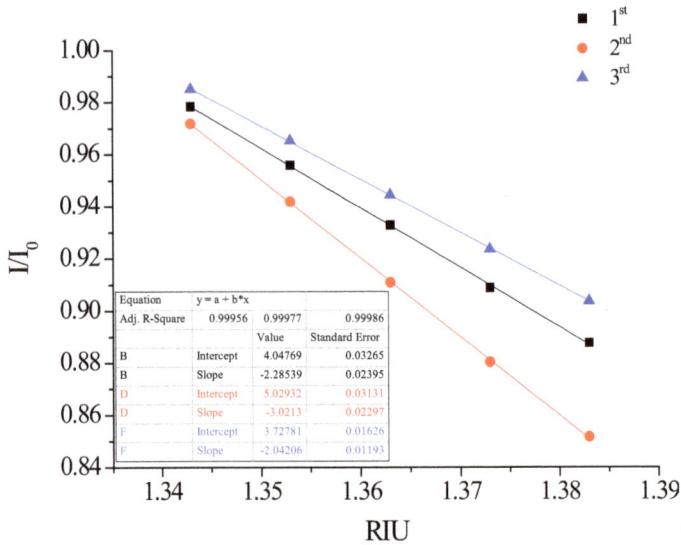

Table 7. Results of the sensing experiment for PPR fiber sensor.

No.	Sensing slope (m)	σ value	RI resolution 3σ/m (RIU)
1	2.285	1.07×10^{-5}	1.41×10^{-5}
2	3.021	6.15×10^{-5}	6.1×10^{-5}
3	2.042	6.6×10^{-5}	9.69×10^{-5}
		Average sensor resolution values	$\mathbf{5.73 \times 10^{-5}}$

Figure 25. Plot of the sensor response *versus* the refractive index of the sucrose solution.

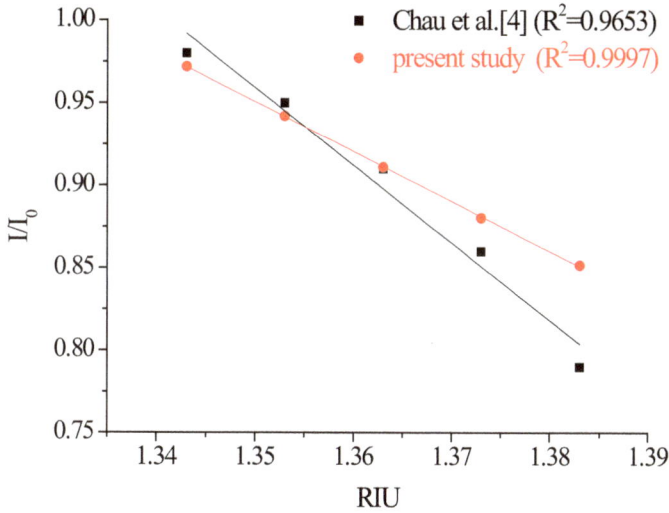

4. Conclusions

The original optical fiber structure of window-type optical fiber sensors has changed according to current sensing needs. The current structure exposes the fiber core to facilitate contact with the sensing environment, and is adopted to measure the environmental refractive index according to the principle of gradually reducing the total reflective sensing. The partial removal method selected for the optical fibers used in this study was the laser thermal removal method. Although adjusting the parameters using this method is extremely easy, because of the thermal effects of removal, if the parameters are not appropriately controlled, defects are likely to result, such as insufficient removal, excessive removal, or property changes. The purpose of removing material is to ensure the core layer of the optical fiber contacts the environment directly. However, if defects exist and the refractive index does not match as expected, the sensitivity of optical fiber sensors may be reduced. The sensing principle employed in this study was the ATR sensing method because the penetration depth of the evanescent wave is extremely limited. In addition, the evanescent wave decays exponentially into the surrounding media because of its distance from the fiber core. However, if other defects exist, the sensing ability is inevitably reduced. In this study, we proposed four methods for analyzing the laser processing quality, that is, size measurements, light transmission defect detection, NA measurement, and a Weibull tensile test. They were employed to verify the quality of the sensing area of optical fiber sensors after laser processing. These methods can also facilitate coating the functional surface, thereby increasing the surface coating rate and sensor sensitivity.

After quality analysis, the remaining processed sensor region measured 400 μm and did not possess light transmission defects. The NA value of the sensing area was 0.52 ± 0.019, with superior processing parameters; that is, a Weibull tensile tested module number $\omega = 0.886$, scale parameter $\eta = 1.186$, minimum destructive pulling force of 0.25 kgf, and maximum destructive pulling force of 4.15 kgf. The processing parameters also include a vertical processing path with a processing power of 4 W and a processing speed of 9.45 cm/s. The parameter-processed optical fiber sensors have ATR refractive index sensing capabilities. The average sensor resolution measured was approximately 1.8×10^{-4} RIU ($R^2 = 0.954$), with the PPR sensing resolution reaching 5.73×10^{-5} RIU ($R^2 = 0.999$).

The above discussion indicates that in this study, we successfully established a quality analysis method for laser-processed optical fiber sensors. The detection method provides superior quality verification of fiber optic sensors before application by including a quantitative analysis of size, NA, and material properties to eliminate any concerns regarding reduce ATR sensing capability or insufficient functional surface coverage.

Acknowledgments

The authors thank Tai-Huei Wei and Lai-Kwan Chau for their laboratory assistance. The partial support of the National Science Council (NSC) of Taiwan under Contract Nos. NSC-99-2120-M-194-005-CC1; NSC101-2120-M-194-001-CC2; NSC-99-2112-M-194-004-MY3; NSC-101-2221-E-020-010-MY3 is gratefully acknowledged.

References

1. Brooks, S.L.; Ashby, R.E.; Turner, A.P.F.; Calder, M.R.; Clarke, D.J. Development of an on-line glucose sensor for fermentation monitoring. *Biosensors* **1987**, *3*, 45–56.
2. Lundstrom, I.; Spetz, A.; Winquist, F.; Albery, W.J.; Thomas, J.D.R. Semiconductor biosensors [and Discussion]. *Philos. Trans. Roy. Soc. Lond. B Biol. Sci.* **1987**, *316*, 47–60.
3. Janata, J. *Principles of Chemical Sensors*; Plenum Press: New York, NY, USA, 1989.
4. Cheng, S.-F.; Chau, L.-K. Colloidal gold-modified optical fiber for chemical and biochemical sensing. *Anal. Chem.* **2002**, *75*, 16–21.
5. Wang, J.-N.; Tang, J.-L. Feasibility of fiber bragg grating and long-period fiber grating sensors under different environmental conditions. *Sensors* **2010**, *10*, 10105–10127.
6. Wei, T.; Han, Y.; Tsai, H.-L.; Xiao, H. Miniaturized fiber inline Fabry-Perot interferometer fabricated with a femtosecond laser. *Opt. Lett.* **2008**, *33*, 536–538.
7. Chen, C.H.; Chao, T.C.; Li, W.Y.; Shen, W.C.; Cheng, C.W.; Tang, J.L.; Chau, L.K.; Wu, W.T. Novel D-type fiber optic localized plasmon resonance sensor realized by femtosecond laser. *J. Laser Micro Nano Eng.* **2010**, *5*, 1–5.
8. Chiu, M.H.; Chiu, P.C.; Liu, Y.H.; Zheng, W.D. Single-mode D-type optical fiber sensor in spectra method at a specific incident angle of 89°. *Sens. Transd.* **2009**, *5*, 41–46.
9. Yoon, H.-J.; Kim, C.-G. The mechanical strength of fiber Bragg gratings under controlled UV laser conditions. *Smart Mater. Struct.* **2007**, *16*, 1315–1319.
10. Chau, L.-K.; Lin, Y.-F.; Cheng, S.-F.; Lin, T.-J. Fiber-optic chemical and biochemical probes based on localized surface plasmon resonance. *Sens. Actuators B Chem.* **2006**, *113*, 100–105.
11. Vengsarkar, A.M.; Lemaire, P.J.; Judkins, J.B.; Bhatia, V.; Erdogan, T.; Sipe, J. E. Long-period fiber gratings as band-rejection filters. *J. Lightwave Technol.* **1996**, *14*, 58–65.
12. Malki, A.; Humbert, G. Investigation of the writing mechanism of electric-arc-induced long-period fiber gratings. *Opt. Lett.* **2003**, *42*, 3776–3779.

13. Chiang, C.-Y.; Hsieh, M.-L.; Huang, K.-W.; Chau, L.-K.; Chang, C.-M.; Lyu, S.-R. Fiber-optic particle plasmon resonance sensor for detection of interleukin-1β in synovial fluids. *Biosens. Bioelectr.* **2010**, *26*, 1036–1042.

14. Zhao, L.; Jiang, L.; Wang, S.; Xiao, H.; Lu, Y.; Tsai, H.-L. A High-Quality Mach-Zehnder Interferometer Fiber Sensor by Femtosecond Laser One-Step Processing. *Sensors* **2010**, *11*, 54–61.

15. Chen, C.-H.; Tsao, T.-C.; Tang, J.-L.; Wu, W.-T A multi-D-shaped optical fiber for refractive index sensing. *Sensors* **2010**, *10*, 4794–4804.

16. Homola, J.; Yee, S.S.; Gauglitz, G. Surface plasmon resonance sensors: review. *Sens. Actuators B Chem.* **1999**, *54*, 3–15.

17. Salamon, Z.; Macleod, H.A; Tollin, G. Surface plasmon resonance spectroscopy as a tool for investigating the biochemical and biophysical properties of membrane protein systems, II: Applications to biological systems. *BBA Rev. Biomembranes* **1997**, *1331*, 131–152.

18. Michael, B.; Eric, W.V.S. *Fiber Devices and Systems for Optical Communications*; McGraw-Hill: Orlando, FL, USA, 2002.

19. Alexey, F.K.; Sergei, L.S.; Alexandr, N.D.; Evgeny, M.D. Mechanical strength and fatigue of microstructured optical fibers. In Proceedings of Optical Fiber Communication Conference (OFC), Anaheim, CA, USA, 25–29 March 2007.

20. Liu, X.Y.; Dai, G.C. Surface modification and micromechanical properties of jute fiber mat reinforced polypropylene composites. *eXPRESS Polym. Lett.* **2007**, *1*, 299–307.

21. Wang, J.-N. A Microfluidic long-period fiber grating sensor platform for chloride ion concentration measurement. *Sensors* **2011**, *11*, 8550–8568.

Reprinted from *Sensors*. Cite as: Montero, D.S.; Vázquez, C. Remote Interrogation of WDM Fiber-Optic Intensity Sensors Deploying Delay Lines in the Virtual Domain. *Sensors* **2013**, *13*, 5870–5880.

Article

Remote Interrogation of WDM Fiber-Optic Intensity Sensors Deploying Delay Lines in the Virtual Domain

David Sánchez Montero * and Carmen Vázquez

Departamento de Tecnología Electrónica, Universidad Carlos III de Madrid, Avda. de la Universidad, 30, 28911 Leganés, Madrid, Spain; E-Mail: cvazquez@ing.uc3m.es

* Author to whom correspondence should be addressed; E-Mail: dsmontero@ing.uc3m.es; Tel.: +34-91-6248-865; Fax: +34-91-6249-430.

Received: 6 March 2013; in revised form: 9 April 2013 / Accepted: 3 May 2013 / Published: 7 May 2013

Abstract: In this work a radio-frequency self-referencing WDM intensity-based fiber-optic sensor operating in reflective configuration and using virtual instrumentation is presented. The use of virtual delay lines at the reception stage, along with novel flexible self-referencing techniques, and using a single frequency, avoids all-optical or electrical-based delay lines approaches. This solution preserves the self-referencing and performance characteristics of the proposed WDM-based optical sensing topology, and leads to a more compact solution with higher flexibility for the multiple interrogation of remote sensing points in a sensor network. Results are presented for a displacement sensor demonstrating the concept feasibility.

Keywords: optical sensing; passive remote sensing; self-referencing; virtual delay line

1. Introduction

Intensity fiber-optic sensors (FOS) provide an optical modulation signal as the measurement and use different self-referencing techniques to avoid noise errors from undesirable intensity fluctuations or variation in losses non-correlated to the sensor modulation. They have been proved to be easily integrated in wavelength-division multiplexing (WDM) networks, including those based on fiber Bragg gratings (FBGs), and have been demonstrated to provide an effective and compact strategy to operate in reflective configuration [1] for exploiting fiber links and for remotely addressing multiple sensing points with a single fiber lead [2,3] instead of spatial multiplexing deployments [4].

Configurations providing self-referencing techniques have been a research topic during the last years. The use of all-optical resonant structures as the basis of a self-referencing intensity type sensor has been widely discussed in the literature. Schemes based on a Fabry-Perot resonant structure [5], Michelson [6], Sagnac [7,8], and ring resonators [9–11], with fiber delay coils [1,12] are reported. For instance, in the work reported in Reference [1] identical fiber coils of 450 m were emplaced at each sensing point, with the requirement of being of identical length to share the two modulation frequencies at the transmission stage for all the sensors, otherwise the point of operation of the measurement technique would be different for each sensor, which is not a desirable situation.

Lately, the long fiber coils were replaced with electrical filters at the reception stage of the remote sensor network [13]. This solution provided arbitrary modulation frequencies, compact sensing points and flexibility in the operation of the sensor network. In addition, the optical power modulation from the remote sensing point could be related to the coefficients of the filter structure thus encoding the filter response either in amplitude or in phase, and then performing self-referencing measurements. Furthermore, a Coarse WDM (CWDM) reflective star sensor network topology for multiplexing and interrogation of N quasi-distributed self-referencing remote sensing points, using two electrical phase-shifts per sensor for flexibility purposes, was recently studied [14]. Two measurement parameters were defined, one based on phase measurements and another based on amplitude measurements. This electro-optical self-referencing solution was verified by modulating the light injected into the network and using a lock-in amplifier and electrical phase-shifters at the reception stage. On the other hand, it was recently reported a virtual delay line deployment [15], but in a Mach-Zehnder interferometric topology in which the optical source was modulated with two different frequencies and based on a power-splitting topology rather than in a WDM approach. The self-referencing parameter was defined as the ratio of voltage values of the optical output sinusoidal wave at a previously defined non-constructive and constructive interference frequency, respectively.

In this work, the feasibility for enhancing the automation of the interrogation, by working in the virtual domain, of remote intensity-based optical sensors operating in reflective configuration and deployed in a radio-frequency (RF) WDM-based passive sensor networks is demonstrated. The proposed solution preserves all the above advantages for sensing interrogation. By using WDM devices with low insertion losses for spectral splitting of a RF modulated broadband light source (BLS) it is possible to enhance the power budget of the network. In Section 2, a brief description of the theoretical background is presented. To test the concept, Section 3 analyzes the performance of an optical intensity sensor system compatible with a CWDM network as well as the self-reference property, thus emulating unexpected losses by means of a variable optical attenuator (VOA). Relative errors of the measurements shown in this section are given. Moreover, a system analysis in terms of optical power budget is studied thus establishing the limits for remote sensor interrogation. Both low-cost Analog-to-Digital converter (ADC) at the reception stage and virtual instrumentation techniques supported on a LabVIEW® platform for developing two virtual delay lines and for controlling the sensor operation are used. Finally, the main conclusions of this work are reported in Section 4.

2. Theoretical Background

The digital filter schematic for a single fiber-optic sensor topology is illustrated in Figure 1, being H the sensor loss modulation. At the reception stage, after signal acquisition, virtual phase-shifts Ω_1 at the reference channel and Ω_2 at the sensing channel are applied, respectively, to the RF modulating signal providing a flexible and easy-reconfigurable operation point of the remote sensor. These delay line filters can be deployed in the optical, electrical or virtual domain but with a coefficient β which depends on the optical sensor loss modulation, H, in the sensing point. In Figure 1 this sensor loss modulation appears squared due to the reflective operation of the sensing structure, as the light crosses twice the sensor thus providing a sensing system with a doubled sensitivity. For a deeper comprehension, the works reported in References [13,14] are recommended.

Figure 1. Filter model of the proposed topology for a generic remote sensing point with two virtual delay lines, after acquisition, at the reception stage. IM: Intensity Modulator, PD: Photodetector.

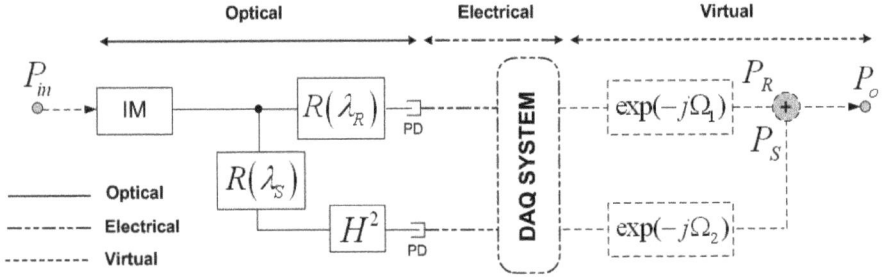

Attending to the digital filter model of the sensor topology, the normalized system output, *i.e.*, the transfer function of the system $H_0 = P_0/P_{in}$, can be directly identified with a digital Finite Impulse Response (FIR) filter in the Z-Transform domain, as shown in Equation (1):

$$H_0 = \frac{P_0}{P_{in}} = \alpha \cdot \left(1 + \beta \cdot z^{-1}\right)$$

(1)

where $z = e^{-j(\Omega_2 - \Omega_1)}$, being α and β defined as follows:

$$\alpha = m_R . R(\lambda_R) . d_R \cdot e^{-j \cdot \Omega_1}$$

(2)

$$\beta = \frac{m_s \cdot R(\lambda_s) \cdot d_s}{m_R \cdot R(\lambda_R) \cdot d_R} H^2$$

(3)

and where m_R, $R(\lambda_R)$ and d_R are the RF modulation index, the reflectivity of the FBG and the photodetector response at the reference wavelength λ_R respectively, and m_S, $R(\lambda_S)$ and d_S are those parameters but for the sensing wavelength λ_S.

Two measurement parameters can then be defined at the remote sensing point. On one hand, the parameter R which is given by the ratio between the voltage values received for different delay configurations. And, on the other hand, the output phase Φ of the acquired electrical signal, also dependant on the delays configured at the reception stage. Those parameters are given by Equations (4) and (5), respectively:

$$R = \frac{V_O(f, \Omega_2)|_{\Omega_1 = 0}}{V_O(f, \Omega_1)_{\Omega_2 = 0}} = \frac{[1 + (\frac{2\beta}{1+\beta^2})\cos\Omega_2]^{1/2}}{[1 + (\frac{2\beta}{1+\beta^2})\cos\Omega_1]^{1/2}} \tag{4}$$

$$\phi = \arctan\left[\frac{-(\sin\Omega_1 + \beta\sin\Omega_2)}{(\cos\Omega_1 + \beta\cos\Omega_2)}\right] \tag{5}$$

For a fixed value of both the modulation frequency and the delays selection, both measurement parameters of the generic remote sensing point depend only on β which is insensitive to external power fluctuations in the optical link. Moreover, both self-referencing parameters can be determined for any pair of values of angular frequencies, *i.e.*, delays, (Ω_1, Ω_2).

3. Experimental Setup and Results

The experimental setup is shown in Figure 2, where the topology is performed partially in the optical domain and partially in the digital electronics and virtual domains.

Figure 2. Experimental setup. Inset: user graphical interface for Φ parameter. BLS: Broadband Light Source, IM: Intensity Modulator, PD: Photodetector.

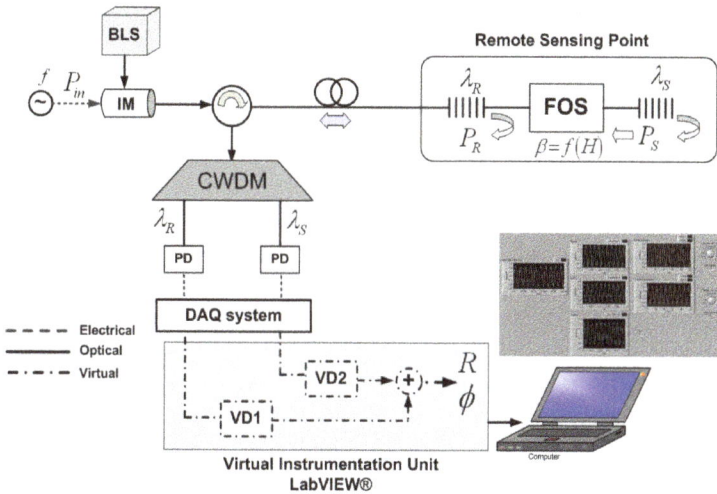

A broadband light source (BLS) modulated at a single frequency $f = 100$ Hz by means of an acousto-optic modulator (IM) is employed to launch optical power into the configuration through the broadband circulator. A 1 km-long SMF feeder fiber connecting the header with the remote sensing point was used for the experiment. A pair of low-cost FBGs is used at the remote sensing point, being placed before and after the FOS, thus obtaining a reference channel and sensing channel, respectively. The latter contains the power modulation induced in the FOS by the measurand. Their central wavelengths are $\lambda_R = 1{,}530.2\ nm$ and $\lambda_S = 1{,}550.1\ nm$ compatible with standard ITU G.694.2 for CWDM networks, see Figure 3. The optical signal is demultiplexed by a CWDM device and delivered to two different switchable gain InGaAs photodetectors. The Data Acquisition (DAQ) board includes an analog/digital converter (ADC) and performs the signal aggregation while delay line functionality and signal processing are achieved with virtual instrumentation techniques. All these elements are located at the reception stage. DAQ card performs a 14-bit Analog-to-Digital conversion for each analog input and a 48 kS/s of maximum aggregate sampling rate. Finally a PC with LabVIEW® software is used to control the system. The LabVIEW® control panel and user graphical interface for the output phase parameter Φ after signal acquisition from the DAQ board can be seen in Figure 2-inset.

Figure 3. Optical spectrum of both the reference (λ_R) and sensing (λ_S) channels before being sliced through the CWDM demultiplexer device, at different β values.

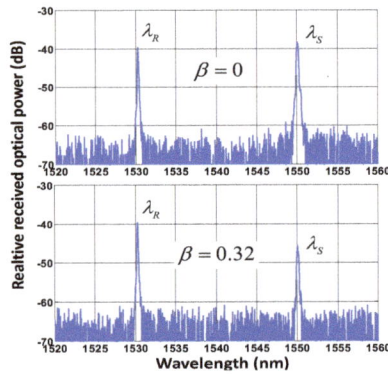

In the reception stage, each signal is compound by 240 samples as a compromise between the performance/limitations of the DAQ board. The displacement sensing system resolution is found to be 14 μm when measuring the output phase, and 2.1 μm (*i.e.*, amplitude resolution of 0.042 dB) when considering the parameter R. As shown in Figure 4, a 700 μm input full range (*i.e.*, Full Scale—F.S.) is considered. A better sensor resolution of 0.025 dB but in the same order of magnitude was obtained in the work reported in [16], where erbium doped fiber (EDF) amplification was used to enhance a frequency modulated continuous wave (FMCW) technique for referencing optical intensity sensors. Nevertheless, since the number of samples and the total sampling time provided by the DAQ board is the same, for a lower modulation frequency a better measurand resolution is expected. And if a greater resolution is required in our passive remote sensing topology, a DAQ system with a higher sampling rate should be used, assuming that there is no limitation in terms of optical power budget in our proposed topology. The 0.3% F.S. measurand resolution obtained is better than the values reported in [15], which were around 1.3% F.S., even though lower frequencies are used, of tenths of Hz. Resolution improvement is mainly due to the reflective operation and better R parameter performance of our sensing structure.

Figure 4. Calibration curve of the sensor loss modulation H for the taper-based displacement sensor.

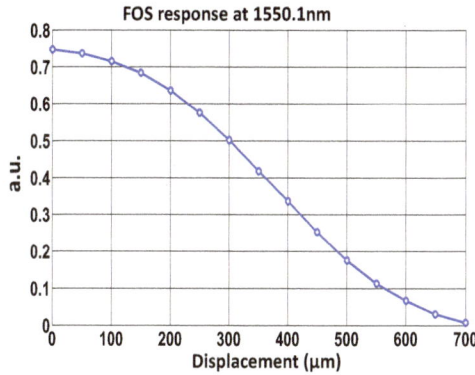

To test the concept, a SMF-based taper operating as a micro-displacement sensor is placed between each FBG with sensor loss modulation H. The taper was obtained by elongation of singlemode fiber during the arc discharge provided by a splicing machine in a semiautomatic fabrication process. Waist cross-section and waist length were 78 μm and 15 mm, respectively. Nevertheless better sensor sensitivities have been demonstrated when decreasing the waist diameter [17], even in combination with FBGs for strain measurements [18]. Once the taper is fixed onto a micro-positioning stage system, the optical loss transmission coefficient of the taper is sensitive to the displacement between the two fiber ends. Figure 4 shows the calibration curve of the sensor loss modulation H *versus* displacement [19].

Different calibration curves *versus* optical loss modulation of the sensor, β, were obtained, for different phase-shifting values. Results showed good agreement between theory and measurements, as can be seen in set of figures comprising Figure 5. Five measurements per each value of β were taken for the different phase-shifts selected thus obtaining the mean of the self-referencing parameters. In all cases, relative errors around 1%–3% were obtained. This fact assures good performance to guarantee the sensor interrogation and the self-referencing property. However, it can be seen that when operating at a phase-shift condition nearly $\Omega_i = \pm\pi$, R measurement stability decreases but not beyond 4%. A similar performance is achieved by using Φ parameter with the exception on those cases in which $\beta \approx 0$. Nevertheless, phase-shifting configurations could be chosen in order to improve any system feature depending on specific requirements. For instance, the blue curve of Figure 5(d) shows a linear regression coefficient of $r = 0.9994$ quite close to the unit with regards to R parameter performance. Phase-shifts requirements for the latter were $\Omega_1 = 0.58\pi$, $\Omega_2 = 0.98\pi$, going beyond previous works [19] in which phase-shifts around π radians caused a malfunction in the automation system.

In addition, the self-reference property of both measurement parameters was tested with regards to power fluctuations along the optical system. A singlemode variable optical attenuator (VOA) was located after the broadband circulator thus emulating unexpected power losses, up to 10 dB, in the fiber lead from the optical source to the remote sensing point. In Figure 6 no correlation between the measurements of both self-referencing parameters and the induced power attenuation is shown, as expected.

Figure 5. Experimental results and theoretical curves for both self-referencing parameters at different virtual phase-shifts, (**a**) and (**c**) R parameter; (**b**) and (**d**) Output phase Φ. Theoretical curves are drawn in solid lines. Figures inset: relative errors (in %) of the measurements taken.

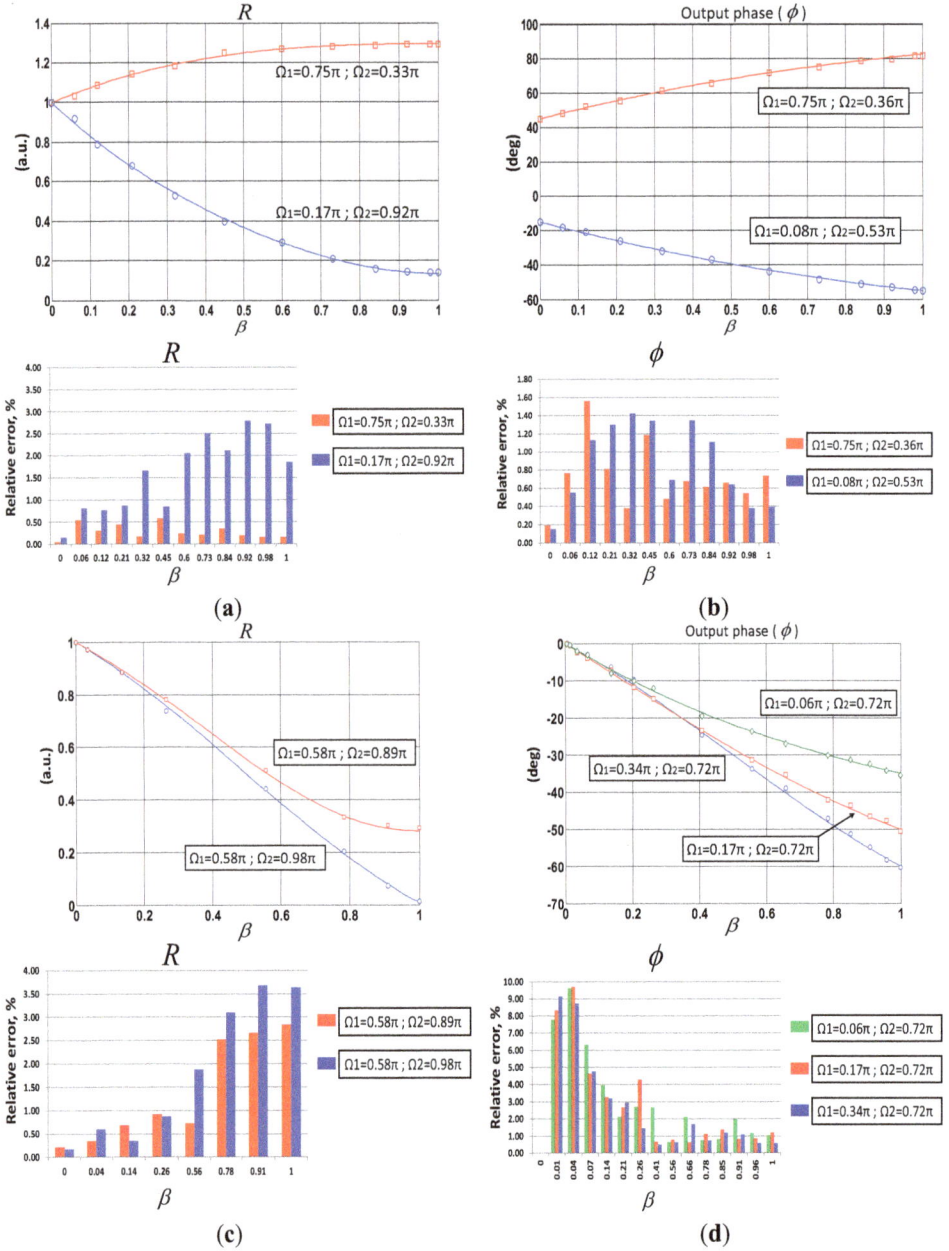

(a)

(b)

(c)

(d)

Figure 6. Self-reference test of both measurement parameters *versus* induced power losses (up to 10 dB) for different values of β. Figure Inset: relative error (%) of the measurements.

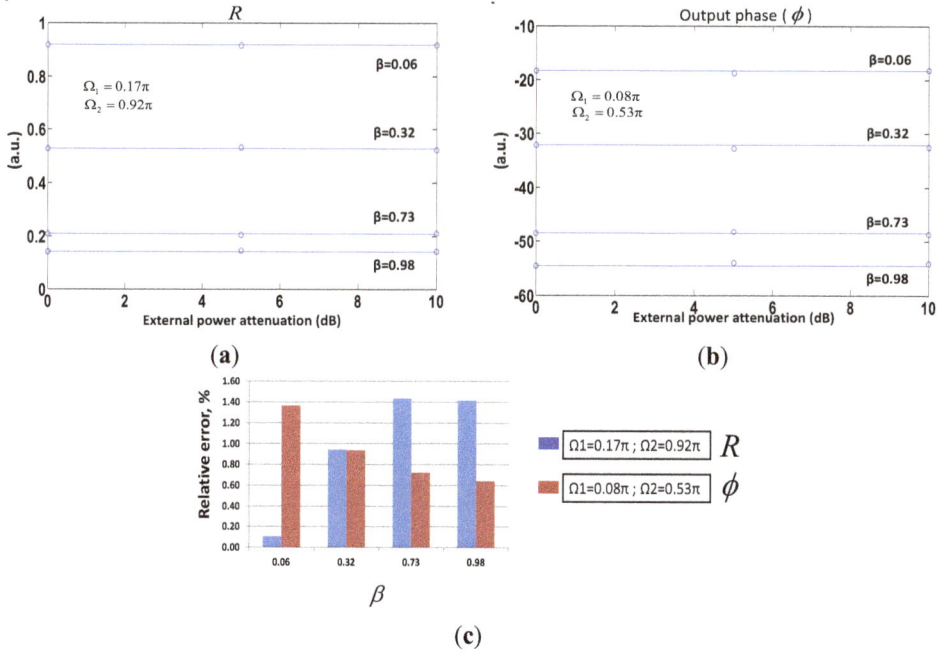

(a)

(b)

(c)

3.1. System Analysis

Considering the measurement results, in the following the optical power budget of the proposed reflective topology is analyzed. Table 1 shows the optical power analysis of the WDM-based remote sensing topology.

Table 1. Optical power budget analysis of the proposed CWDM-based remote sensing scheme.

BLS output power, P_{in}	PD sensitivity (dBm)	Device, Insertion Loss					
		IM, Att_{AOM} (dB)	Circulator, Att_{cir} (dB)	CWDM demux, Att_{CWDM} (dB)	Adapters *, Att_{con} (dB)	FBG (reflectivity), R_{FBG-i} (dB)	FOS [1], Att_{FOS} (dB)
1,530 nm (ref)	−15 dBm	10	0.6	1.8	0.5	3.4 (46%)	17
1,550 nm (sens)	−17.5 dBm	−60.5				3 (50%)	

* losses per connector; [1] measured at maximum input Full-Scale (F.S), worst case (β = 1).

The power budget of the system can be calculated using Equation (6):

$$P_{in}(dBm) = Att_{AOM}(dB) + Att_{cir}(dB) + \alpha(dB/km) \cdot L(km) + R_{FBG-ref} + Att_{FOS} + R_{FBG-sens}$$
$$+ \alpha(dB/km) \cdot L(km) + Att_{CWDM}(dB) + Att_{con}(dB) + P_{out}(dBm) \tag{6}$$

where $P_{in}(dBm)$ is the optical power launched into the system, $L(km)$ is the total length of the fiber lead connecting the remote sensing point, and $P_{out}(dBm)$ is the photodiode (PD) sensitivity.

The latter is directly related to the PD noise-equivalent power (NEP) figure of merit and the value provided in Table 1 refers to the worst case, in terms of both NEP and bandwidth.

Computing Equation (6) at both reference and sensing channels, and assuming an optical attenuation in singlemode fiber of 0.2 dB/km, $\alpha(dB/km)$, in the C-band as well as seven adapters within the link, a maximum length of 16 km could be obtained. Worst case was obtained when evaluating the sensing channel ($\lambda = 1,550$ nm). It is worth noticing that with 99% reflectance FBG at 1,550 nm, the reach could be extended up to 23 km.

4. Conclusions

In this work, a single radio-frequency self-referenced CWDM intensity-based fiber-optic sensor using virtual delay lines at the reception stage is presented. The sensor topology operates in reflective configuration and takes advantage of the use of FBGs and CWDM devices. It also allows a high scalability and an enhancement of the power budget as CWDM devices with low insertion losses are used for spectral splitting. With the proposed topology, and considering the available optical power provided by the optical source, a remote sensing distance of 16 km could be obtained. Improving the reflectivity of the FBG at the sensing channel a maximum reach of 23 km is estimated. The proposed virtual configuration avoids the deployment of physical delay lines, either in the optical and electrical domains, thus allowing an even more compact, cost-effective and easy-reconfigurable solution, through two delays, for sensor operation while keeping all the advantages of optical sensing. In addition, there is no need for a lock-in amplifier at the reception stage. Assuming no optical power budget limitation, system resolution and number of channels to be multiplexed are limited by DAQ board sampling rate and its ADC resolution, but easily improved. A micro-displacement sensor is used and two different self-referencing parameters have been tested. This solution can lead either to a higher automation in the interrogation of future WDM-based remote sensor networks or to a lower cost. It can also improve scalability and automation in monitoring system [20] of fault location or drop fibers optical power losses in WDM passive optical network (PON) architectures.

Acknowledgments

This work has been supported by Spanish CICyT projects TEC2012-37983-C03-02 and TEC2009-14718-C03-03 from the Spanish Ministry of Science. Authors want to thank Eng. Laura Valverde for her technical support with the virtual instrumentation software.

Conflicts of Interest

The authors declare no conflict of interest.

References

1. Montalvo, J.; Frazao, O.; Santos, J.L.; Vazquez, C.; Baptista, J.M. Radio-frequency self-referencing technique with enhanced sensitivity for coarse WDM fiber optic intensity sensors. *J. Lightwave Technol.* **2009**, *27*, 475–482.

2. Leandro, D.; Ullán, A.; Loayssa, A.; López-Higuera, J.M.; López-Amo, M. Remote (155 km) fiber bragg grating interrogation technique combining Raman, Brillouin, and erbium gain in a fiber laser. *IEEE Photon. Tech. Lett.* **2011**, *23*, 621–623.

3. Saitoh, T.; Nakamura, K.; Takahashi, Y.; Iida, H.; Iki, Y.; Miyagi, K. Ultra-long-distance fiber bragg grating sensor system. *IEEE Photon. Tech. Lett.* **2007**, *19*, 1616–1618.

4. Murtaza, G.; Senior, J.M. Methods for providing stable optical signals in dual wavelength referenced LED based sensors. *IEEE Photon. Tech. Lett.* **1994**, *6*, 1020–1022.

5. Wang, A.; Xiao, H.; Wang, J.; Wang, Z.; Zhao, W.; May, R.G. Self-calibrated interferometric intensity-based optical fiber sensors. *J. Lightwave Technol.* **2001**, *19*, 1495–1501.

6. Baptista, J.M.; Abad, S.; Rego, G.M.; Ferreira, L.A.; Araújo, F.M.; Santos, J.L.; Lage, A.S. Wavelength multiplexing of frequency-based self-referenced fiber optic intensity sensors. *Opt. Eng.* **2004**, *43*, 702–707.

7. Baptista, J.M.; Santos, J.L.; Lage, A.S. Self-referenced fibre optic intensity sensor based on a multiple beam Sagnac topology. *Opt. Commun.* **2000**, *181*, 287–294.

8. Dong, X.; Tam, H.Y.; Shum, P. Temperature-insensitive strain sensor with polarization maintaining photonic crystal fiber based Sagnac interferometer. *Appl. Phys. Lett.* **2007**, *90*, doi:10.1063/1.2722058.

9. Vázquez, C.; Montalvo, J.; Montero, D.S.; Pena, J.M.S. Self-referencing fiber-optic intensity sensors using ring resonators and fiber bragg gratings. *IEEE Photon. Tech. Lett.* **2006**, *18*, 2374–2376.

10. Spillman, W.B.; Lord, J.R. Self-referencing multiplexing technique for fiber-optic intensity sensors. *J. Lightwave Technol.* **1987**, *LT-5*, 865–869.

11. Caucheteur, C.; Mussot, A.; Bette, S.; Kudlinski, A.; Douay, M.; Louvergneaux, E.; Mégret, P.; Taki, M.; González-Herráez, M. All-fiber tunable optical delay line. *Opt. Express* **2010**, *18*, 3093–3100.

12. Abad, S.; López-Amo, M.; Araújo, F.M.; Ferreira, L.A.; Santos, J.L. Fiber Bragg grating-based self-referencing technique for wavelength-multiplexed intensity sensors. *Opt. Lett.* **2002**, *27*, 222–224.

13. Montalvo, J.; Araujo, F.M.; Ferreira, L.A.; Vazquez, C.; Baptista, J.M. Electrical FIR filter with optical coefficients for self-referencing WDM intensity sensors. *IEEE Photon. Tech. Lett.* **2008**, *20*, 45–47.

14. Montero, D.S.; Vázquez, C.; Baptista, J.M.; Santos, J.L.; Montalvo, J. Coarse WDM networking of self-referenced fiber-optic intensity sensors with reconfigurable characteristics. *Opt. Express* **2010**, *18*, 4396–4410.

15. Fernandes, A.J.G.; Jesus, C.; Jorge, P.A.S.; Baptista, J.M. Fiber optic intensity sensor referenced with a virtual delay line. *Opt. Commun.* **2011**, *284*, 5665–5668.

16. Pérez-Herrera, R.A.F.; Frazao, O.; Santos, J.L.; Araújo, F.M.; Ferreira, L.A.; Baptista, J.M.; López-Amo, M. Frequency modulated continuous wave system for optical fiber intensity sensors with optical amplification. *IEEE Sens. J.* **2009**, *9*, 1647–1653.

17. Arregui, F.J.; Matías, I.R.; López-Amo, M. Optical fiber strain gauge based on a tapered single-mode fiber. *Sens. Actuators A Phys.* **2000**, *79*, 90–96.

18. Frazão, O.; Silva, S.O.; Guerreiro, A.; Santos, J.L.; Ferreira, L.A.; Araújo, F.M. Strain sensitivity control of fiber Bragg grating structures with fused tapers. *Appl. Opt.* **2007**, *46*, 8578–8582.

19. Montero, D.S.; Vázquez, C. Interrogation of remote intensity-based fiber-optic sensors deploying delay lines in the virtual domain. *Proc. SPIE* **2012**, *8421*, doi: 10.1117/12.968580.

20. Montalvo, J.; Montero, D.S.; Vázquez, C.; Baptista, J.M.; Santos, J.L. Radio-frequency self-referencing system for monitoring drop fibres in wdm passive optical networks. *IET Optoelectron.* **2010**, *4*, 226–234.

Reprinted from *Sensors*. Cite as: Chang, Y.-T.; Yen, C.-T.; Wu, Y.-S.; Cheng, H.-C. Using a Fiber Loop and Fiber Bragg Grating as a Fiber Optic Sensor to Simultaneously Measure Temperature and Displacement. *Sensors* **2013**, *13*, 6542–6551.

Article

Using a Fiber Loop and Fiber Bragg Grating as a Fiber Optic Sensor to Simultaneously Measure Temperature and Displacement

Yao-Tang Chang [1], Chih-Ta Yen [2], Yue-Shiun Wu [3] and Hsu-Chih Cheng [3],*

[1] Department of Information Technology, Kao Yuan University, Kaohsiung City 821, Taiwan;
E-Mail: t10066@cc.kyu.edu.tw

[2] Department of Electrical Engineering, National Formosa University, Yunlin 632, Taiwan;
E-Mail: ctyen@nfu.edu.tw

[3] Department of Electro-Optical Engineering, National Formosa University, Yunlin 632, Taiwan;
E-Mail: chenghc@nfu.edu.tw

* Author to whom correspondence should be addressed; E-Mail: chenghc@nfu.edu.tw;
Tel.: +886-5-6315-657; Fax: +886-5-6329-257.

Received: 8 April 2013; in revised form: 18 April 2013 / Accepted: 8 May 2013 /
Published: 16 May 2013

Abstract: This study integrated a fiber loop manufactured by using commercial fiber (SMF-28, Corning) and a fiber Bragg grating (FBG) to form a fiber optic sensor that could simultaneously measure displacement and temperature. The fiber loop was placed in a thermoelectric cooling module with FBG affixed to the module, and, consequently, the center wavelength displacement of FBG was limited by only the effects of temperature change. Displacement and temperature were determined by measuring changes in the transmission of optical power and shifts in Bragg wavelength. This study provides a simple and economical method to measure displacement and temperature simultaneously.

Keywords: fiber Bragg grating; fiber loop; temperature; displacement

1. Introduction

Recently, with the rapid development of optical communications, numerous active and passive optical components have been developed, including wavelength division multiplexers (WDMs) and fiber Bragg gratings (FBGs). Fiber-optic sensors possess several advantages, including small size, light weight, high bandwidth, the ability to measure multiple points, the ability to withstand electromagnetic interference, and high operating temperature range. Fiber optic sensors have been used for various applications, such as medical endoscopes, chemical solution concentrations, and the measurement of the area of important components inside the mechanical precision measurement system.

A fiber optic sensor employs light to carry messages; this light passes through an optical fiber to a sensing area. The resulting changes in physical quantities in the sensing area (e.g., changes in strain, temperature, or refractive indices) can cause optical signals to transform, including the wavelength, power, and phases of these signals. Transformations in optical signals, in turn, can be employed to measure physical quantities. FBG is often used in fiber optic sensors and is developed by shining UV laser light on the fiber core with sub-micron modulations of the refractive index. This establishment of FBG can be achieved using various technologies, among which the phase masking technique is the most popular. Because of the invention of Bragg grating, fiber optic sensors can be developed for more numerous applications. For example, sensor heads can be used to measure a wide variety of physical quantities (e.g., strain, temperature, vibration, pressure, acceleration, and voltage) and can be applied for sensor multiplexing and sensor signal processing. In addition, multi-parameter fiber-sensing techniques are attractive for various reasons, including low loss, high bandwidth, immunity to electromagnetic interference (EMI), small size, light weight, safety, comparatively lower cost, and low maintenance requirements [1–7].

Although FBGs possess numerous advantages and wide applications, FBG measurements encounter a problem in that reflection wavelength displacement FBG is caused in the because of changes in strain and temperature. Thus, when the FBG is used as a sensor head, frequent and simultaneous changes in temperature and strain can lead to the integration of temperature and strain in the FBG, causing the two factors to become indistinguishable to the grating. The easiest way to overcome this problem and separately measure temperature is to employ special packaging around the sensing element; in this way, the inside of FBG does not cause wavelength drift because of changes in external strains. Patrick *et al.* [2] placed an FBG in the hollow of a steel casing or silica tube, with both sides of the FBG affixed to the casing using high-temperature resistant epoxy glue; as a result, FBG was isolated from strain. However, if both temperature and pressure were to be determined simultaneously, two FBGs with special settings or other auxiliary components were required. Lo [3] attached a single FBG to a cantilever by first pulling on both ends of the grating before releasing them after the FBG had been affixed to the cantilever. Reflection wavelength from non-adhered FBG was used to measure temperature change, and adhered grating was used to measure strain as well as temperature change. Temperature change measured using non-adhered grating was subtracted from the center wavelength displacement measured using the adhered grating to achieve a simultaneous measurement of temperature and strain. Jung *et al.* [4] proposed that FBG could be used with an erbium-doped fiber amplifier (EDFA) to simultaneously measure temperature and strain. This system employed declines in linear changes of spontaneous radiated power from an EDFA and increases in temperature to verify temperature measurements. Strain was verified by subtracting the temperature-affected grating wavelength displacement. In the study by [5], a special distributed-feedback fiber laser with four modes was used to measure strain and temperature simultaneously, which is smart and compact. However, this system had several drawbacks; for example, a fiber laser with four modes is difficult to fabricate, and an expensive high-frequency photodetector (PD) and electrical spectrum analyzer (ESA) are required. In a later study, the authors extended the basic concept to the simultaneous measurements of strain and temperature distributions using four FBGs [6]. Although the methods proposed in [6] proved successful, the computational time of the genetic algorithm was lengthy because examining the reflection intensity spectra of two or more FBGs was necessary to obtain the arbitrary strain and temperature profiles. Lai *et al.* proposed obtaining simultaneous measurements of the level and specific gravity of a liquid by using a dual-optical-fiber-sensor system comprising a fiber Bragg grating (FBG) level sensor and a Fabry-Pérot (FP) pressure sensor [7]. The experimental results showed that the average measurement errors of the dual-sensor system for the liquid level and specific gravity were 0.0323 and 0.0528 m, respectively.

Polymer FBGs have recently been widely used in fiber strain-sensor systems because of their low Young modulus and the high elasticity limit of the polymer fibers [8]. A simple fiber-strain sensor was presented using two cascaded polymer FBGs to cancel the thermal effect in the experiment; it successfully obtained the temperature-free strain recovery results and demonstrated the high wavelength tune range (12 nm with 2.25% strain) of the polymer FBGs [8]. The wavelength-based detection method offers various advantages, including simplicity, more economical polymer FBGs, and robustness in the optical-source power variation. In [9], the authors compared the polymer FBG accelerometer and silica-based FBG accelerometer. They showed that the polymer FBG accelerometer has superior sensitivity to the silica-based FBG. A temperature-compensated strain sensor can also be devised based on the wavelength detection methods by using two cascaded Sagnac interferometers and solid birefringent hybrid photonic crystal fibers (PCFs) [10]. The experimental results showed the high-strain and low-temperature sensitivities. However, the optical path difference of the two Sagnac interferometers must be well aligned, and using PCFs is more costly compared to previous methods employing FBGs.

In this study, a fiber loop, manufactured by using commercial fiber (SMF-28, Corning), and FBG were combined to form a fiber-optic sensor capable of the simultaneous measurement of displacement and temperature. The principle of optical fiber bending loss and temperature-affected FBG wavelength, which possess excellent linear change characteristics (as shown in the spectrogram), were used to measure changes in optical power loss and wavelength displacement and produce a simultaneous measurement of displacement and temperature. Compared to [8], the proposed method used a cascaded fiber loop and the FBG to achieve the simultaneous measurement of displacement and temperature; this is more economical and simpler because a fiber loop (manufactured using commercial fiber) is used. Moreover, the architecture of the proposed method is less complex for the fiber-strain sensor of the cascaded Sagnac interferometers [11].

2. Sensing Principles and Architecture

In this study, we employed a combination of FBG and a fiber loop to achieve a simultaneous measurement of temperature and displacement. The grating period of FBG was changed using temperature and strain, which caused displacement of the reflected Bragg wavelength. The equation demonstrating this relationship is as follows:

$$\Delta\lambda_B = 2n\Delta\Lambda \tag{1}$$

where $\Delta\lambda_B$ is the change in Bragg wavelength, n is the effective refractive index, and $\Delta\Lambda$ is the change in grating period. Because this study employed FBG to measure temperature, Bragg wavelength was affected by the thermal expansion coefficient of grating and the thermal breakdown of optical fiber [3]. The equation demonstrating this relationship is as follows:

$$\lambda_B = \lambda_{B0} + \lambda_{B0}(\xi + \alpha_F)\Delta T \tag{2}$$

where ξ is the thermo-optic coefficient of fiber material, α_F is the thermal expansion coefficient of FBG, and ΔT is changing temperature.

To test the practicality of our sensors, we affixed the FBG to a thermoelectric cooling module to isolate the effect of FBG displacement, as the architecture diagram in Figure 1 shows. Because experimental settings were employed, factors affecting the temperature sensitivity coefficient for FBG varied. If FBG was not affixed or tied to a particular object, factors affecting the temperature sensitivity coefficient included the thermo-optic coefficient (ζ) and the fiber thermal expansion coefficient (α_F). Because, during the experiment, both ends of the FBG were affixed to the thermoelectric cooling module, the thermal expansion coefficient (α_S) for the thermoelectric cooling module was also a factor affecting the temperature sensitivity coefficient for FBG. Therefore, the relational equation for temperature and FBG wavelength can be obtained by modifying Equation (2) to become the following equation:

$$\lambda_B = \lambda_{B0} + \lambda_{B0}(\xi + \alpha_F + \alpha_S)\Delta T \tag{3}$$

Figure 1. Experimental architecture diagram.

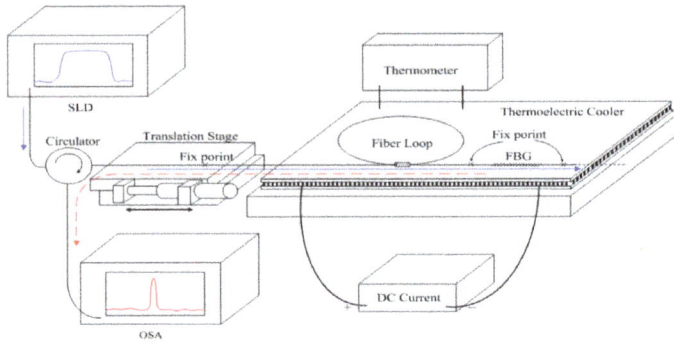

Figure 2 shows center FBG wavelengths of 1,547.78 nm, 1,547.88 nm, 1,548.1 nm, and 1,548.31 nm (shown from left to right), which correspond to temperatures of 20 °C, 30 °C, 50 °C, and 70 °C, respectively. The fiber loop used in this study was composed of single-mode bare optical fiber with a circumference of 9 cm, which was wrapped in a 3 mm to 5 mm long plastic sleeve at the overlap area of the loop. Consequently, when either end of the fiber loop was subjected to tension, the loop maintained its shape, despite changes in size. The fiber loop was installed in a thermoelectric cooling module. One end of the loop was affixed to a mobile platform with a resolution of 10 μm, and one end was affixed to the thermoelectric cooling module. When the fiber in a loop is bent to a certain degree, the amount of signal lost increases in conjunction with increases in the degree to which a light signal is bent. This is the principle used when measuring displacement for a fiber loop. Assuming that light traveling through the bent section of the fiber loop causes optical power loss, ideal single-mode optical fiber output power is defined as follows [11]:

$$P_{out} = P_{in} \exp(-2\alpha_B l^e) \tag{4}$$

In this equation $l^e = 2\pi R^e$, and R^e is the effective bending radius that accounts for bending stress in the reflection coefficient of the fiber core material; therefore, R^e is different from the actual diameter of the fiber loop. $2\alpha_B$ represents the bending loss coefficient per unit length. When displacement is very small, the exponential function of displacement (Δd) in Equation (5) can be considered a linear function of Δd. Therefore, the relationship between optical power change and displacement can be rewritten as the following equation:

$$\Delta P = -2\alpha_B P_{in} \Delta d \tag{5}$$

The results of Equations (4) and (5) show that the optical power change (ΔP) of the fiber loop is proportionally to the displacement (Δd).

Fiber must bend only to a certain degree before significant optical power loss occurs. However, this does not imply that further bending of fiber will lead to greater optical loss because light refracted in the fiber structure may be reflected back to the core layer, leading to increases in measured optical power. Therefore, linear optical power loss for a fiber loop must occur in a certain range.

Figure 2. Center wavelength of the FBG at various temperatures.

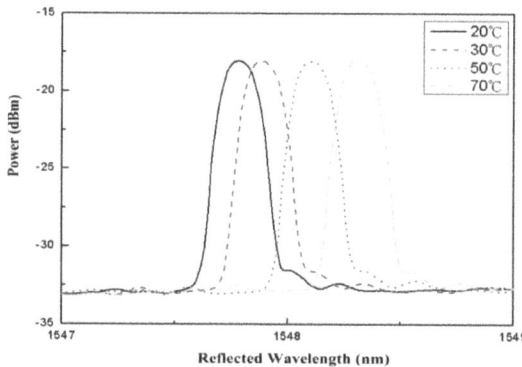

The fiber loop was manufactured using commercial fiber (SMF-28, Corning). Figure 3 shows the decrease, at room temperature, in the circumference of a fiber loop from 9 cm to 4.5 cm at a displacement increment of 0.5 mm as changes in optical power for the loop are measured and observed. The figure shows that fiber loop circumference in the ranges of approximately 6.35 cm–6.75 cm and 4.9 cm–5.8 cm coincide with locations in which power loss occurs in Figure 3. However, optical power loss in the second range is much greater because of temperature change, and, consequently, temperature and displacement are difficult to distinguish at this range. Thus, circumferences in the range of approximately 6.35 cm–6.75 cm were chosen as the application range for this study. Matrices were employed to represent the effect of grating wavelengths on changes in temperature and the effect of optical power on changes in displacement and temperature, as shown in the following equation:

$$\begin{bmatrix} \lambda_B \\ P \end{bmatrix} = \begin{bmatrix} k_{T1} & k_{d1} \\ k_{T2} & k_{d2} \end{bmatrix} \begin{bmatrix} \Delta T \\ \Delta d \end{bmatrix} + \begin{bmatrix} \lambda_0 \\ P_0 \end{bmatrix} \tag{6}$$

where λ_B and P represent the measured Bragg wavelength and optical power, respectively; k_{T1} and k_{T2} represent the temperature change sensitivity for FBG and the fiber loop, respectively; k_{d1} and k_{d2} represent the displacement change sensitivity for FBG and the fiber loop, respectively; and λ_{B0} and P_0 represent the grating reflection wavelength and reflected power, respectively, before any change in temperature or displacement occurs. Because of the setup for the experiment, FBG was not affected by displacement; consequently, k_{d1} can be reduced to 0. Therefore, Equation (6) can be converted to the following:

$$\begin{bmatrix} \lambda_B \\ P \end{bmatrix} = \begin{bmatrix} k_{T1} & 0 \\ k_{T2} & k_{d2} \end{bmatrix} \begin{bmatrix} \Delta T \\ \Delta d \end{bmatrix} + \begin{bmatrix} \lambda_{B0} \\ P_0 \end{bmatrix} \tag{7}$$

Figure 3. Changes in optical power for a fiber loop at various circumferences.

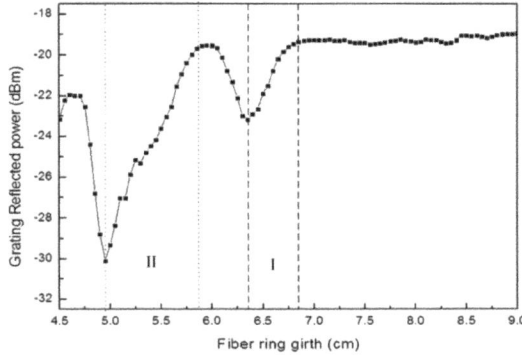

If temperature and displacement sensitivity coefficients are known, Equation (7) can be employed to calculate amounts of temperature change and displacement as well as measure temperature and displacement. The relationship matrix for change in wavelength and power loss and temperature and displacement is expressed in Equation (8), where $D = k_{T1}k_{d2}$:

$$\begin{bmatrix} \Delta T \\ \Delta d \end{bmatrix} = \begin{bmatrix} k_{d2} & 0 \\ -k_{T2} & k_{T1} \end{bmatrix} \begin{bmatrix} \Delta \lambda \\ \Delta P \end{bmatrix} \tag{8}$$

3. Experimental Results

During the experimental measurements, a superluminescent diode (SLD) (ASLD155-200, Amonics) was used as the light source, and an Anritsu MS9710C was used as the spectrometer. The measured spectral range was 600 nm to 1,750 nm, and 0.1 nm was chosen as the resolution setting for the spectrometer. We first re-evaluated the temperature sensitivity coefficient for FBG, and ensured that both ends of the FBG were fixed to the thermoelectric cooling module and not affected by displacement effects. After these procedures, the temperature sensitivity coefficient for

FBG was relatively easy to obtain. Power supply was used to provide currents to the thermoelectric cooling module, and current size was used to control temperature. The experiment employed 1,547.81 nm as the grating center wavelength and 0.21 nm as the bandwidth.

The dots in Figure 4 represent various FBG wavelengths within a temperature range of approximately 20 °C to 80 °C. This 60-°C change in temperature resulted in a wavelength shift from 1,547.74 nm to 1,548.46 nm. The line portion represents the use of linear equations to simulate the curves closest to the wavelength displacements. Most wavelength displacements obtained 0.0117 nm/°C as the temperature sensitivity coefficient (k_{T1}) for FBG. After temperature coefficients for FBG were obtained, temperature and displacement sensitivity coefficients for the fiber loop were determined. To obtain temperature sensitivity coefficients for different temperatures, displacement was rendered constant. By contrast, to obtain displacement sensitivity coefficients, temperature was rendered constant, with only displacement being allowed to change. Figure 5 shows optical power change at a displacement increment of 50 μm for various temperature environments for the fiber. The displacement sensitivity coefficient (k_{d2}) for the fiber loop was −0.00274 dBm/μm. Figure 6 shows optical power change for the fiber loop at various temperatures, with displacement amounts maintained at 150 μm, 300 μm, and 500 μm. The obtained temperature sensitivity coefficient (k_{T2}) was 0.0044 dBm/°C.

Figure 4. Changes in the center wavelength of FBG at differing temperatures.

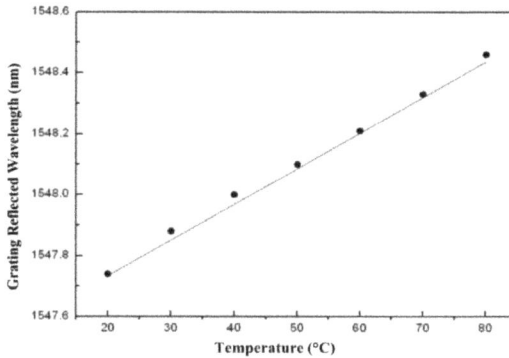

Figure 5. Changes in optical power as displacement increased at 50-μm increments at an FBG temperature of 70 °C.

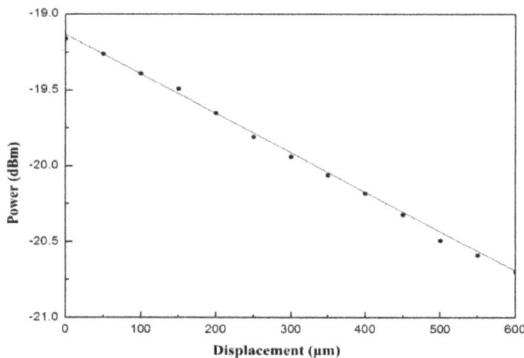

94

Figure 6. Changes in sensor optical power with displacement maintained at 150 μm, 300 μm, and 500 μm as different temperatures were applied.

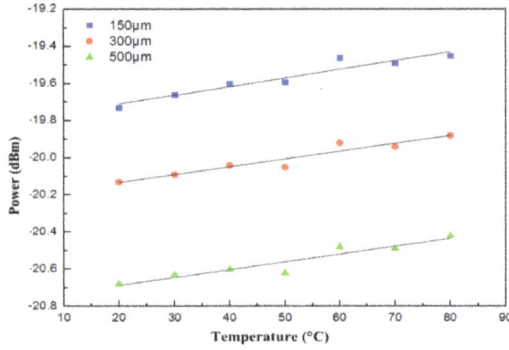

Displacement and temperature sensitivity coefficients were obtained for the fiber loop and FBG based on the experiment conducted in this study. We then used these coefficients to create the relationship matrix for optical power change and wavelength displacement and displacement and temperature change:

$$\begin{bmatrix} \Delta T \\ \Delta d \end{bmatrix} = \begin{bmatrix} 0.0117 & 0 \\ 0.0044 & -0.00274 \end{bmatrix} \begin{bmatrix} \Delta\lambda \\ \Delta P \end{bmatrix} \tag{9}$$

$$\begin{bmatrix} \Delta T \\ \Delta d \end{bmatrix} = \frac{1}{-3.21\times10^{-5}} \begin{bmatrix} -0.00274 & 0 \\ -0.0044 & 0.0117 \end{bmatrix} \begin{bmatrix} \Delta\lambda \\ \Delta P \end{bmatrix} \tag{10}$$

In the present study, we simultaneously measured temperature and displacement by first measuring the center wavelength displacement of FBG to determine changes in temperature. The effects of both temperature and displacement on the fiber loop caused it to lose optical power. However, displacement amounts can be measured by subtracting detected temperature change from center wavelength displacement. Therefore, the experiment simultaneously measured temperature and displacement through measurements of optical power and wavelength displacement. Figure 7 shows a comparison between the measured and applied values of temperature and displacement for the sensor at various temperature conditions.

Figure 7. Comparison of actual displacement and temperature and measured displacement and temperature at various temperatures.

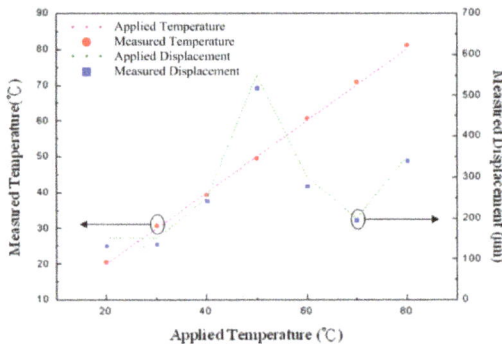

When varying amounts of displacement for each temperature level were exerted, the root mean square errors for temperature and displacement for temperatures ranging from 20 °C to 80 °C were ±10.57 μm and ±0.78 °C, respectively. These results demonstrate that both temperature and displacement measurements were consistent with the actual temperatures and displacements applied to a sensor. Therefore, the experimental setup used to simultaneously measure temperature and displacement is feasible for real situations.

4. Conclusions

This study introduced the basic principles of FBG and the reasoning behind wavelength drift caused by the effects of temperature and strain on the center wavelength of FBG. The drifting of the center wavelength results from the influence of changes during the grating period, or from changes in temperature that lead to transformations during the grating period because of heating or cooling as well as changes in the effective refractive index at the core of FBG. Although change to FBG caused by temperature and strain is linear, other settings are required to distinguish the effect of temperature on FBG. Thus, in the present study, we combined a fiber loop and FBG in a sensor head to simultaneously measure temperature and displacement. We first measured changes in temperature using the displacement of the center wavelength of FBG. Factors that caused optical power loss included temperature and displacement. By simply subtracting the effect of temperature, which is measured using center wavelength displacement, from optical power loss, the sensitivity coefficient for optical power loss in the fiber loop and displacement could be obtained. Therefore, changes in temperature and displacement can be determined simultaneously by measuring center wavelength displacement and optical power loss. Finally, we verified that sensors demonstrate varying amounts of displacement at different temperatures. Results from comparing measured values against actual values showed that our proposed sensor setting is feasible for practical application to simultaneously measure temperature and displacement. The root mean square error for temperature and displacement was ± 0.78 °C and ± 10.57 μm, respectively.

Conflicts of Interest

The authors declare no conflict of interest.

References

1. Frazão, O.; Ferreira, L.A.; Araújo, F.M.; Santos, J.L. Applications of fiber optic grating technology to multi-parameter measurement. *Fiber Integrated Opt.* **2005**, *24*, 227–244.
2. Patrick, H.; Williams, G.M.; Kersey, A.D.; Pedrazzani, J.R.; Vengsarkar, A.M. Thermally compensated bending gauge using surface-mounted fiber gratings. *Int. J. Optoelectronics* **1994**, *9*, 281–283.
3. Lo, Y.L. Using in-fiber Bragg-grating sensors for measuring axial strain and temperature simultaneously on surface of structures. *Opt. Eng.* **1998**, *37*, 2272–2276.
4. Jung, J.; Nam, H.; Lee, J.H.; Park, N.; Lee, B. Simultaneous measurement of strain and temperature by use of a single-fiber Bragg grating and an erbium-doped fiber amplifier. *Appl. Opt.* **1999**, *38*, 2749–2751.

5. Hadeler, O.; Ibsen, M.; Zervas, M.N. Distributed-Feedback fiber laser sensor for simultaneous strain and temperature measurements operating in the radio-frequency domain. *Appl. Opt.* **2001**, *40*, 3169–3175.

6. Cheng, H.C.; Huang, J.F.; Lo, Y.L. Simultaneous strain and temperature distribution sensing using two fiber Bragg grating pairs and a genetic algorithm. *Opt. Fiber Technol.* **2006**, *12*, 340–349.

7. Lai, C.W.; Lo, Y.L.; Yur, J.P.; Chuang, C.H. Application of fiber Bragg grating level sensor and Fabry-Perot pressure sensor to simultaneous measurement of liquid level and specific gravity. *IEEE Sens. J.* **2012**, *12*, 827–831.

8. Yuan, W.; Stefani, A.; Bang, O. Tunable polymer fiber Bragg grating (fbg) inscription: fabrication of dual-fbg temperature compensated polymer optical fiber strain sensors. *IEEE Photon. Technol. Lett.* **2012**, *24*, 401–403.

9. Stefani, A.; Andresen, S.; Yuan, W.; Herholdt-Rasmussen, N.; Bang, O. High sensitivity polymer optical fiber-Bragg-grating-based accelerometer. *IEEE Photon. Technol. Lett.* **2012**, *24*, 763–765.

10. Gu, B.; Yuan, W.; He, S.; Bang, O. Temperature compensated strain sensor based on cascaded Sagnac interferometers and all-solid birefringent hybrid photonic crystal fibers. *IEEE Sens. J.* **2012**, *12*, 1641–1646.

11. Nguyen, N.Q.; Gupta, N. Power modulation based fiber-optic loop-sensor having a dual measurement range. *J. Appl. Phys.* **2009**, *106*, doi:10.1063/1.3187917.

Reprinted from *Sensors*. Cite as: García, I.; Beloki, J.; Zubia, J.; Aldabaldetreku, G.; Illarramendi, M.A.; Jiménez, F. An Optical Fiber Bundle Sensor for Tip Clearance and Tip Timing Measurements in a Turbine Rig. *Sensors* **2013**, *13*, 7385–7398.

Article

An Optical Fiber Bundle Sensor for Tip Clearance and Tip Timing Measurements in a Turbine Rig

Iker García [1,*], **Josu Beloki** [2], **Joseba Zubia** [1], **Gotzon Aldabaldetreku** [1], **María Asunción Illarramendi** [3] and **Felipe Jiménez** [4]

[1] Department of Communications Engineering, University of the Basque Country, Alda. Urquijo s/n Bilbao 48013, Spain; E-Mails: joseba.zubia@ehu.es (J.Z.); gotzon.aldabaldetreku@ehu.es (G.A.)

[2] CTA, Centro de Tecnologías Aeronáuticas, Parque tecnológico de Bizkaia, Edif. 303, Zamudio 48170, Spain; E-Mail: josu.beloki@ctabef.com

[3] Department of Applied Physics I, University of the Basque Country, Alda. Urquijo s/n Bilbao 48013, Spain; E-Mail: ma.illarramendi@ehu.es

[4] Department of Applied Mathematics, University of the Basque Country, Alda. Urquijo s/n Bilbao 48013, Spain; E-Mail: felipe.jimenez@ehu.es

* Author to whom correspondence should be addressed; E-Mail: iker.garciae@ehu.es; Tel.: +34-946-017-305; Fax: +34-946-014-259.

Received: 9 April 2013; in revised form: 16 May 2013 / Accepted: 29 May 2013 / Published: 5 June 2013

Abstract: When it comes to measuring blade-tip clearance or blade-tip timing in turbines, reflective intensity-modulated optical fiber sensors overcome several traditional limitations of capacitive, inductive or discharging probe sensors. This paper presents the signals and results corresponding to the third stage of a multistage turbine rig, obtained from a transonic wind-tunnel test. The probe is based on a trifurcated bundle of optical fibers that is mounted on the turbine casing. To eliminate the influence of light source intensity variations and blade surface reflectivity, the sensing principle is based on the quotient of the voltages obtained from the two receiving bundle legs. A discrepancy lower than 3% with respect to a commercial sensor was observed in tip clearance measurements. Regarding tip timing measurements, the travel wave spectrum was obtained, which provides the average vibration amplitude for all blades at a particular nodal diameter. With this approach, both blade-tip timing and tip clearance measurements can be carried out simultaneously. The results obtained on the test turbine rig demonstrate the suitability and reliability of the type of sensor used, and

98

suggest the possibility of performing these measurements in real turbines under real working conditions.

Keywords: optical fiber bundle; real operating conditions measurement; tip clearance; tip timing; turbine

---

1. Introduction

Blade-tip timing (BTT) and tip clearance (TC) are two critical parameters for turbine engineering, since their measurement and optimization lead to more effective, secure and reliable engines [1]. BTT is a technique for blade vibration measurements that uses the differences between real and theoretical blade arrival times to calculate the vibration amplitude of the blade. Pioneering works in the BTT technique were performed in the 70 s and during the last forty years ample research has been developed and published related to this technique. Some examples of these works, chronologically ordered, can be found in References [2–11]. Blade vibration measurements are crucial to assess turbine operation and predict blade failures due to fatigue [12]. Blade vibrations can occur due to different causes. For instance, combustors or stators can produce synchronous responses in the rotor blades, or even irregularities in the casing or in the intake geometry can produce non-regular pressure distributions that lead to synchronous responses in the rotor blades. In contrast, rotating stall or adverse flow-blade interaction with negative aero-damping can cause non-synchronous responses, such as flutter, which consists of a self-sustained aerodynamic instability. In order to predict the lifetime of blades and to prevent damages that can lead to huge repairing costs or even to engine destruction, a low-cost and effective blade vibration system is needed. The BTT technique fulfils both requirements detecting all blade vibrations.

Blade vibrations are usually measured with strain gauges. These devices can be used provided that the influence of their mass is negligible. Results are provided only for the blades on which gauges are mounted and the signals from the gauges have to be transmitted by telemetry or a slip ring. Despite their proven suitability, strain gauges require considerable instrumentation, their use is restricted to a few blades of the turbine and they are in physical contact with the blades. A second standard method is the frequency modulated grid system [13]. This method is based on small permanent magnets attached to the tips of some blades that induce an electromagnetic voltage in a wire installed in the casing. Its disadvantages are similar to those of the strain gauges [14].

As to the TC, it is defined as the distance between the blade-tip and the engine casing, and its usual values are 2–3 mm. This distance is one of the factors on which engine efficiency depends, as the latter increases as TC decreases. A high TC allows an amount of air to flow without generating a useful work, whereas a lack of clearance can put engine integrity at risk. A 0.25-mm-TC reduction is equivalent to a reduction of 1% in specific fuel consumption, and of 10 °C in exhaust gas temperature [15]. As a consequence, the engine works at lower temperatures and the life cycle of its components is increased. In addition to the economical benefit, aircraft noise and emissions are also reduced, implying additional environmental advantages.

For the measurement of TC, traditional methods make use of capacitive, eddy current and discharging probe sensors. Capacitive sensors are simple and inexpensive but they have a poor frequency response and require iron blades. Eddy current sensors provide non-contact measurements, but the magnetic disturbance of the turbine engine can interfere with their output. In addition, they need to be calibrated in advance, because they are highly dependent on the tip shape and temperature [16]. Finally, discharging probes, in the same way as eddy current sensors, require conducting blades and they only measure the shortest clearance. On the contrary, optical fiber sensors provide small size and simplicity, non-contact measurements and simple instrumentation, high sensitivity, resolution and bandwidth, insensitivity to electromagnetic interference and measurement of every blade [17]. In this paper we will show the results obtained for the BTT and TC measurement using an optical fiber bundle sensor.

2. Experimental Section

All tests and measurements took place in the Aeronautical Technologies Center (CTA) facilities and were performed in a turbine rig with a rotor of 146 blades. The main component of the sensor is a trifurcated optical fiber bundle (manufactured *ad-hoc* by Fiberguide Industries, Stirling, NJ, USA, according to our requirements). The bundle structure is depicted in Figure 1: its length is 3 m and it consists of multimode glass optical fibers. Optical fibers temperature operating range goes from −190 to 350 °C, whereas the maximum temperature reached during the measurements is 88.1 °C.

Figure 1. Trifurcated optical fiber bundle and legs cross-section.

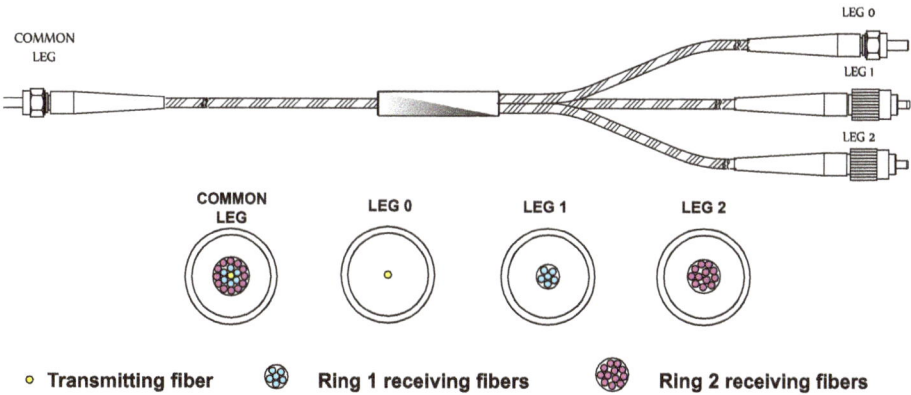

Figure 2. Microscope image of the cross-section of the common leg.

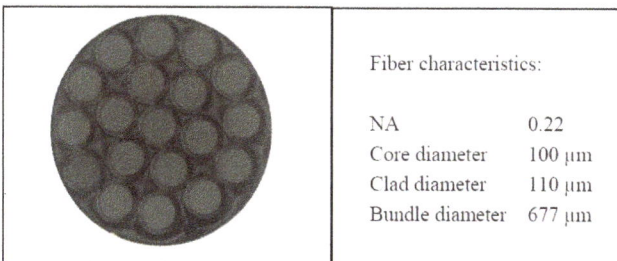

Fiber characteristics:

NA	0.22
Core diameter	100 μm
Clad diameter	110 μm
Bundle diameter	677 μm

Figure 2 shows a microscope image of the cross-section of the common leg. The central fiber is the transmitting fiber which guides the light from the laser to the probe end in order to illuminate the blade. The reflected light is collected by two receiving fiber rings around the transmitting fiber. The inner ring is formed by six fibers that are gathered in leg 1 and the outer ring is composed of twelve fibers gathered in leg 2.

A trifurcated bundle is chosen to eliminate the effects of variations in the light source, reflectivity of the blade surface, and optical losses and misalignments between the probe and the target surface [16,18]. This is possible because the distance to the target is obtained as a function of the quotient of two photodetector voltages (V_1 and V_2), so the influence of any previous disturbance is cancelled. To evaluate the reflected light irradiance collected by each ring of receiving fibers, let us call I_0 the light irradiance leaving the transmitting fiber in the common leg of the bundle. Then the optical irradiance at the end of legs 1 and 2 can be expressed as:

$$I_1 = K_0 R I_0 K_1 F_1(d) \tag{1}$$

$$I_2 = K_0 R I_0 K_2 F_2(d) \tag{2}$$

where R represents the reflectivity of the blade, the coefficients K_1 and K_2 account for the losses in the corresponding receiving fibers, each function $F_1(d)$ and $F_2(d)$ represents the relationship between the collected irradiance and the target distance for each group of fibers considered as a bifurcated bundle [19–24], and K_0 is a factor that accounts for laser fluctuations.

Dividing both equations we obtain:

$$I_2 / I_1 = K_2 F_2(d) / K_1 F_1(d) = K F(d) \tag{3}$$

Therefore, the quotient of the irradiances only depends on a constant related to the losses in the optical fibers and it is a function of the distance to the illuminated blade.

We employ a 655-nm-wavelength laser (FP-65 7FE-SMA, Laser Components, Olching, Germany) as the light source with a typical power of 7 mW. The laser is coupled to leg 0. Two identical photodetectors (PDA100A-EC, Thorlabs, Dachau, Germany) are connected to the end of legs 1 and 2, which convert the reflected light collected by each ring into a voltage signal. The photodetectors consist of a reverse-biased PIN photodiode and a switchable gain transimpedance amplifier. The adjustable gain range goes from 0 to 70 dB, and the bandwidth decreases from 1.5 MHz to 2 kHz as the gain increases. The maximum output signal amplitude for a 50 Ω load is 5 V through a BNC connector. To acquire the output signal an Agilent Technologies (Santa Clara, CA, USA) Infinium MSO9104A oscilloscope was used. It has four analog channels of 1 GHz and a maximum sample rate of 20 Gsa/s. All these components are depicted in Figure 3a with the exception of the oscilloscope, which was placed at the control room due to the high temperatures reached. In order to carry the signal from the photodetectors to the oscilloscope and reduce the noise of the engine, two 25 m long double-shielded coaxial cables were used.

The first problem we encountered during the assembly was how to couple the common leg SMA connector to the casing of the turbine. It was solved using a threaded tip in order to fix the SMA connector to a coupler which was inserted in a probe hole, shown in Figure 3b, that placed the bundle tip at a distance of 0.45 mm from the inner face of the casing.

Figure 3. (a) Experimental setup for the tip clearance and tip timing measurements in the turbine rig at the CTA facilities. (b) Coupler to fix the bundle tip to the turbine casing.

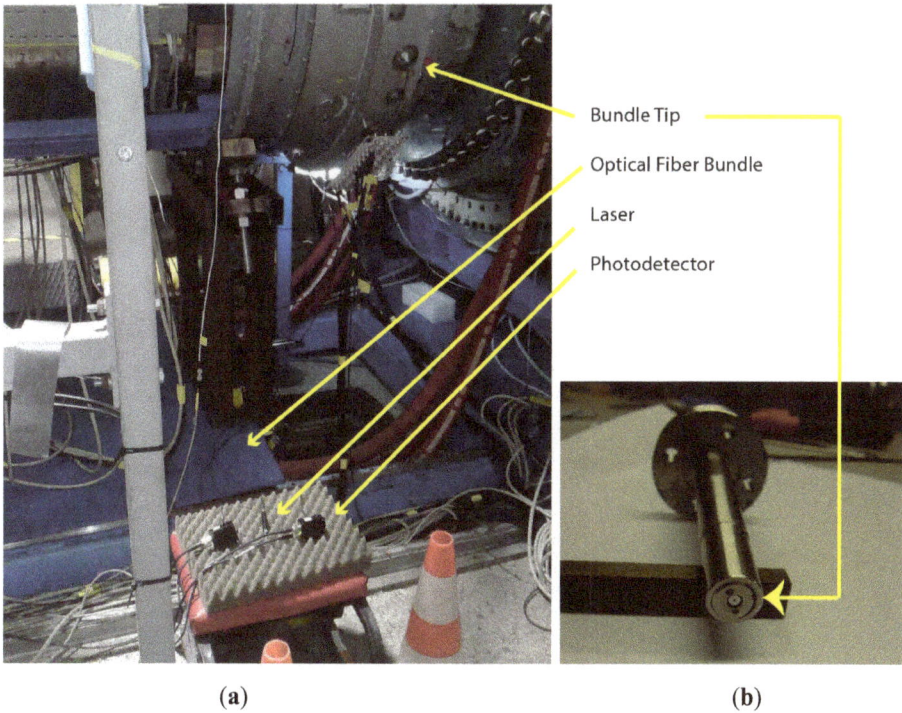

Bundle Tip

Optical Fiber Bundle

Laser

Photodetector

(a)   (b)

In Figure 4 the blade shape can be observed. We decided to illuminate the flat platform of the blade, so that the whole laser spot illuminates a flat surface and, therefore, most of the reflected light returns to the bundle. Since the difference between the flat platform and the nearest part of the blade to the casing is 1.3 mm, 1.75 mm should be subtracted from the measured distance to obtain the real TC.

Figure 4. Blade profile image.

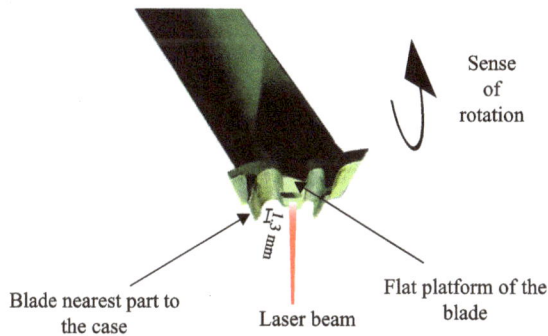

Sense of rotation

1.3 mm

Blade nearest part to the case

Laser beam

Flat platform of the blade

The experimental set-up is depicted in Figure 5. In addition to the sensor components, the inset shows the operational principle of the sensor. The light from the illuminating fiber is reflected by the blade and collected by the two rings of receiving fibers. Afterwards, the photodetectors connecting to each receiving leg of the optical fiber bundle carry out the photoelectric conversion. Finally, the oscilloscope acquires and saves the signals from both photodetectors. Additionally, the inset clarifies the tip timing and TC parameters.

Figure 5. Experimental set-up and operational principle. Legend: DO: Digital Oscilloscope; PD1: Photodetector 1; PD2: Photodetector 2; OFB: Optical Fiber Bundle; LD: Laser Diode; TC: Tip Clearance; BTT: Blade Tip Timing; RB: Rotor Blade; L0: Leg 0; L1: Leg 1; L2: Leg 2; CL: Common Leg.

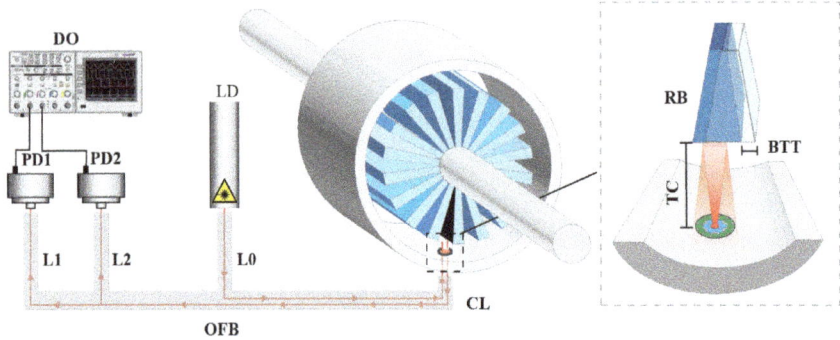

During the calibration we used the same components except for the oscilloscope, which was replaced by two digital multimeters (notice that during the calibration process a high acquisition rate is not necessary and that multimeters are much easier to automate). The bundle tip was fixed to a self-made SMA adapter. The blade was installed on a linear stage and it was longitudinally moved away from the common leg of the bundle in 10-μm steps. Using this procedure we obtained the calibration curve of the sensor. This calibration was carried out both in darkness and in the presence of light, and the differences between them were found negligible. The calibration curve is shown in Figure 6.

Figure 6. Calibration curve of the optical fiber sensor: quotient of photodetector voltages (V_2/V_1) as a function of the blade distance (d). The red line represents the fit to the calibration curve of a straight line in the range 3–7 mm. The shaded area corresponds to usual values of the TC (without correction).

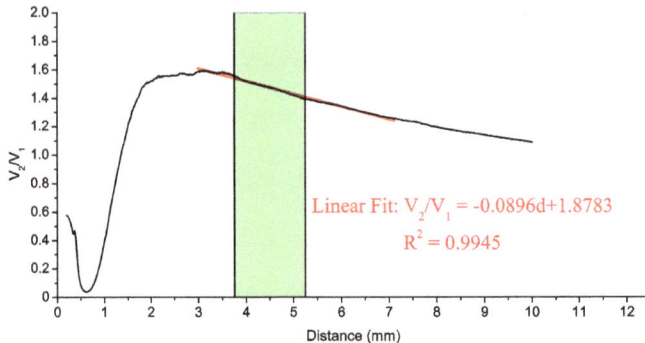

Taking into account that typical TC values for our turbine range between 2 and 3.5 mm, and that we have to add the 1.75 mm offset distance, the measurement range should be between 3.75 mm and 5.25 mm (represented by the shaded area). A straight line has been fitted to the calibration curve, as shown in Figure 6, by the least-square method for the 3–7 mm interval, with a coefficient of determination $R^2 = 0.9945$.

3. Results and Discussion

An important restriction of our acquisition system is that the oscilloscope has only 8 bits for vertical resolution. Therefore, we performed an oversampled acquisition to obtain a better-detailed waveform. For the case of TC measurements the sampling frequency was 250 Msa/s, and the oscilloscope memory depth allowed a maximum acquisition time of 82 ms from both photodetectors. In BTT measurements the accuracy in the amplitude of the signals is not as critical as in the former case, so the sampling frequency was set to 250 Ksa/s, obtaining 82 s of acquisition time. Both measurements were performed for 84 different working conditions (working points) of the engine. To identify each revolution, a blade with a particular reflection pattern has been used. The data from the sensor are post-processed after a low-pass filtering (cut-off frequency 50 KHz). Once the sensing principle is validated, a real time implementation will be possible. In Figure 7(a), we can observe the post-filtering signal from photodetector 2 for a specific working point where the turbine rotates at 2,400 rpm.

Figure 7. (**a**) Raw signal of photodetector 2. (**b**) Second derivative of the raw signal. (**c**) Detected blade events.

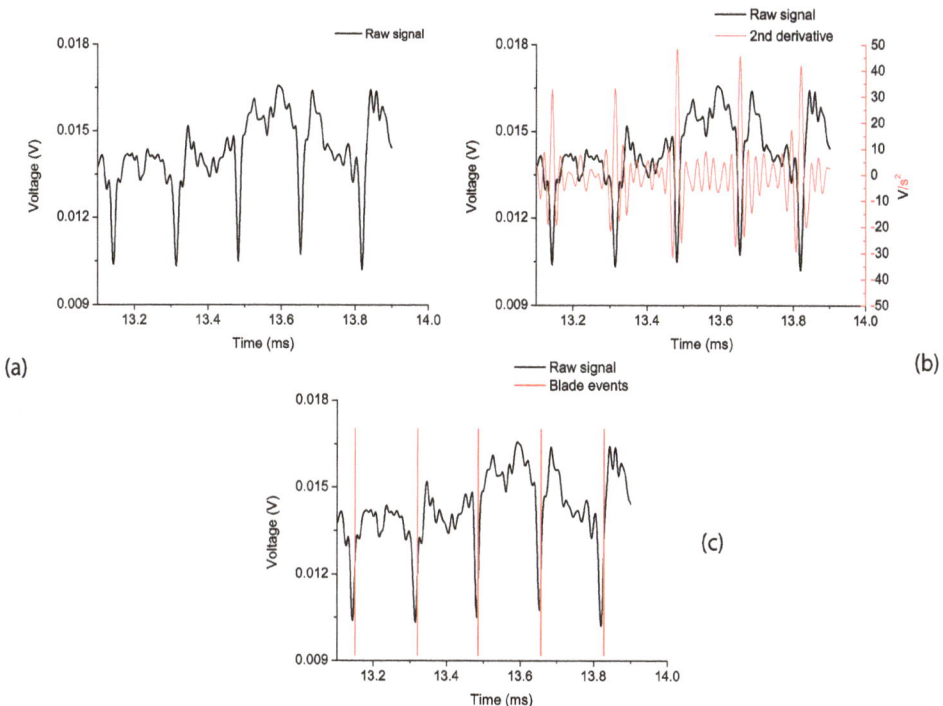

3.1. Tip Timing Measurements

The BTT technique is based on the measurement of the arrival times of all the blades. If the blades do not vibrate, the theoretical arrival times are obtained from a simple relationship involving the number of blades, the rotation frequency and the blade radius. However, if the blades do vibrate, their arrival times precede or succeed the theoretical non-vibrating arrival times. The difference between the theoretical and the real arrival time is related to blade deflections and it is used in the post-processing section of the system [25].

Furthermore, by placing probes at different circumferential positions, a sine wave model can be fitted to the blade deflections measured at each position. From these sinusoidal fits, we will be able to obtain the frequencies and amplitudes of the vibrations.

In our tests, a single probe was used to measure the blade arrival times. With this single probe application it is not possible to carry out an appropriate model fitting and a complete modal analysis cannot be achieved. Nevertheless, it is still possible to perform a travelling wave analysis in order to obtain the average amplitude of the blade tips at a particular nodal diameter (*ND*) (note that the minimum value of *ND* is 0, which is obtained when all the blades vibrate with the same phase, whereas *ND* reaches its maximum value (half of the number of blades) when each blade is out of phase with its neighbour). This travelling wave analysis can be used for monitoring the integrity of the blades against flutter, crack propagation or foreign object damage (*FOD*).

The travelling wave mode is the vibration condition of the blades. In a bladed disk system, the blades vibrate at the same amplitude but with a phase lag between them. This is known as the inter blade phase angle (*Ibpa*). This phase lag between the blades is related to the nodal diameter (*ND*) of the bladed disk mode according to:

$$ND = \frac{2\pi n}{Ibpa} \qquad (4)$$

where *n* is the number of blades.

As a consequence the frequency detected by the probe (f_{probe}) is the frequency of the blade (f_{blade}) plus the nodal diameter multiplied by the rotation frequency (ω), *i.e.*, [26]:

$$f_{probe} = f_{blade} + \omega ND \qquad (5)$$

Let us define the engine order (*EO*) as:

$$EO = \frac{f_{blade}}{\omega} \qquad (6)$$

Dividing Equation (5) by the rotation frequency and substituting Equation (6):

$$f_{probe} / \omega = f_{blade} / \omega + ND = EO + ND \qquad (7)$$

Therefore, a probe placed in the casing that measures the frequency of a rotating part detects not only the frequency of the blade, but also the phase lag that a blade has with its neighbour, so that it is not possible to discriminate *EO* from *ND*, because the probe detects the arrival time of each blade as a superposition of both terms.

The travelling wave analysis is mainly used for non-synchronous responses, such as flutter, where *EO* + *ND* has non-zero values. For synchronous responses it turns out that *EO* + *ND* = 0, since the excitation and the response are in phase and, therefore, synchronous responses are more difficult to detect with the travelling wave analysis.

Finally, the working point of the turbine is defined by two parameters: the rotational velocity N and the work per unit mass flow Ws. The specific work is defined as [27]:

$$Ws = C_p * \Delta T_0 \tag{8}$$

where C_p is the air specific heat capacity at a constant pressure and ΔT_0 is the total temperature drop within the stage.

Figure 7a shows the raw signal obtained with the optical probe for a test performed at nominal working conditions (2,400 rpm). Notice that we can estimate the arrival time of each blade by calculating the second derivative of the signal. This derivative gives the change in concavity/convexity of the raw signal as can be seen in Figure 7b. Choosing a threshold value for the second derivative of the raw signal, the blade arrival event can be obtained for every blade. In Figure 7c, the raw signal together with the blade arrival events can be observed.

From the arrival times we can obtain the deflection or deviation of each blade from its theoretical equilibrium position. The deviations from the theoretical equilibrium position in one revolution for every blade are shown in Figure 8. This deviation provides us with useful information for health monitoring to predict possible damages in blades. By plotting the deviations in real time, flutter or crack propagations can be detected as the deviation of a certain blade increases in time and gets close to predefined pre-alarm values.

Figure 8. Deviation of each blade from the equilibrium position in a complete revolution.

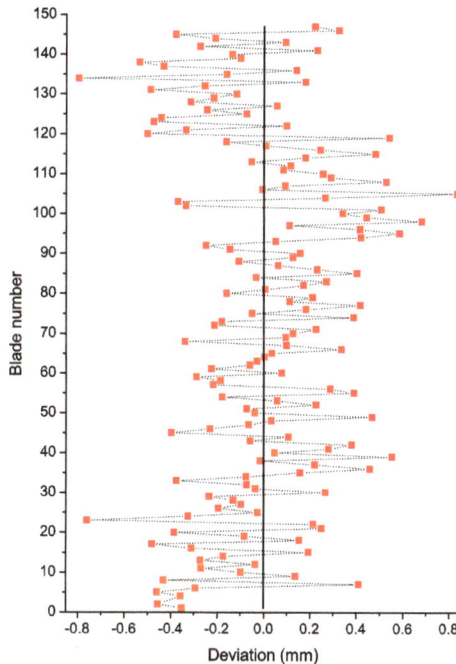

With the deflection values of the blades a fast Fourier transform can be performed to obtain the travelling wave spectrum of the system, which gives the average tip amplitude of all the blades, as shown in Figure 9. The travelling wave spectrum gives an average value of all the blade vibration amplitudes at a certain nodal diameter and it can also be used to monitor the system in real time, in order to check that the average blade tip amplitude does not exceed a predefined maximum value.

Figure 9. Travelling wave frequency spectrum of the system.

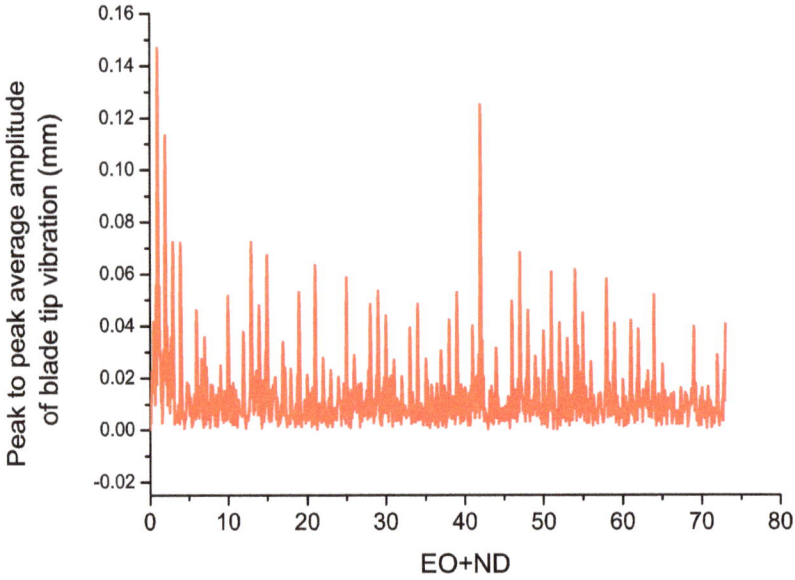

Figure 9 depicts the peak to peak average amplitude of blade tip vibration. There is a dominant peak of 0.15 mm at $EO + ND = 1$. This is probably a solid rigid rotation ($EO = 1$, $ND = 0$) due to some unbalancing of the rotor or the facility instead of a blade vibration mode.

There is another important peak at $EO + ND = 42$ of 0.13 mm that could be due to a low amplitude non-synchronous response, such as a flutter. When a blade starts moving, the surrounding flow exerts an aerodynamic force on it, and the direction and phase of this force can dampen or speed up the motion of the blade, leading to flutter. The *Ibpa* determines the phase between the local unsteady flow and local blade motion and this phase affects the unsteady aerodynamic work done on the blades. Unfavorable phase angles can lead to positive work being performed on the blades, which results in flutter.

All in all, the vibration amplitudes are always below 0.2 mm in the frequency spectrum of Figure 9 which is indicative of the stability of the rotor. More tests were performed for different rotation frequencies, leading to similar conclusions.

In order to discriminate the frequency of the blade and the nodal diameter, more probes should be placed in the casing in other circumferential positions. In the future, more probes are planned to be mounted in order to perform more detailed modal analysis of the rotors.

3.2. Tip Clearance Measurements

As we have already mentioned, TC represents the gap between the blade tip and the engine casing. This distance is obtained as a function of the quotient of two photodetector voltages, and the calibration curve of Figure 6 is used to relate both quantities. Figure 10 shows the two filtered photodetector signals and its quotient during the acquisition of a working point at 3,148 rpm. In this figure, 15 blades can be observed. If we pay attention to the minimum values that limit each blade, we can notice that each minimum has a different level (dashed line in Figure 10). This offset is probably due to the vibrations of the rotor-shaft assembly, so the first step in the signal processing, after the low pass filtering, is to remove this offset so that the signals start at the same level for all blades. Afterwards V_2 and V_1 are divided and its maximum value is found for a complete revolution. By definition, the TC is the distance corresponding to the maximum value of the quotient.

Figure 10. Filtered signals from the photodetectors and their quotient at a working point of 3,148 rpm.

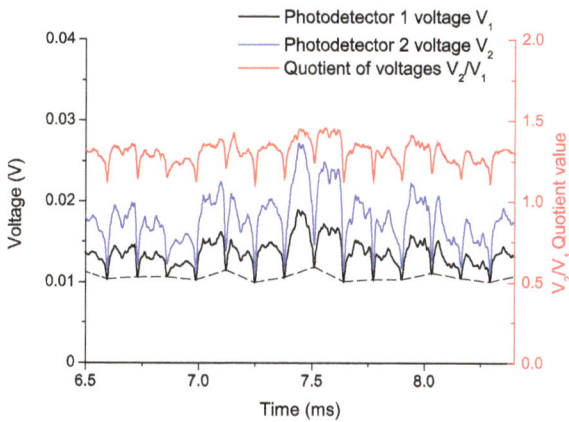

As we have already mentioned, 84 different working points were acquired during turbine assessment. Since the TC value depends mainly on the number of rpm, for a first evaluation of the sensor operation, all working points with the same rpm were averaged to obtain a TC for each turning speed. The optical sensor response was compared with a discharging probe sensor (Rotatip RCSM4 from Rotadata, Derby, UK) used by CTA. After appropriate data processing, the results for both sensors are shown in Table 1.

Whereas the optical sensor can obtain the distance to the casing for each blade at any instant, the discharging probe sensor gets only the smallest clearance of all the blades in an unknown revolution. The measurement instant is, therefore, not the same for both sensors, and even though measurements are taken in a steady state, they are measuring different distances (different vibration modes of the blades in different revolutions). It must be also taken into account that the standard deviation in the data from the sensor before speckle correction was about 150 µm. However, since our aim is to compare the behaviour of both sensors, we believe it is useful to show the obtained results. The latest statement is demonstrated by the fact that the differences between the two sensors are on the order of some tens of microns and the relative difference is less than 3%, as it can be observed in Table 1.

Table 1. Tip clearance measurements obtained during turbine tests.

Working Point (rpm)	Discharging Probe Tip Clearance (mm)	Optical Sensor Tip Clearance (mm)	Difference (%)
3148.52	2.890	2.954	2.22
3390.71	2.919	2.961	1.42
3632.91	2.893	2.843	1.72
3875.10	2.851	2.840	0.38

4. Conclusions

An optical sensor based on a trifurcated optical fiber bundle has been used to carry out BTT and TC measurements using a reflected intensity-modulation technique. Despite the limitation of using a unique probe for BTT measurements, the travelling wave spectrum has been obtained and with it the average vibration amplitude for all blades. With respect to TC measurements, a good approximation to a commercial sensor has been achieved, with differences lower than 3% for average TC values at each turning speed.

The great potential of optical fiber bundles for this kind of measurements has been demonstrated. The main novelties of this work with respect the previous ones are the possibility to simultaneously carry out both BTT and TC measurements with the same probe, and the reporting of these results as obtained from a real test of a turbine rig. In addition, the measurement system is a non-contact one, which allows to obtain information from all the blades with very short instrumentation times and relatively low cost. In future tests, several probes will be installed in the casing to make the most of the BTT measurements.

Acknowledgments

This work has been sponsored by the institutions Ministerio de Economía y Competitividad under projects TEC2009-14718-C03-01, TEC2012-37983-C03-01, Gobierno Vasco/Eusko Jaurlaritza under projects AIRHEM-II, S-PE12CA001, IT664-13 and by the University of the Basque Country (UPV/EHU) through program UFI11/16.

Conflicts of Interest

The authors declare no conflict of interest.

References

1. Ye, D.C.; Duan, F.J.; Guo, H.T.; Li, Y.; Wang, K. Turbine blade tip clearance measurement using a skewed dual-beam fiber optic sensor. *Opt. Eng.* 2012, doi:10.1117/1.OE.51.8.081514.
2. Zablotskii, I.E.; Korostelev, Y.A.; Sviblov, L.B. *Contactless Measuring of Vibrations in the Rotor Blades of Turbines*; Defense Technical Information Center: Fort Belvoir, VA, USA, 1974.

3. Nieberding, W.C.; Pollack, J.L. Optical Detection of Blade Flutter. In Proceedings of the Gas Turbine Conference and Products Show, Philadelphia, PA, USA, 27–31 March 1977.

4. Roth, H. Vibration Measurements on Turbomachine Rotor Blades with Optical Probes. In *Measurement Methods in Rotating Components of Turbomachinery*, Proceedings of the Joint Fluids Engineering Gas Turbine Conference and Products Show, New Orleans, LA, USA, 10–13 March 1980; pp. 215–224.

5. McCarthy, P.E.; Thompson, J.W., Jr. Development of a Noninterference Technique for Measurement of Turbine Engine Compressor Blade Stress; Arnold Engineering Development Center, US Air Force: Arnold AFB, TN, USA, 1980.

6. Watkins, W.B.; Robinson, W.W.; Chi, R.M. Noncontact engine blade vibration measurements and analysis. In Proceedings of the AIAA/SAE/ASME/ASEE 21st Joint Propulsion Conference, Monterey, CA, USA, 8–10 July 1985.

7. Watkins, W.B.; Chi, R.M. Noninterference blade-vibration measurement system for gas turbine engines. *J. Propul. Power* 1989, *5*, 727–730.

8. Kawashima, T.; Iinuma, H.; Wakatsuki, T.; Minagawa, N. Turbine Blade Vibration Monitoring System. In Proceedings of the 37th International Gas Turbine and Aeroengine Congress and Exposition, Cologne, Germany, 1–4 June 1992; p. 6.

9. Bloemers, D.; Heinen, M.; Krämer, E.; Wüthrich, C. Optische Überwachung von Turbinenschaufelschwingungen in Bettrieb. *VGB Kraftwerkstechnik*, 1993, *73*, 29–33 (in German).

10. Krämer, E.; Plan, E. Optical vibration measuring system for long, free-standing LP rotor blades. *ABB Rev.* 1997, *5*, 4–9.

11. Heath, S.; Imregun, M. A survey of blade tip-timing measurement techniques for turbomachinery vibration. *J. Eng. Gas Turbines Power* 1998, *120*, 784–791.

12. Georgiev, V.; Holík, M.; Kraus, V.; Krutina, A.; Kubín, Z.; Liška, J.; Poupa, M. The Blade Flutter Measurement Based on the Blade Tip Timing Method. In Proceedings of the 15th WSEAS International Conference on Systems, Corfu Island, Greece, 14–16 July 2011; pp. 270–275.

13. Zielinski, M.; Ziller, G. Optical Blade Vibration Measurement at MTU. In Proceedings of the AGARD conference, Brussels, Belgium, 20–24 October 1998.

14. Zielinski, M.; Ziller, G. Noncontact vibration measurements on compressor rotor blades. *Meas. Sci. Technol.* 2000, *11*, 847–859.

15. Wiseman, M.W.; Guo, T.-H. An Investigation of Life Extending Control Techniques for Gas Turbine Engines. In Proceedings of the American Control Conference, Arlington, VA, USA, 25–27 June 2001; pp. 3706–3707.

16. Cao, S.Z.; Duan, F.J.; Zhang, Y.G. Measurement of rotating blade tip clearance with fibre-optic probe. *J. Phys. Confer. Ser.* 2006, *48*, 873–877.

17. López-Higuera, J.M. *Handbook of Optical Fibre Sensing Technology*; Wiley: New York, NY, USA, 2002.

18. Ma, Y.; Li, G.; Zhang, Y.; Liu, H. Tip Clearance Optical Measurement for Rotating Blades. In Proceedings of the International Conference on Management Science and Industrial Engineering (MSIE), Harbin, China, 8–11 January 2011; pp. 1206–1208.

19. Dhadwal, H.S.; Kurkov, A.P. Dual-laser probe measurement of blade-tip clearance. *J. Turbomach.* 1999, *121*, 481–485.

20. Jia, B.; Zhang, X. An optical fiber blade tip clearance sensor for active clearance control applications. *Procedia Eng.* 2011, *15*, 984–988.

21. Tong, Q.; Ma, H.; Liu, L.; Zhang, X.; Li, G. Measurements of radiation vibrations of turbomachine blades using an optical-fiber displacement-sensing system. *J. Russ. Laser Res.* 2011, *32*, 216–229.

22. Faria, J.B. A theoretical analysis of the bifurcated fiber bundle displacement sensor. *IEEE Trans. Instrum. Meas.* 1998, *47*, 742–747.

23. Cao, H.; Chen, Y.; Zhou, Z.; Zhang, G. Theoretical and experimental study on the optical fiber bundle displacement sensors. *Sens. Actuators A Phys.* 2007, *136*, 580–587.

24. Cook, R.O.; Hamm, C.W. Fiber optic lever displacement transducer. *Appl. Opt.* 1979, *18*, 3230–3241.

25. Non-Intrusive Stress Measurement Systems. Available online: http://agilis.com/documents/NSMS.pdf (accessed on 11 March 2013).

26. Washburn, R.; Kim N.-E.; Brouckaert, J.-F. Hardware transmitted excitation sources and the associated blade responses using tip timing instrumentation. In *VKI Lecture Series*; von Karman Institute: Rhode-Saint-Genése, Belgium, 2007.

27. Cohen, H.; Rodgers, G.F.C.; Saravanamuttoo, H.I.H. *Gas Turbine Theory*, 4th ed.; Pearson Education Limited: Essex, UK, 1996; p. 273.

Reprinted from *Sensors*. Cite as: Rota-Rodrigo, S.; Pinto, A.M.R.; Bravo, M.; Lopez-Amo, M. An In-Reflection Strain Sensing Head Based on a Hi-Bi Photonic Crystal Fiber. *Sensors* **2013**, *13*, 8095–8102.

Article

An In-Reflection Strain Sensing Head Based on a Hi-Bi Photonic Crystal Fiber

Sergio Rota-Rodrigo *, Ana M. R. Pinto, Mikel Bravo and Manuel Lopez-Amo

Department of Electric and Electronic Engineering, Universidad Pública de Navarra, Pamplona 31006, Spain; E-Mails: anamargarida.rodrigues@unavarra.es (A.M.R.P.); mikel.bravo@unavarra.es (M.B.); mla@unavarra.es (M.L.-A.)

* Author to whom correspondence should be addressed; E-Mail: sergio.rota@unavarra.es; Tel.: +34-948-169-841; Fax: +34-948-169-720.

Received: 13 May 2013; in revised form: 13 June 2013 / Accepted: 17 June 2013 / Published: 25 June 2013

Abstract: A photonic crystal fiber-based sensing head is proposed for strain measurements. The sensor comprises a Hi-Bi PCF sensing head to measure interferometric signals in-reflection. An experimental background study of the sensing head is conducted through an optical backscatter reflectometer confirming the theoretical predictions, also included. A cost effective setup is proposed where a laser is used as illumination source, which allows accurate high precision strain measurements. Thus, a sensitivity of ~7.96 dB/mε was achieved in a linear region of 1,200 $\mu\varepsilon$.

Keywords: optical fiber sensor; photonic crystal fiber; strain sensor; interferometry

1. Introduction

There are a number of applications of practical interest in which the monitoring of strain-induced changes is important, such as experimental mechanics, aeronautics, metallurgy and health monitoring of complex structures, among others. These applications need continuous monitoring, aiming to control and prevent accidents or abnormal states early in time. Through the monitoring of structures, maintenance and rehabilitation advice can be provided, opening the possibility to avoid casualties [1]. In order to meet the increasing measurement requirements of modern industry, different types of strain sensors based on fiber-optic techniques have been developed. Fiber-optics have a number of characteristics that make them very appealing for sensing purposes, such as immunity to electromagnetic interference, light weight, remote sensing ability, multiplexing capability, and the ability for continuous *in situ* measurement [2]. Photonic crystal fibers (PCFs) are a recently developed class of optical fibers [3], which present a geometry characterized by a periodic arrangement of air-holes running along the entire length of the fiber, centered on a solid or hollow-core. The major difference between PCFs and single mode fibers (SMFs) relies on the fact that the waveguide properties of photonic crystal fibers are not due to spatially varying glass composition, as in conventional fibers, but from an arrangement of very tiny and closely spaced air-holes which go through the whole fiber length. In contrast with standard optical fibers, photonic crystal fibers can be made of a single material and have several geometric parameters that can be manipulated offering great flexibility of design. As such, PCFs present a diversity of new and improved features when compared to common SMFs, introducing innovative solutions in the sensing field [4].

Several strain sensors based on PCFs have been developed. Some modal interferometers have been accomplished using PCFs to measure strain or displacement: by tapering solid-core silica PCFs [5]; or by constructing a sensing head with a sensitivity of ~2.8 pm/με through splicing a piece of PCF to a SMF and interrogating it with a LED and a miniature spectrometer [6]; or even throughout a core offset at one of the joints of a SMF-PCF-SMF structure with 0.0024 dB/μm of sensitivity [7]. Other authors reported strain sensors that used highly birefringent (Hi-Bi) PCF based Sagnac interferometers showing temperature insensitivity, using a wavelength based measurement (~1.11 pm/με) [8] and a power based measurement (~2.7 dB/mε to 3.2 dB/mε of sensitivity) [9]. Displacement sensors were also reported using a Hi-Bi PCF in a Sagnac interferometer with a sensitivity of 0.283 nm/mm [10], and using a three-hole suspended-core fiber in a high precision Sagnac configuration (~0.45 μm) [11]. In addition, a strain sensor based in a Mach-Zehnder interferometer was accomplished by splicing a short length of PCF between two SMFs with collapsed air holes over a short region in the two splicing points (sensitivity of ~0.21 μs^{-1}/mε) [12]. A miniature in-line Fabry-Perot interferometer was as well obtained for strain sensing by splicing a small length of hollow-core photonic bandgap fiber between two SMFs in order to obtain a strain sensitivity of 1.55 pm/με [13]. Furthermore, a strain sensor was obtained based in a birefringent interferometer fabricated by an all-silica Hi-Bi PCF in transmission with a sensitivity of 1.3 pm/με [14].

In this work, an in-reflection interferometric Hi-Bi PCF sensing head for strain measurement is proposed. A study of the sensing head characteristics is shown, where a theoretical study is in accordance with the experimental data obtained through a high resolution optical backscatter reflectometer. Strain sensing is carried out using an accessible setup, where the interference signal is obtained through an in-line fiber polarizer.

2. Operation Principle

The operation principle of the proposed sensing system is based on two main properties of the Hi-Bi PCF: high birefringence and low temperature sensitivity. When light is launched into a highly birefringent fiber the difference in velocities between the two birefringent axes causes the resultant polarization state to vary along the length of the fiber in a controlled manner. The beat length (L_B) is a measure of the birefringence, or ability to preserve polarization. The beat length is defined as the distance over which the polarization rotates through 360 degrees:

$$L_B = \frac{\lambda}{b} \tag{1}$$

where λ is the wavelength at which the beat length is measured and b is the birefringence of the fiber. Since the sensing head works in reflection, the interferometric signal is proportional to twice the fiber length and its wavelength dependence can be expressed by Equation (2), where l is the fiber length, A is the amplitude and ϕ is the total phase:

$$R(\lambda)\big|_{dB} = 10Log\left[A \cdot Cos\left(2\pi \cdot \frac{2l \cdot b}{\lambda} - \phi \right) \right]^2 \tag{2}$$

The total phase is defined as $\phi = \phi_0 + \Delta\phi$. Where ϕ_0 is the initial phase and $\Delta\phi$ is the phase change induced by external perturbations. When strain is applied to the fiber, the phase variation will be given by:

$$\Delta\phi(\lambda) = \frac{2\pi}{\lambda} \cdot (\Delta b \cdot 2l + 2\Delta l \cdot b) \tag{3}$$

where Δl and Δb are the length and birefringence variations, respectively.

The low temperature sensitivity characteristic is a direct consequence of single material fabrication. Conventional optical fibers contain two different materials with different thermal (thermal expansion coefficient) and mechanical properties (Young's modulus and Poisson's ratio), which will generate high thermal stress when the fiber is subjected to temperature variations, consequently changing the birefringence of the fundamental mode. Since the Hi-Bi PCF is made of a single material, it will not present thermal stress, and thus, it is not surprising that Hi-Bi PCF temperature associated variations were experimentally measured to be negligible [15].

3. Sensing Head Characterization

The Hi-Bi PCF sensing head was obtained by splicing one end of ~20.8 cm Hi-Bi PCF to a SMF (maximum loss of 2 dB) and cleaving the other end. The Hi-Bi PCF is a polarization maintaining photonic crystal fiber (PM-1550-01 from NKT Photonics, Birkerød, Denmark) with a beat length of ~3.65 mm at 1,550 nm and an attenuation of 1.0 dB/Km (a cross section photograph can be seen in the inset of Figure 1). Figure 1 presents the characterization setup using an optical backscattering reflectometer (OBR), a linear polarizer, a polarization controller (PC) and the Hi-Bi PCF sensing head. The OBR used was developed by Luna Technologies and presents characteristics such as high spatial resolution (up to ~10 μm) for different measurable magnitudes, such as amplitude, polarization states and return loss in time and frequency domains.

In the experimental setup depicted in Figure 1, the linear polarizer converts the polarization state of the source light into a linear one, while the polarization controller allows one to adjust the alignment angle with the PCF. When the light propagates along the PCF, a phase shift is generated between the two birefringent axes due to its own birefringence.

Figure 1. (a) Schematic of the experimental setup used to characterize the Hi-Bi PCF sensing head and **(b)** optical microscopic picture of the Hi-Bi PCF cross-section.

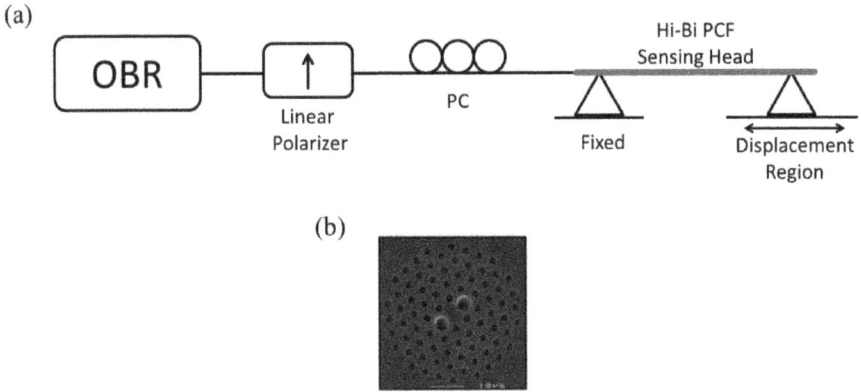

(a)

(b)

The reflected light passes again through the linear polarizer producing the interference between the retarded component signals. The interferometric signal obtained for the sensing head illustrated above (when no external forces act on it) is presented in Figure 2.

Figure 2. Reflected spectrum of the Hi-Bi PCF interferometer in a relaxed position, when no external force is induced on the sensing head.

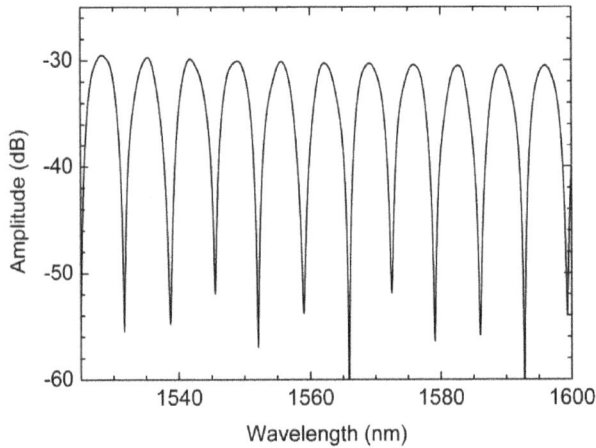

Since the sensing head is based on a Hi-Bi PCF, it will be sensitive to the angle between the input polarized light and the birefringent axes of the fiber. It is expected from theory that if this angle is 0° or 90° there is no interference signal, however if the angle is 45° both components will have the same optical input and the interference will be maximum. Figure 3 presents the interferometric spectra obtained for different angles between the input light and the birefringent axes. These results were obtained using the setup illustrated in Figure 1.

Figure 3. Interferometer spectra obtained for different polarization controller positions.

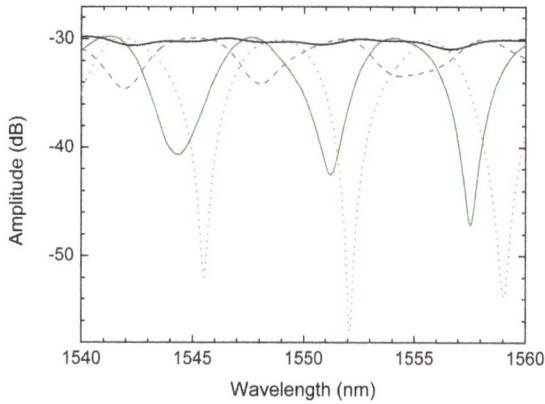

When strain variations are imparted to the PCF sensing head, the interference output signal (presented in Figure 2) will shift in wavelength. Figure 4 displays the experimental and theoretical results obtained for three different strain variations (0 με, 500 με and 1,000 με). As it can be seen in Figure 4, the interferometric spectrum presents a wavelength shift when strain variations are forced into the sensing head, which is quite in agreement with the simulations presented. Based on this characterization, strain measurement can be achieved by monitoring the interference wavelength shift, which presents a proportionality behavior with strain variations.

Figure 4. Experimental results (solid line) and theoretical simulations (dash line) of the reflected output signal for three different strain induced variations.

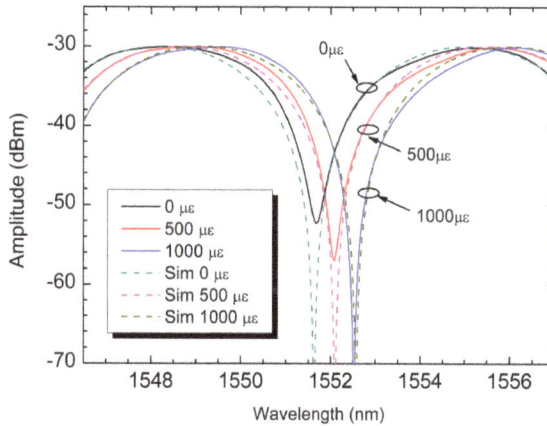

4. Sensor System and Results

After the characterization of the Hi-Bi PCF sensing head, its response to strain variations was measured through a more accessible, intensity based, setup which is presented in Figure 5. The experimental configuration consisted of a laser working at 1,554 nm (Ando AQ8201-13), a circulator, a linear polarizer, a polarization controller, the Hi-Bi PCF sensing head, and an optical spectrum analyzer (OSA) with a maximum resolution of 10 pm. After passing through the circulator, the laser light is linearly polarized and the polarization angle optimized before reaching the Hi-Bi PCF sensing head. The interferometric reflected signal will make a pass again through the circulator before reaching the OSA.

Figure 5. Experimental setup for strain measurement with an interferometric in-reflection Hi-Bi PCF sensing head.

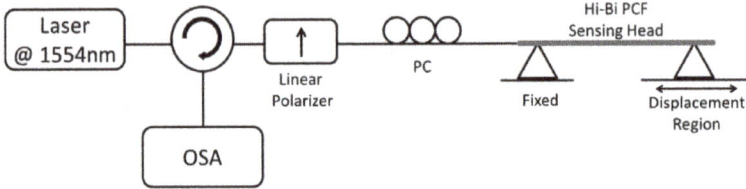

When strain changes are imparted to the Hi-Bi PCF sensing head, its output signal shifts in wavelength (see Figure 4). Using the cost effective system illustrated in Figure 5, the sensing head interrogation will be made through the laser, and as so in intensity. Since its interrogation is now made with a peak laser the output signal due to strain changes will present power shifts. Since the Hi-Bi PCF sensing head is illuminated by the peak laser, the output signal will be obtained only in the part of the interferometric signal that is in its cone of illumination. As so, if the sensing head signal is in an interferometric minimum the output peak power will be at its minimum value, meanwhile if it is at an interferometric maximum the output peak power will be at its maximum value. This will provide a visual sensation that the laser line is sweeping the interferometric signal, as the output peak power varies between a maximum and a minimum. The observed power shift with strain induced variations is depicted in Figure 6, using a stepper motor with increments of 22.2 $\mu\varepsilon$. The Hi-Bi PCF sensing head response showed a quadratic behavior followed by a linear one; this last with a sensitivity to strain variations of 7.96 dB/mε in an operational region of 1,200 $\mu\varepsilon$. The rupture point of the sensor head was found to be close to 5,000 $\mu\varepsilon$.

Figure 6. Measured optical power variation with strain.

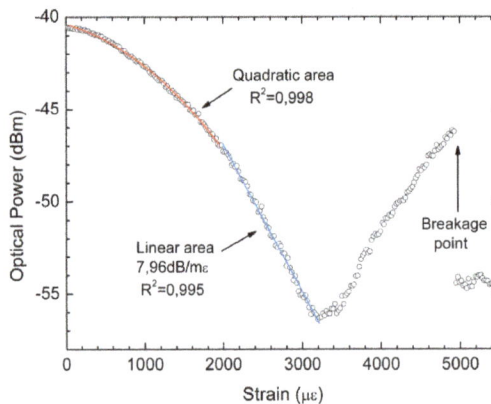

The presented sensing head sensitivity to strain induced variations is higher than other developed structures based on this fiber. For instance, when using the Hi-Bi PCF as the sensing element in a fiber loop mirror a sensitivity to strain that varied from 2.7 dB/mε at 1,530 nm to 3.2 dB/mε at 1,545 nm was obtained [9]. This interferometric sensing head shows appropriate response to strain variations, opening the possibility to obtain even better performance with a proper auto-referenced interrogation scheme such as a highly stable in-quadrature dual-wavelength fiber laser [16]. Also, the use of this system in-reflection is an attractive choice as a basic sensing element since it is simple, compact and presents the ability for remote sensing and multiplexing. Even more, reflective sensors enable the possibility for interrogation from a network header using a single fiber, as done in OTDR interrogation systems [17].

5. Conclusions

A simple configuration for an interferometric fiber optic strain sensor was presented and experimentally demonstrated. The sensing head is achieved by using a Hi-Bi PCF in-reflection. An experimental characterization of this sensing head was made using an optical backscatter reflectometer, which was in accordance with the presented theoretical simulations. Using a more cost-effective setup, strain variations could be accurately retrieved. The in-reflection sensing head presented a sensitivity of ~7.96 dB/mε to strain induced variations. Due to the demonstrated strain sensitivity, this interferometric sensing head is a very attractive solution for applications such as strain measurement in hazard environments and health monitoring of complex structures.

Acknowledgments

The authors are grateful to the Spanish government projects TEC2010-20224-C02-01 and Innocampus.

Conflicts of Interest

The author declares no conflict of interest.

References

1. López-Higuera, J.M.; Cobo, L.R.; Incera, A.Q.; Cobo, A. Fiber optic sensors in structural health monitoring. *J. Lightwave Technol.* **2011**, *29*, 587–608.
2. Li, H.N.; Li, D.S.; Song, G.B. Recent applications of fiber optic sensors to health monitoring in civil engineering. *Eng. Struct.* **2004**, *26*, 1647–1657.
3. Knight, J.C.; Birks, T.A.; Russell, P.S.; Atkin, D.M. All-silica single-mode optical fiber with photonic crystal cladding. *Opt. Lett.* **1996**, *21*, 1547–1549.
4. Pinto, A.M.R.; Lopez-Amo, M. Photonic crystal fibers for sensing applications. *J. Sens.* **2012**, *2012*, doi:10.1155/2012/598178.
5. Villatoro, J.; Minkovich, V.P.; Monzón-Hernández, D. Temperature-independent strain sensor made from tapered holey optical fiber. *Opt. Lett.* **2006**, *31*, 305–307.
6. Villatoro, J.; Finazzi, V.; Minkovich, V.P.; Pruneri, V.; Badenes, G. Temperature-insensitive photonic crystal fiber interferometer for absolute strain sensing. *Appl. Phys. Lett.* **2007**, *91*, 091109.

7. Dong, B.; Hao, E.J. Temperature-insensitive and intensity-modulated embedded photonic-crystal-fiber modal-interferometer-based microdisplacement sensor. *J. Opt. Soc. Am. B* **2011**, *28*, 2332–2336.

8. Frazao, O.; Baptista, J.M.; Santos, J.L. Temperature-independent strain sensor based on a Hi-Bi photonic crystal fiber loop mirror. *IEEE Sens. J.* **2007**, *7*, 1453–1455.

9. Qian, W.W.; Zhao, C.L.; Dong, X.Y.; Jin, W. Intensity measurement based temperature-independent strain sensor using a highly birefringent photonic crystal fiber loop mirror. *Opt. Commun.* **2010**, *283*, 5250–5254.

10. Zhang, H.; Liu, B.; Wang, Z.; Luo, J.; Wang, S.; Jia, C.; Ma, X. Temperature-insensitive displacement sensor based on high-birefringence photonic crystal fiber loop mirror. *Opt. Appl.* **2010**, *40*, 209–217.

11. Bravo, M.; Pinto, A.M.R.; Lopez-Amo, M.; Kobelke, J.; Schuster, K. High precision micro-displacement fiber sensor through a suspended-core Sagnac interferometer. *Opt. Lett.* **2012**, *37*, 202–204.

12. Zhou, W.; Wong, W.C.; Chan, C.C.; Shao, L.Y.; Dong, X.Y. Highly sensitive fiber loop ringdown strain sensor using photonic crystal fiber interferometer. *Appl. Opt.* **2011**, *50*, 3087–3092.

13. Shi, Q.; Lv, F.Y.; Wang, Z.; Jin, L.; Hu, J.J.; Liu, Z.Y.; Kai, G.Y.; Dong, X.Y. Environmentally stable Fabry–PÉrot-type strain sensor based on hollow-core photonic bandgap fiber. *IEEE Photonic. Technol. Lett.* **2008**, *20*, 237–239.

14. Han, Y.G. Temperature-insensitive strain measurement using a birefringent interferometer based on a polarization-maintaining photonic crystal fiber. *Appl. Phys. B* **2009**, *95*, 383–387.

15. Kim, D.H.; Kang, J.U. Analysis of temperature-dependent birefringence of a polarization-maintaining photonic crystal fiber. *Opt. Eng.* **2007**, *46*, 075003.

16. Pinto, A.M.R.; Frazão, O.; Santos, J.L; Lopez-Amo, M.; Kobelke, J.; Schuster, K. Interrogation of a suspended-core Fabry-Perot temperature sensor through a dual wavelength Raman fiber laser. *J. Lightwave Technol.* **2010**, *28*, 3149–3155.

17. Bravo, M.; Baptista, J.M.; Santos, J.L.; Lopez-Amo, M.; Frazão, O. Ultralong 250 km remote sensor system based on a fiber loop mirror interrogated by an optical time-domain reflectometer. *Opt. Lett.* **2011**, *36*, 4059–4061.

Reprinted from *Sensors*. Cite as: Xu, G.; Liang, C.; Chen, X.; Liu, D.; Xu, P.; Shen, L.; Zhao, C. Investigation on Dynamic Calibration for an Optical-Fiber Solids Concentration Probe in Gas-Solid Two-Phase Flows. *Sensors* **2013**, *13*, 9201–9222.

Article

Investigation on Dynamic Calibration for an Optical-Fiber Solids Concentration Probe in Gas-Solid Two-Phase Flows

Guiling Xu, Cai Liang *, Xiaoping Chen, Daoyin Liu, Pan Xu, Liu Shen and Changsui Zhao

Key Laboratory of Energy Thermal Conversion and Control of Ministry of Education,
School of Energy and Environment, Southeast University, Nanjing 210096, China;
E-Mails: guilingxu@seu.edu.cn (G.X.); xpchen@seu.edu.cn (X.C.); dyliu@seu.edu.cn (D.L.);
230099066@seu.edu.cn (P.X.); 220120371@seu.edu.cn (L.S.); cszhao@seu.edu.cn (C.Z.)

* Author to whom correspondence should be addressed; E-Mail: liangc@seu.edu.cn;
Tel./Fax: +86-25-8379-5652.

Received: 3 April 2013; in revised form: 4 July 2013 / Accepted: 15 July 2013 / Published: 17 July 2013

Abstract: This paper presents a review and analysis of the research that has been carried out on dynamic calibration for optical-fiber solids concentration probes. An introduction to the optical-fiber solids concentration probe was given. Different calibration methods of optical-fiber solids concentration probes reported in the literature were reviewed. In addition, a reflection-type optical-fiber solids concentration probe was uniquely calibrated at nearly full range of the solids concentration from 0 to packed bed concentration. The effects of particle properties (particle size, sphericity and color) on the calibration results were comprehensively investigated. The results show that the output voltage has a tendency to increase with the decreasing particle size, and the effect of particle color on calibration result is more predominant than that of sphericity.

Keywords: optical-fiber probe; solids concentration; calibration method; gas-solid two-phase flows

1. Introduction

The measurements of solids concentration are essential to understand the gas-solid flow behavior in fluidized beds, blow tanks, pneumatic conveying lines, and other multiphase flow systems. A detailed knowledge of solids concentration profile is critical to the accurate design and valid modeling of these systems. Many techniques have been carried out for measuring solids concentration: X-ray or γ-ray absorption, laser Doppler anemometry, acoustics methods, capacitance probes, optical-fiber probes, and so on [1–7]. Optical-fiber probes have been widely used in recent years for the determination of velocity and concentration of particles in gas-solid flow systems [1–3,8–17], which have the advantages of high sensitivity, fast response, high signal-to-noise ratio, large dynamic range, small volume and light weight, fire and shock resistance and corrosion proof, freedom from disturbance by temperature, humidity, electrostatics and electromagnetic fields, and suitability for remote transmission and multi-channel detection. The application of optical-fiber probes to the solids concentration measurement is based on the principle that the particles in the fluid produce scattering of incident light [18]. There are two different arrangements of optical-fiber probes [19]: transmission-type probe and reflection-type probe, as shown in Figure 1.

Figure 1. Two different types of optic fiber probes [19] (**a**) transmission-type; (**b**) reflection-type.

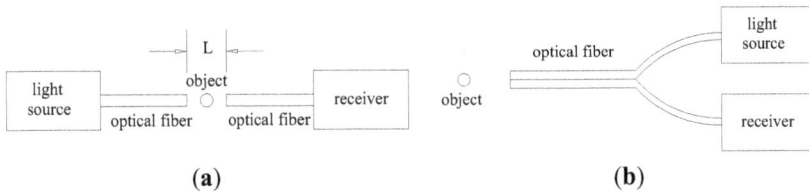

(**a**)   (**b**)

As for the transmission-type probe, which is based on the forward scattering of particles against the incident light, the object to be measured is located between the two probe tips and the light input and output are coaxial. The effective measurement volume is dependent on the distance between the two probe tips L, diameter and numerical aperture of the probe. The output signal is independent of the chromaticness of particles, which means that a single white or black particle may produce the same output signals. As for the reflection-type probe, which is based on the back scattering of incident light by particles, it has only one tip, where the projecting and receiving fibers are intermingled or in rows, in parallel or crossed. The effective measurement volume is dependent on the diameter, numerical aperture, overlap region of the capture angles and the optic sensitivity of the photoelectric converter. The output signals depend on the chromaticness and reflectivity of the particles [18]. The transmission-type probe is restricted to relatively low solids concentration and considered to cause strong disturbance of the flow structure compared to the reflection-type probe [13,20]. The reflection-type probe may be used over the entire range of particle concentration, from extremely dilute flows to the fixed bed state [21]. Thus, more attention has been paid to the reflection-type optical-fiber probe.

According to the ratio of particle diameter to fiber diameter, Matsuno *et al.* [19] classified the reflection-type probes into two categories. Figure 2a shows the optical-fiber probe with fiber diameter larger than particle diameter, the output signals are all generated by the reflected light from the particles existing within the measuring area. The integrated values of the output signal can be correlated with the particle concentration by any calibration method, and the instantaneous concentration can be obtained. Figure 2b shows the optical-fiber probe with fiber diameter smaller than the particle diameter. The output signals from the light receiver are converted into pulses at some threshold level V_s, and the pulse count corresponds to the number of particles. The particle velocity must be known in order to convert pulses to concentration. Only the average concentration can be measured if the flow field fluctuates.

Figure 2. Two categories of reflection-type probe [19,22].

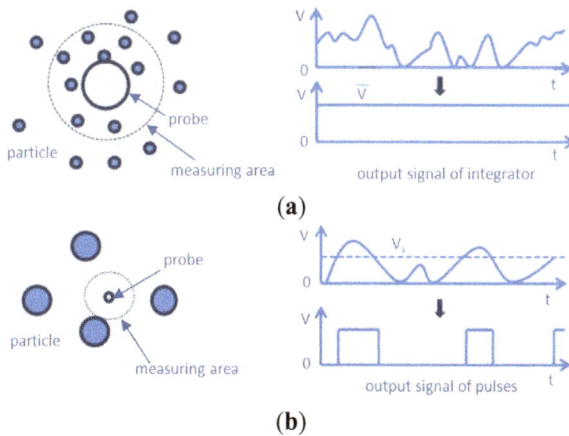

(a)

(b)

For the measurement of solids concentration, the emitted light reflected by the moving particles is magnified by a photo-multiplier and converted into voltage signals. The optical-fiber solids concentration probe should be calibrated to obtain a relationship between the output voltages and the solids concentration, which depends on the particle size distribution, concentration and optical properties of the solids and the surrounding fluid. The reliability of the measurements is strongly affected by the accuracy of the calibration method [9,16,21,23]. Calibration of a probe is finding a calibration curve with experiments and computations to convert the voltage time series into solids concentration time series [24]. The appropriate calibration method must be suitable for a wide range of solids concentrations from very low concentrations up to the solids concentrations of a packed bed [14]. However, it is difficult to generate a reference suspension of particles with a given concentration, and it is nearly impossible to produce a homogeneous gas-solid flow. Thus, calibration has become a problem.

The output signals of the reflection-type probe depend, to a large extent, on the aforementioned chromaticness of the particles, and the intensity of reflected light of white and black particles of the same diameter may differ by several times, and the roughness of particle surfaces may also have a similar influence [18]. Consequently, a calibration is required for each kind of particle. Very few studies have been carried out with regard to the effects of particle properties on the calibration curve of an optical-fiber solids concentration probe systematically and comprehensively, hindering the progress on the accurate calibration of the optical-fiber solids concentration probes.

The objective of this paper is to carry out investigation on dynamic calibration for an optical-fiber solids concentration probe in gas-solid two-phase flows. The remainder of this paper is organized as follows: a brief review of existing techniques for the calibration of optical-fiber solids concentration probes has been presented in Section 2. From the literature survey, it can be concluded that satisfactory calibration procedures are lacking in the literature and the calibration methods mainly focus on the low solids concentration. Thus, a calibration method, which is capable of calibrating the optical-fiber at nearly full range of the solids concentrations from 0 to packed bed concentration, is proposed. We combined the calibration method proposed by Hong et al. [23] with the calibration method used by Qi et al. [24]. The detailed experimental setup and calibration procedures are described in Section 3. With this combined method, the probe was uniquely calibrated with two kinds of powders. Finally, the calibration results are discussed and summarized in Section 4. The effects of particle properties on the calibration curve were also investigated.

2. Literature Survey

Both the transmission-type probe and the reflection-type probe needed to be calibrated for their measuring range before using them for the solids concentration measurement. Several techniques for the calibration of optical-fiber probes have been developed.

Matsuno et al. [19] calibrated the optical-fiber probe containing a pair of bundles of two small plastic optical fibers by employing the free-falling particles at their terminal velocities after having travelled a certain distance. The particles were poured through a vibrating sieve at a sufficient height to fall at a uniform velocity. The particle concentration was varied by changing the weight of particle on the sieve which was located at the top of the system and also by using sieves of different apertures. The probe was set at a height sufficiently below the sieve. The solids concentration was calculated with the following equation:

$$\varepsilon_s = \frac{\Delta W}{Au_t \Delta t} \tag{1}$$

where ΔW is the cumulative weight of particles on the cross-sectional area A within time Δt, and u_t is the calculated terminal velocity of the particles. An approximately linear relationship between the output voltage and the solids concentration was obtained. This method is usable for measuring the solids concentration certainly with a maximum concentration of 0.001. The manual vibration of the sieve and the fouling of particles will produce some erroneous data. The method is not only unsuitable to the system in which particles have high velocities but also to fast fluidized beds where the clusters of particles are formed.

Cutolo et al. [15] calibrated his probe with an apparatus consisted of a solids feed hopper and a 41 mm i.d. and 1 m high Plexiglas pipe. The particles were charged into the hopper and fell downward through a series of nets which acted as a solids distributor. The solids flow rate was regulated by the number of nets and their meshes. Measurement was made throughout the pipe core at a distance from the axis from −15 mm to +15 mm. A smaller tube with 33 mm i.d. was coaxially positioned at the bottom of the high pipe, which allowed the separation of solids falling along the walls of the larger tube. A collecting vessel placed on a balance was located below the apparatus. The average solids concentration $\varepsilon_{s,v}$ in the pipe of cross section A could be calculated with the following expression:

$$\varepsilon_{s,v} = \frac{\Delta W}{\rho AU(\varepsilon_{s,v})\Delta t} \tag{2}$$

where ΔW is the weight of the particles collected during the time internal Δt, ρ is the particle density, A is the cross-section area of the pipe, U is the average falling velocity of the particles, and $U(\varepsilon_{s,v})$ is a function involved of gas pressure and composition, particle size, concentration and weight. The relationship between the output voltages and solids concentration shows good linearity for the condition that the solids concentration is below 0.1. The maximum solids concentration achieved by using this method is 0.16.

Lischer and Louge [20] calibrated the optical-fiber probe used for measuring the particle volume fraction in dense suspensions against a quantitative capacitance probe. The material used was poured randomly along the probe assembly, which was mounted flush with the inside wall of a pipe that had a 15 cm i.d. They found that in a practical system, such regular calibration may be mandated by long-term variations of the average backscattered signal caused by subtle changes in the optical alignment or the quality of the fiber tip exposed to the particle suspension. However, the values from the optical-fiber probe and the capacitance probe didn't agree well.

Yamazaki *et al.* [25] carried out calibration experiments in a flat-bottomed cylindrical tank of 6.0 cm diameter. A known mass of solids was charged into a tank filled with water. After a steady state suspension was reached, the intensity of reflected light from the solids particles was measured by immersing the probe at various angular positions in the stirred tank. The light reflected from the solid particles in the solids-liquid mixtures had been measured in the range of solids concentration from 5% to 40% by volume. The axial concentration profiles at different radial positions agreed well, which implied that a homogeneous suspension of solids in the radial direction could be obtained by this method. It was found that measureable range for solids concentration was affected by particle species, particle diameter, particle shape, and particle color, variations in the refractive index of particles and continuous phase.

Zhou *et al.* [26] calibrated their optical-fiber solids concentration probe in two liquid-solid systems. For voidages less than 0.8, the calibration was carried out in a liquid-solid fluidized bed for the reason that particles were quite uniformly distributed in such a system. The solids concentration could be obtained with the following expression:

$$\varepsilon_s = \frac{H_0}{H}(1 - \varepsilon_0)$$
(3)

where H_0 is the height of a packed bed and ε_0 is the voidage of the packed bed. Concentration data were obtained by changing the fluid velocity to vary the bed height, H. For voidages larger than 0.8, the calibration was carried out in a beaker, where a certain volume of solid particles was put into and mixed with water of known volume. The liquid-solid mixture was stirred until the particles were uniformly distributed in the water. The calibration was very nearly linear over the entire solids concentration range obtained. Different concentrations were achieved by mixing different volumes of particles into water. However, it must be pointed out that water calibration may cause measurement errors while using the probe in a gas-solid flow due to the refraction index difference between liquid and gas. The index of refraction of a liquid is greater than that of a gas, and the probes should be calibrated in the same medium for which they will be utilized [27]. Furthermore, calibrations should also be performed accordingly for each kind powders or a certain kind of powders with different diameter, different color and particle shape based on the research of Yamazaki *et al.* [25].

Herbert *et al.* [21] calibrated a single-fiber optical reflection-type probe with a technique similar to that used by Cutolo *et al.* [15]. They established a stable downward flow of particles at a known velocity and in a column small enough so that a local measurement could yield a cross-sectional average value. The particles flowed from a fluidized bed feeder with a diameter of 12.5 cm, through an orifice in the center of the porous metal grid into a tube of square cross-section 8 × 8 mm, then fell 2.5 m into a collection pot. The particle flow rate from the fluidized bed could be kept stable by changing the orifice diameter. The average solids concentration could be calculated with the following expression:

$$\varepsilon_{s,v} = \frac{m_p}{u_p A \rho_p}$$ (4)

where m_p is the particle mass flow rate, u_p is the particle velocity measured with an optical-fiber velocity probe, A is the tube cross-sectional area, ρ_p is the particle density. During the experiments, it was found that electrostatic forces caused fine particles to cling to the probe tip and obscured light transmission, which was very undesirable since the reflected light intensity was no longer a function of the particle concentration in front of the probe, but also related to the degree of coverage of the fiber which could not be known. To overcome the electrostatic force effect, a fine-particle additive Larostat 519 was mixed with the particles.

Hong *et al.* [23] developed a method for calibrating the optical-fiber probe using a polynomial regression to correlate the output signal with the solids volume concentration in the fluidized calibration vessel. Solid particles of mass W_s were placed in the fluidized vessel made of transparent glass and then fluidized to the full vessel volume after closing the top end of the vessel with a filter screen. The overall volume concentration of solids within the full vessel could be calculated with the following expression:

$$\varepsilon_s = \frac{W_s}{\rho_s A H}$$ (5)

where ρ_s is the solids density, A is the cross-sectional area of the vessel and H is the vessel height. They found that simply assuming a linear relationship between the output signal and solids concentration may cause a significant deviation from experiment in the range of concentration above 0.05. They also pointed out that a powder with very homogeneous fluidized behavior would have a more reliable calibration curve using their method. The disadvantage of this method was that the formation of bubbles in the bed during calibration which would cause some scatter of the calibrating points.

San José *et al.* [28] calibrated the optical-fiber probe used for voidage measurement of a conical spouted bed. For the spout zone, the calibration was carried out in a 60 mm i.d. column, where the probe had been introduced at a given level. The solid was fed by a hopper to the column. A linear relationship between the intensity of the reflected light and the volume fraction of the bed occupied by the particles, $1 - \varepsilon$, was obtained. Bed voidage, ε, which was changed by adjusting the solid mass flow rate in the feed, Q, could be calculated with the following expression:

$$Q = \rho A v_z (1 - \varepsilon) \tag{6}$$

where ρ is the particle density, A is the column cross sectional area, v_z is the velocity of the particles along the longitudinal direction at the probe zone, which can be calculated as follows:

$$v_z = \frac{d_e}{\tau} \tag{7}$$

where d_e is the effective distance between the two receiving fibers, τ is the delay time between the two signals. For the annular zone, the calibration consisted of measurement in moving beds, in a column of 60 mm i.d., using different particle sizes. Three kinds of different experiments were carried out: packed beds, beds loosened to the maximum and partially loosened beds. They found that the position of the probe in the bed did not affect the resulting calibration curve. This method has two disadvantages that the solids concentration along the cross-section of the downer column is mal-distribution and using the measurement of particle velocity with optical-fiber to evaluate solids concentration is not absolutely a good measurement technique.

Zhang et al. [27] proposed a back pressure control method to calibrate a multi-fiber optical reflection probe in a downer to obtain quantitatively precise solids concentration. The calibration apparatus was similar to that of Herbert et al. [21], and the significant improvement was adding a back pressure control system by sealing the bottom collection vessel. By using quick closing valves [1], the particles were trapped to determine various solids concentrations to compare with the data obtained with optical-fiber probe. The solids concentration could be obtained as large as 0.56 because the increase of the back pressure decreased the particle velocity and increased the solids concentration. Meanwhile, an iteration procedure was employed to modify the initial calibration curves. They also found that the probes were sensitive to minor variations of particle color and reflective properties. However, the flow was not entirely radial uniform even by using a vibrator to distribute the particles in the downer, which would cause calibration errors.

Johnsson et al. [29] carried out calibration experiments of an optical-fiber probe in a cold CFB riser, and compared the output signals with those obtained by an optical reference probe, which was calibrated with a guarded capacitance probe. The guarded capacitance probe was calibrated from measurements in a packed bed. They found that the reference probe and the optical-fiber probe gave similar response in amplitude and frequency with respect to variations in solids volume-fraction. The shape of the calibration function of the reference probe was also valid for the optical-fiber probe. The calibration and the reference measurement were carried out at ambient conditions, while the measurements were done at 850 °C with the assumption that the shape of the calibration function obtained under ambient condition was valid at elevated temperatures. They believed that it was a reasonable assumption because the shape of the calibration function depended on the optical properties of the probe and not on the temperature of the gas and particles.

Cui et al. [30] suggested a novel calibration method and correlation for different optical-fiber probes. They made a series of uniform mixture samples of FCC and amorphous transparent polystyrene by blending the two materials in a Brabender mixer at 200 °C, shaped the samples into a cubic form in a size of 20 mm, and finally polished them to attain good optical properties. For different samples, the FCC particle concentration varied from 0 to that of the minimum fluidization state. The mixtures were used to simulate gas-solid flow systems with different solids concentrations, in which the transparent polystyrene was seen as air. The transparent polystyrene had refractive index higher than that of air. They found that the cross probe and parallel probes with glass window had similar calibration curves. The output voltages of the probe increased sharply with increasing solids concentration at low concentration but increased slowly at high solids concentration. This method is very interesting and novel. The probes were calibrated with a homogeneous dispersion of solids in the polymer-solid cubes with different and exact solids concentrations which is representative of different solids concentrations inside the gas-solid fluidized bed.

Rundqvist *et al.* [10] proposed an improvement in the design of a dual optical-fiber probe as well as a general calibration theory. They calibrated the probes in a small stirred tank filled with water, and a controlled mass of particles was added. The results of their calibration experiments showed some discrepancies relative to theoretical calibrations. A disproportional fraction of the particles were observed to reside close to the walls and the bottom of the calibrating vessel, which hindered the calibration experiments, leading to lower volume fractions than expected and offered a possible explanation to the above mentioned discrepancies. It was concluded that the calibration theory scaled well with probe size, particle size and particle volume fraction, although the exact shape of the calibration function could not be verified exactly.

Liu *et al.* [22,31] calibrated three-fiber optical probes with two units. The first was a dropping/trapping technique [32]. Particles fall into a collection vessel from an incipiently fluidized bed through a short tube located in the center of a punched plate distributor covered with fine wire mesh into a 12 mm i.d. tube. After a steady flow was obtained, two slide plates located 32 mm apart were closed quickly and simultaneously to trap the particles in a section of pipe where the probe inserted. By using different sizes of feeding tube and the flow rate of aeration air, different solids concentration could be obtained. The second was to obtain a water-FCC suspension in a well-stirred beaker. The probe with and without the quartz glass window was calibrated using both the two calibration units, and the results were compared with simulation predictions. The simulation results were in good agreement with the calibration results. They found that without the glass window, the calibration curves were highly nonlinear, which meant that the protective window could improve the linearity of the calibration curve. They also suggested that one should not directly apply calibrations obtained in liquids to calibrate probes for gas-solids systems.

Magnusson *et al.* [33] calibrated a dual fiber-optical probe based on the calibration theory proposed by Rundqvist *et al.* [10] in a circulating fluidized bed, for the reason that circulating fluidized beds provide a wide range of flow conditions and solids volume fractions. The particle volume fractions measured by the optical-fiber probe were compared to the pressure drop measured for a range of operation conditions. The relation between the pressure drop and the solids volume fraction could be expressed as follows:

$$\Delta P = \varepsilon_s (\rho_s - \rho_g)gz + p_{acc} \tag{8}$$

where ε_s is the solids volume fraction, ρ_s and ρ_g are the solids and gas phase densities, g is the acceleration due to gravity and z is the distance between the pressure measuring taps, p_{acc} is the pressure drop due to acceleration of the particles, which may be neglected. They found that glare points on the particles and the beam length from the probe to the particle determines the curvature of the calibration curve. The values obtained with pressure drop measurements and the values of optical-fiber probe measurements showed poor agreement. The measurement volumes for the optical probe and the pressure drop are different from each other. Thus, application of this method for calibration of optical-fiber solids concentration probe is not satisfactory.

As can be seen from the literature review above, most of the researchers performed their calibration procedure and obtained specific kind of calibration curve, either linear or non-linear. All the calibration procedures were established by comparing the optical-fiber probe output signals to the concentration values obtained by the traditional methods for direct measurement of solids concentration, which could be mainly divided into three categories. The first is to use quick closing valves [1], with which the column to be studied can be positioned in sections of suitable length. The two valves were used to trap the solids within the desired section of the downer. After a certain period of time, the valves were closed simultaneously. The solids contained in the section can be collected and weighed, thus the solids concentration may be determined. This method was used by Zhang et al. [27] and Issangya et al. [32]. The second is to measure the pressure drop over a certain section of a riser tube, and the Bernoulli equation neglecting wall friction and acceleration forces was solved for solids concentration. This method was adopted by Magnusson et al. [33]. The third is to build a stable downward flow system with particle density deduced from mass flux of particles and measurement where phase velocities were nearly equal [18]. This method was widely used by researchers with appropriate modifications, like Matsuno et al. [19], Cutolo et al. [15], Lischer and Louge [20], Herbert et al. [21] and so on. In addition, liquid fluidized beds had been used by some investigators with the purpose to overcome the difficulty of obtaining stable suspensions of solids in gases, like Yamazaki et al. [25], Zhou et al. [26], Rundqvist et al. [10], and Liu et al. [22,31]. However, it is worthwhile to mention that the validity of such a calibration in gaseous suspensions is questionable due to the differences in the refractive index between gases and liquids [13], and the probes should be calibrated in the same medium for which they will be utilized. Different calibration methods used by researchers are summarized in Table 1.

The accuracy of solids concentration measurement by using an optical-fiber solids concentration probe is strongly dependent on the precision of the calibration technique utilized. The calibration of optical-fiber solids concentration probe in gas-solid environment is challenging due to the heterogeneity and instability of gas-solids flow. It is difficult to offer a series of standard gas-solid flow covering solids concentrations from 0 to packed bed. Because it is rather difficult to maintain a homogeneous gas-solids flow at high solids concentrations, the majority of calibration methods developed to date are focused on obtaining relatively homogeneous gas-solids flow at low solids concentrations. When applying them to the practical measurements, the calibration methods may be inaccurate or problematic. There are no widely-accepted calibration methods which cover a wide range of solids concentrations up to now. Thus, a feasible and simple method for the calibration of optical-fiber solids concentration probe which could solve the above mentioned questions of the existing calibration methods is urgently needed to be developed. More research on the calibration of optical-fiber solids concentration probe, especially experimental, is still required.

Table 1. Different calibration methods.

Year	Author	Probe Type	Material	Particle Diameter	Particle Density	Calibration Apparatus	Verification Method	Linear/ Non-Linear Calibration Curve	Remarks	Measuring System
1983	Matsuno et al.	Reflection type	Glass beads	56.5 μm	2520 kg/m³	vibrating sieves	Theoretical calculation	linear	Limited to low solids concentration	Gas-fluidized bed
1990	Cutolo et al.	Transmission-type	Glass beads	90 μm	-	A solids feed hopper with a 41 mm i.d. and 1 m high Plexiglas pipe	Theoretical calculation	linear	Good linearity when the volume concentration was below 0.1	High concentration (up to 0.16) gas-solid suspension
1992	Lischer and Louge	Reflection type	Glass beads	70 μm 210 μm	-	15 cm i.d. pipe	capacitance probe/simulation calculation	non-linear	Simple construction	-
1992	Yamazaki et al.	Reflection type	Glass beads / Toyoura sands / PVC powders	225 μm, 131 μm, 42 μm, 163 μm, 164 μm	2,490 kg/m³, 2,490 kg/m³, 2,350 kg/m³, 2,650 kg/m³, 1,500 kg/m³	a flat-bottomed cylindrical tank of 6.0 cm diameter	Theoretical calculation	non-linear	The refractive index of liquids is different from that of gases	Slurry mixing tank
1994	Zhou et al.	Reflection type	Ottawa sand	213μm	2,640 kg/m³	a liquid-solid fluidized bed/ well-stirred water-sand beaker	Theoretical calculation	near-linear	The refractive index of liquids is cross-section	CFB of square cross-section
1994	Herbert et al.	Reflection type	FCC particles	0.78 mm	1,630 kg/m³	A fluidized feeder with a 2.5 m long and 8 × 8 mm square cross-section tube	Theoretical calculation	non-linear	The volume fraction range calibrated was only 0.01 to 0.1	0.05 m diameter downflow CFB reactor
1994	Zhou et al.	Reflection-type	Ottawa sand	213 μm	2,640 kg/m³	a liquid-solid fluidized bed/ well-stirred water-sand beaker	Theoretical calculation	near-linear	The refractive index of liquids is different from that of gases	CFB of square cross-section

Table 1. *Cont.*

Year	Author	Probe Type	Material	Particle Diameter	Particle Density	Calibration Apparatus	Verification Method	Linear/ Non-Linear Calibration Curve	Remarks	Measuring System
1995	Hong et al.	Reflection-type	Limestone	0.124 mm	2,170 kg/m³	fluidized vessel	Theoretical calculation	non-linear	A powder with very homogeneous fluidized behavior is more suitable	Horizontal pneumatic piplies
1998	San José et al.	Reflection-type	Glass beads	3 mm 4 mm 5 mm	2,420 kg/m³	moving bed/ 60 mm i.d. column	image treatment system	linear	-	Conical spouted beds
1998	Zhang et al.	Reflection-type	FCC particles	49.4 μm 59.0 μm	1,500 kg/m³ 1,420 kg/m³	An incipiently fluidized bed and vibrating solids feeder with a 3.81 m downer	quick closing valves	non-linear	The solids concentration range calibrated could be from 0 to about 0.56	-
2001	Johnsson et al.	Reflection-type	Silica sand	0.30 mm	-	cold CFB riser	Optical reference probe	non-linear	-	Electrically heated fluidized bed/CFB boiler
2001	Cui et al.	Reflection-type	Sand	70 μm 385 μm	1,673 kg/m³ 2,650 kg/m³	Mixtures of FCC and amorphous transparent polystyrene	Theoretical calculation	non-linear	Use known solids concentration mixture to simulate gas-solid flow	Air-fluidized bed
2003	Rundqvist et al.	Reflection-type	Silica sand	0.15 mm 0.20 mm	-	small stirred tank with water	Theoretical calculation	near-linear	-	-
2003	Liu et al.	Reflection-type	FCC particles	70 μm	-	3-D gas-solid suspension/ well-mixed water-FCC tank	quick-closing valves/ Theoretical calculation	non-linear	-	High-density CFB riser
2005	Magnusson et al.	Reflection-type	Silica sand	0.08 mm 0.46 mm	2,600 kg/m³	circulating fluidized bed	pressure drop measurement	non-linear	-	CFB

3. Experiment Setup and Calibration Procedure

This study utilized a model PC6M optical-fiber solids concentration probe which was developed by the Institute of Process Engineering, Chinese Academy of Science, Beijing, China. This measurement system was composed of PC6M concentration measurement main unit, optical-fiber probes, and the signal cable, A/D converter and application software. The probe tip is 4 mm in diameter and contains approximately 8,000 emitting and receiving quartz fibers, each of a diameter of about 25 μm. These fibers are arranged in an alternating array, corresponding to emitting and receiving layers of fibers. The active area, where the fibers are located, is approximately 2 mm × 2 mm. The fiber tips are protected from the material by a glass window with thickness of 0.2 mm. The received light reflected by the particles is multiplied by the photo-multiplier and converted into a voltage signal. The voltage signal is further amplified and fed into a computer. The high voltage adjustment is used to adjust the upper measuring limit or full scale of the instrument. There is a zero voltage potentiometer which adjusts the output signal to zero when no powder is on the tip of the probe, thus the offsets of PC6M were set at zero with empty black box and the gains roughly at 4.5 V with packed box (less than the full range of 5 V), making the calibration procedure respond to most of possible particle concentrations. In order to make the day-to-day measurement comparable, the lamp voltage which sets the power voltage of the light source, and the gain factor should be kept constant during the calibration procedures [27]. Figure 3 shows the schematic diagram of the optical-fiber probe system.

Figure 3. The optical-fiber probe system [16].

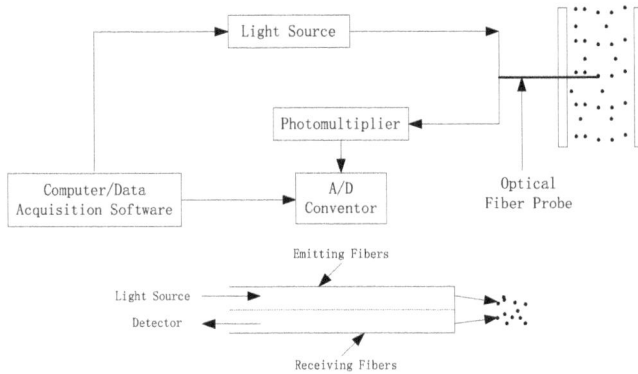

A Plexiglas column was used to calibrate the probes. As shown in Figure 4, the column is 50 mm in diameter and 300 mm high. A perforated Plexiglas plate covered with fine screens was employed as the gas distributor. Optical-fiber probes were installed along the column with their tips located on the axis of the column. Two pressure taps, one at the wind room and the other at the exit section of the column were provided to record the pressure drops. Pressure drop was measured by a 1 m long U-tube manometer, and water was used as the manometric fluid. The fluidizing gas was supplied by a nitrogen cylinder to the wind room which is 50 mm in diameter and 100 mm in length below the gas distributor. Bubble suppressor was installed in the calibration column, which contained a set of metal meshes with 25 mm axial pitches, the diameter of the metal mesh was 45 mm and the opening of each metal mesh was 3 mm × 3 mm with 0.5 mm steel wire. The top end of the column was covered with a filter screen to prevent the fine particles escaping from the column. Meanwhile, all metal portions of the whole calibration apparatus were carefully connected and grounded to eliminate static electricity effect.

Figure 4. Calibration apparatus.

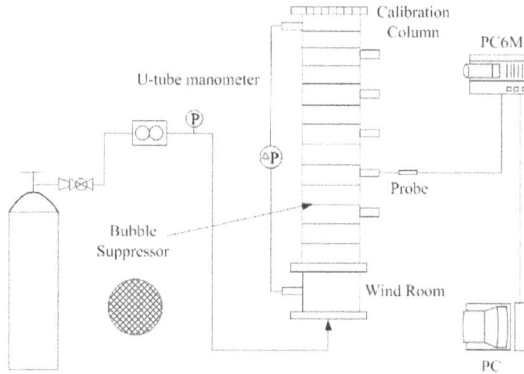

Glass beads and quartz sand were used to investigate the particle properties. Both of the two powders were sieved into three narrow distribution parts, and their physical properties are listed in Table 2. The particle size distributions were measured with a laser particle analyzer (LS, Beckman Coulter Inc., Brea, CA, USA), as shown in Figure 5. Because the particle diameter is much smaller than the bundle diameter, light is reflected by the particles in the measurement volume, allowing the probe to detect to measure the solids concentration.

Table 2. Physical properties of powders.

Powder		Average Diameter (μm)	Particle Poured Bulk Density (kg/m³)	Particle Density (kg/m³)	Color
Glass beads	GB 1#	60.29	1,390		
	GB 2#	104.3	1,400	2,650	grey
	GB 3#	166.8	1,430		
Quartz sand	QS 1#	66.53	1,200		
	QS 2#	78.85	1,300	2,610	white
	QS 3#	192.8	1,320		

The novel calibration procedure in the present study are consisted of two parts: one for high solids concentration (Method I) and the other for low solids concentration (Method II). The calibration procedure for high solids concentrations was similar to that of Qi [24] used a Pseudo Bubble-Free Fluidized Bed (PBFF). The powder was filled into the column with an initial bed height of about 100 mm, then fluidized by the fluidizing gas and was better distributed throughout the column with the assistance of bubble suppressors compared to without the installing of bubble suppressors.

132

Figure 5. Particle size distribution: (**a**) Glass beads; (**b**) Quartz sand.

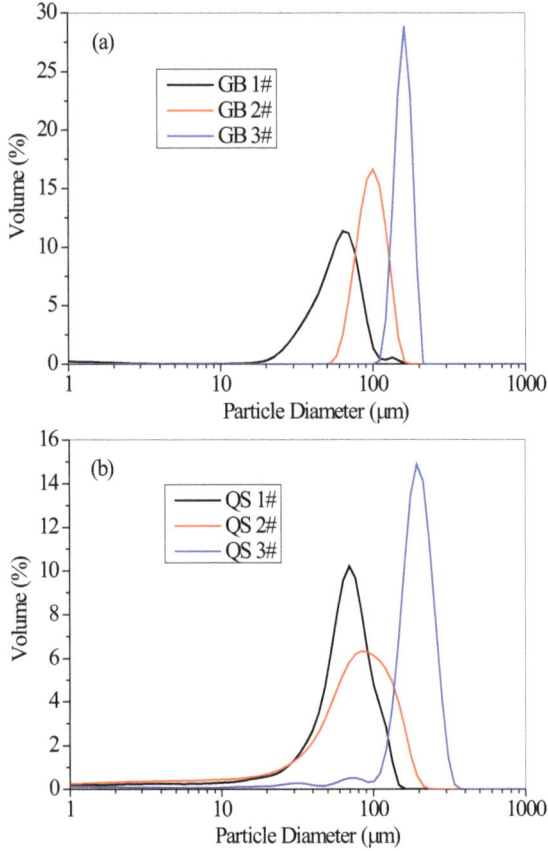

During the calibration process, the optical-fiber probe was inserted into the column center between two successive bubble suppressor plates. The experiments were carried out under different ratio of H/H_{mf}, where H_{mf} is the bed height at minimum fluidization state and H is the bed height under other fluidizing gas flow rate. The solids concentration could be calculated according to the following equation:

$$\frac{\varepsilon_s}{\varepsilon_{s,mf}} = \frac{H}{H_{mf}} \tag{9}$$

From the results obtained by this paper, this kind of calibration procedure was limited to solids concentration higher than 0.2. Thus, for the low solids concentration, the calibration data was obtained with the method similar to Hong *et al.* [23] and the detail procedures were with reference to Wang *et al.* [34]. All the calibration experiments were repeated 6 times under the same condition for the accuracy of the reproducibility. To minimize the influence of fine particles adhering to the tip surface of the probe on the calibration, particularly at the relatively high concentration, the optical-fiber probe was removed for cleaning of its surface for every test. In order to avoid the effect of any outside light source, the calibration column was covered with black cloth after fluidization of the solids particles. The sampling rate of the optical-fiber probe was 1 kHz and the sampling time approximately 4 s.

4. Results and Discussion

Due to that the complexity of the heterogeneous gas-solids flow manifests in the irregular, non-periodic variation of solids concentration with time, the typical time-resolved signals from the probe exhibit sharp spikes that correspond to the passage of individual particles in the near vicinity of the probe [20]. Different from pressure fluctuation signals, the solids concentration signals show binary behavior [35]. The peaks of the signals stand for high solids concentration and the valleys represent low solids concentration. During the calibration experiments, after reaching steady state, the mean output voltages of the probe were measured by immersing the probe at four radical positions: r/R = 0, r/R = 0.2, r/R = 0.4, r/R = 0.8. The radial voltage values of the probe are plotted in Figure 6. It can be seen that the output voltages of the probe at different radial positions basically agree well. The output voltages of the probe recorded near the wall is higher than those of the other three radial positions, which indicates higher solids concentration near the wall due to the wall effect. It can be inferred that relatively homogeneous suspension of solids was obtained.

Figure 6. The probe output voltages of GB 3# at four different radial positions.

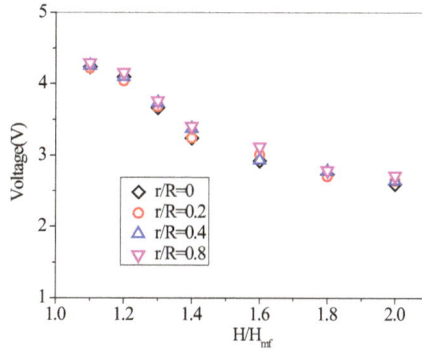

In order to ensure that day-to-day measurements could be compared on an equal basis and to eliminate any parasitic effects caused by this particular apparatus which may prevent the use of the calibration curve in other systems, the output voltages were normalized according to the following expression [21]:

$$v = \frac{V - V_0}{V_p - V_0} \tag{10}$$

where V_0 is the output voltage when the solids concentration is zero, and V_p is the output voltage when the solids concentration is equal to packed bed concentration. According to the research of Cui et al. [30], the calibration correlations of the normalized output voltages and normalized solids concentration can be expressed as follows:

$$\frac{\varepsilon_s}{\varepsilon_{s,p}} = \frac{av}{(1+a)-v} \tag{11}$$

134

where $\varepsilon_{s,p}$ is the packed bed solids concentration, $\varepsilon_{s,p}/\varepsilon_{s,p}$ is normalized solids concentration, and a is the calibration coefficient.

The relationships between normalized output voltages and normalized solids concentration are shown in Figures 7 and 8. From Figures 7 and 8, it can be seen that the output voltages increases sharply with increasing solids concentration at low solids concentrations, while increases slowly at high solids concentrations, which is consistent with the variation trend obtained by Cui *et al.* [30]. It may due to that for higher solids concentrations, the probe has a smaller measuring volume and lead to a more 'local' measurement because particles in the measuring volume woule be blocked by particles in front of them. Meanwhile, it can also been seen that the calibration coefficient a increases with the increasing of particle size.

Figure 7. Calibration curves for Glass beads: (**a**) GB 1#; (**b**) GB 2#; (**c**) GB 3#.

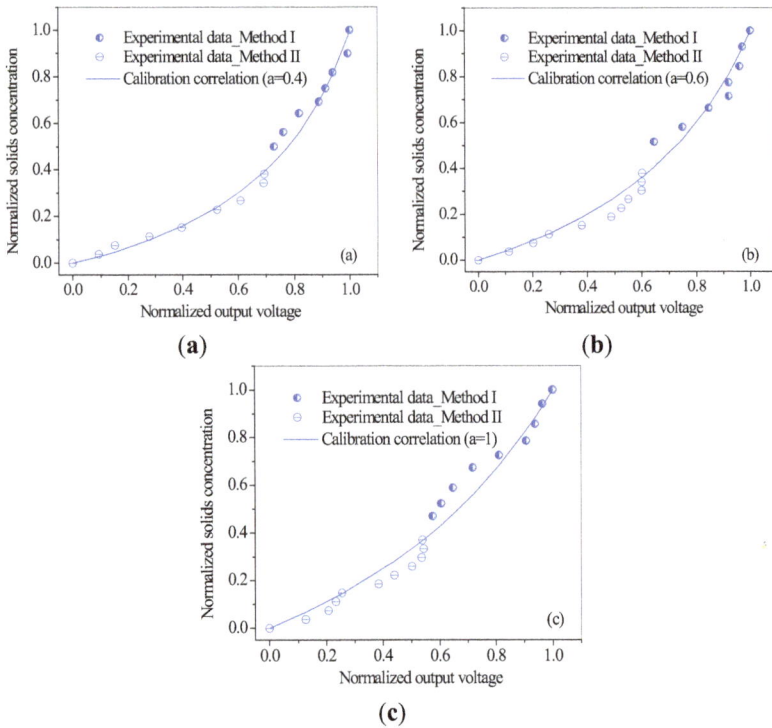

(a)

(b)

(c)

Method II was used to calibrate the probe with a solids concentration range from 0.02 to 0.2 for both of the two powders. Method I worked at the solids concentrations from 0.26 to 0.53 for glass beads and from 0.22 to 0.5 for quartz sand. From Figures 7 and 8, it can also be seen that method I and method II together are capable of calibrating the optical-fiber probe at nearly full range of the solids concentrations from 0 to packed bed concentration.

Figure 8. Calibration curves for Quartz sand: (**a**) QS 1#; (**b**) QS 2#; (**c**) QS 3#.

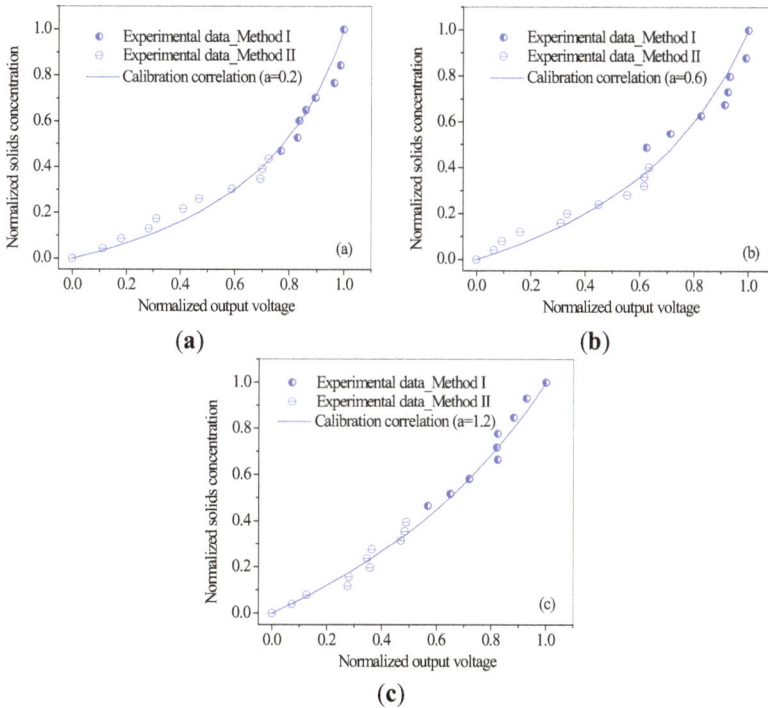

(**a**)

(**b**)

(**c**)

The suspensions encountered in practical problems of chemical engineering usually consist of particles with statistically varying irregular shapes, size and surface properties, locations and spatial orientations which are altogether unknown and for practical problems also may not be exactly describable [36]. The effect of particle size on calibration results is shown in Figure 9. It can be inferred that the output voltage of the optical fiber probe has a tendency to increase with decreasing particle size for the whole range of solid concentrations, but the increase tendency is not so obvious when the solids concentration less than 0.1. The intensity of the backscattered light is a function of the solids concentration and the mean particle size [25,37]. Yamazaki *et al.* [25] found that the intensity of the reflected light is affected by the particle diameter and the back-scattered light reach the probe decreases as the particle diameter increases. This may be attributed to the increase in the average path length of the light beam, which is caused by an increased particle diameter. Bos *et al.* [37] indicated that the intensity of the reflected light increases as the particle diameter decreases and the solids concentration increases. The present finding is consistent with their results. Meanwhile, Qi [24] also obtained similar conclusions that the output voltage decreases substantially with increasing particle size. According to Amos *et al.* [38], the relationship between the probe output and the solids concentration and the variation of this relationship seemed to be strongly dependent on the relationship between the particle size and the fiber diameter. With fixed fiber diameter, the relationship between the probe output and the solids concentration seemed to have the same slop at the low solids concentrations. In our study, the offset of the probe were adjusted to nearly zero with empty black box for each power with different particle diameter. Thus, the difference of calibration curves within the low solids concentration range between different particle sizes is not obvious. Amos *et al.* [38] also found that particle size effect was greater at higher solids concentrations, and that increasing the particle diameter while keeping constant the solids concentration and fiber diameter would lead to more light penetrate into the solid suspension. The light penetrating into the bed deeper than one particle would never be reflected out of the solids suspension. Meanwhile, the discrepancy of calibration results with different particle size exists at moderate solids concentration may also have some relationship with some heterogeneity effect in the gas-solid system due to that the inherent fluctuations are unable to be completely avoided in a gas-solid flow system. During the calibration experiments, it was found that the bubbles are more likely to form and expand at moderate solids concentrations. The bubble suppressor can reduce the generation of bubbles and avoid inherent fluctuations inside the calibration column to a certain extent, but it can't suppress the generation of bubbles totally. The degree of heterogeneity was relatively larger at moderate solids concentrations. Thus, a powder with very homogeneous fluidized behavior would be expected to have a more reliable calibration curve with this applied methodology.

Figure 9. Effect of particle size on the calibration results: (**a**) Glass beads; (**b**) Quartz sand.

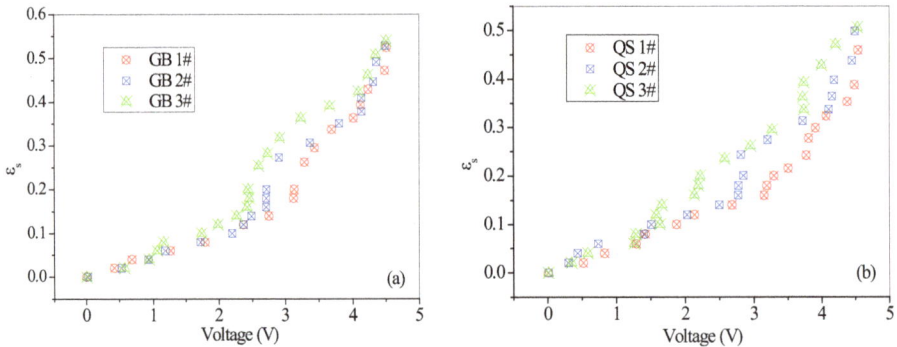

GB 1# and QS 1# are similar in particle size, but different in sphericity and color. Thus, comparisons are difficult to be made directly between these two powders, as shown in Figure 10. A batch of fresh glass beads with the same particle size as GB 1# but has a color of white, which was named GB 0#, was used to compare with the above mentioned two powders separately. The effect of particle color on calibration results is shown in Figure 11. The difference between GB 0# and GB 1# was color, and GB 1# was darker than GB 0#. It can be seen from Figure 11 that GB 0# has higher output voltages than those of GB 1#, which is due to that dark powder reflects less light [27]. The effect of particle sphericity on calibration results is shown in Figure 12. The calibration curves of GB 0# and QS 1# overlapped together. For sand particles, the sphericity is in the region of 0.8~0.98 [10] which is usually less than that of glass beads. Rundqvist *et al.* [10] pointed out that the sampling light scattered from irregular particles will approximate the light scattered from the same number of spherical particles if a sufficient number of particles are considered, as the particles are oriented randomly in the suspension. The irregular object will appear more spherical as angular velocity increases, which is equivalent to sampling the same object from different angles. The impact of sphericity on the calibration theory was neglected in their research. Thus, it can be inferred that the effect of particle color on calibration result is more predominant than that of sphericity, Qi [24] also obtained the same conclusion. The calibration curves in Figure 10 are very close and overlapped at solids concentrations less than 0.2, which means that the calibration curve is less sensitive to the particle color when the solids concentration is low.

Figure 10. The compare of calibration results: GB1# and QS1#.

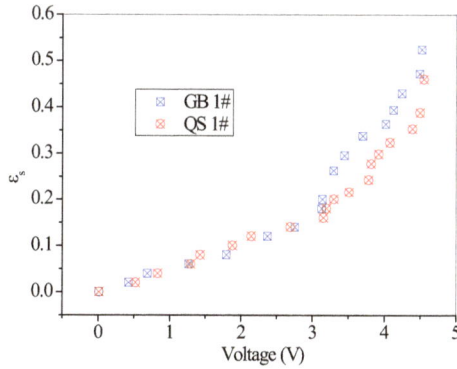

Figure 11. Effect of particle color on the calibration results.

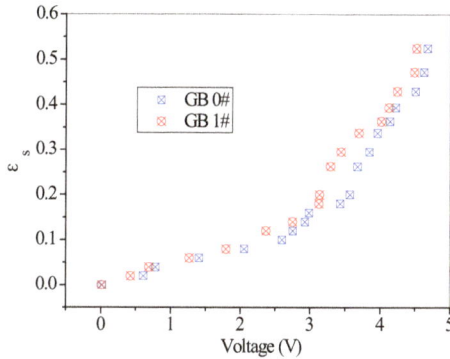

Figure 12. Effect of particle sphericity on the calibration results.

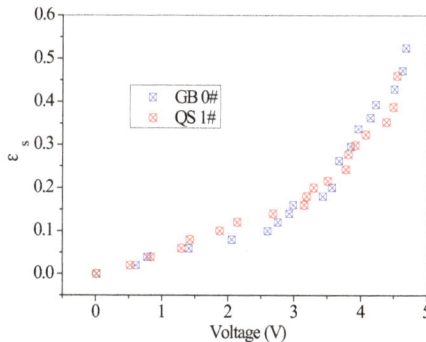

138

The difference of the calibration curve at solids concentration more than 0.2 may due to the color difference. The GB 1# are darker in color than QS 1#. Grey glass beads have the tendency to absorb more light compared to white quartz sand. It can be seen that under the same solids concentration, the output voltages of QS 1# are higher than that of GB 1#.

5. Conclusions

Different calibration methods of optical-fiber solids concentration probes reported in the literature were reviewed in this paper. Satisfactory calibration procedures are lacking in the literature, and the exact shape of the calibration function has not been verified exactly. A combined calibration method, which is capable of calibrating the optical-fiber probe at nearly full range of the solids concentrations from 0 to packed bed concentration, is proposed. With this combined method, the probe was uniquely calibrated with two kinds of powders. The effects of particle properties (particle size, sphericity and color) on the calibration results were comprehensively investigated. From the experiments carried out here, it can be concluded that the output voltage has a tendency to increase with the decreasing particle size, and the effect of particle color on the calibration curve is more predominant than that of sphericity.

Acknowledgments

The authors acknowledge the financial supports of the National Natural Science Foundation of China (51276036), Special Funds of National Key Basic Research and Development Program of China (2010CB227002), and the anonymous reviewers are thanked for their insightful and detailed comments.

Conflicts of Interest

The authors declare no conflict of interest.

References

1. Nieuwland, J.J.; Meijer, R.; Kuipers, J.A.M.; van Swaaij, W.P.M. Measurements of solids concentration and axial solids velocity in gas-solid two-phase flows. *Powder Technol.* **1996**, *87*, 127–139.
2. Wiesendorf, V.; Werther, J. Capacitance probes for solids volume concentration and velocity measurements in industrial fluidized bed reactors. *Powder Technol.* **2000**, *110*, 143–157.
3. Werther, J. Measurement techniques in fluidized beds. *Powder Technol.* **1999**, *102*, 15–36.
4. Van Ommen, J.R.; Mudde, R.F. Measuring the gas-solids distribution in fluidized beds—A review. *Int. J. Chem. Reactor Eng.* **2008**, *6*, doi:10.2202/1542-6580.1796.
5. Makkawi, Y.; Ocone, R. Integration of ECT measurements with hydrodynamic modelling of conventional gas—Solid bubbling bed. *Chem. Eng. Sci.* **2007**, *62*, 4304–4315.
6. Rautenbach, C.; Mudde, R.F.; Yang, X.; Melaaen, M.C.; Halvorsen, B.M. A comparative study between electrical capacitance tomography and time-resolved X-raytomography. *Flow Meas. Instrum.* **2013**, *30*, 34–44.

7. Wang, F.; Marashdeh, Q.; Fan, L.-S.; Warsito, W. Electrical capacitance volume tomography: Design and applications. *Sensors* **2010**, *10*, 1890–1917.

8. Xu, J.; Zhu, J. Effects of particle properties on flow structure in a 2-D circulating fluidized bed: Solids concentration distribution and flow development. *Chem. Eng. Sci.* **2011**, *66*, 5064–5076.

9. Saberi, B.; Shakourzadeh, K.; Guigon, P. Local solid concentration measurement by fibre optics: Application to circulating fluidized beds. *Chem. Eng. Res. Des.* **1998**, *76*, 748–752.

10. Rundqvist, R.; Magnusson, A.; van Wachem, B.G.M.; Almstedt, A.E. Dual optical fibre measurements of the particle concentration in gas/solid flows. *Exp. Fluids* **2003**, *35*, 572–579.

11. Zhang, H.; Zhu, J.X.; Bergougnou, M.A. Hydrodynamics in downflow fluidized beds (1): Solids concentration profiles and pressure gradient distributions. *Chem. Eng. Sci.* **1999**, *54*, 5461–5470.

12. Zhu, H.; Zhu, J. Characterization of fluidization behavior in the bottom region of CFB risers. *Chem. Eng. J.* **2008**, *141*, 169–179.

13. Amos, G.; Rhodes, M.J.; Benkreira, H. Calculation of optic fibres calibration curves for the measurement of solids volume fractions in multiphase flows. *Powder Technol.* **1996**, *88*, 107–121.

14. Link, J.M.; Godlieb, W.; Tripp, P.; Deen, N.G.; Heinrich, S.; Kuipers, J.A.M.; Schönherr, M.; Peglow, M. Comparison of fibre optical measurements and discrete element simulations for the study of granulation in a spout fluidized bed. *Powder Technol.* **2009**, *189*, 202–217.

15. Cutolo, A.; Rendina, I.; Arena, U.; Marzocchella, A.; Massimilla, L. Optoelectronic technique for the characterization of high concentration gas-solid suspension. *Appl. Opt.* **1990**, *29*, 1317–1322.

16. Li, X.; Yang, C.; Yang, S.; Li, G. Fiber-optical sensors: Basics and applications in multiphase reactors. *Sensors* **2012**, *12*, 12519–12544.

17. Ximei, Z. Measurements on the Local Solids Concentration in the Lower Part of a Circulating Fluidized Bed Riser. In Proceedings of the International Conference on Energy and Environment Technology (ICEET 2009), Guilin, China, 16–18 October 2009; pp. 728–732.

18. Soo, S.L. *Instrumentation for Fluid-Particle Flow*; Noyes Publications: Park Ridge, NJ, USA, 1999.

19. Matsuno, Y.; Yamaguchi, H.; Oka, T.; Kage, H.; Higashitani, K. The use of optic fiber probes for the measurement of dilute particle concentrations: Calibration and application to gas-fluidized bed carryover. *Powder Technol.* **1983**, *36*, 215–221.

20. Lischer, D.J.; Louge, M.Y. Optical fiber measurements of particle concentration in dense suspensions: Calibration and simulation. *Appl. Opt.* **1992**, *31*, 5106–5113.

21. Herbert, P.M.; Gauthier, T.A.; Briens, C.L.; Bergougnou, M.A. Application of fiber optic reflection probes to the measurement of local particle velocity and concentration in gas—Solid flow. *Powder Technol.* **1994**, *80*, 243–252.

22. Liu, J.; Grace, J.R.; Bi, X. Novel multifunctional optical-fiber probe: I. Development and validation. *AIChE J.* **2003**, *49*, 1405–1420.

23. Hong, J.; Tomita, Y. Measurement of distribution of solids concentration on high density gas-solids flow using an optical-fiber probe system. *Powder Technol.* **1995**, *83*, 85–91.

24. Maozhan, Q. Hydrodynamics and Micro Flow Structure of Gas-Solid Circulating Turbulent Fluidized Beds. Ph.D. Thesis, The University of Western Ontario, London, ON, Canada, 2012.

25. Yamazaki, H.; Tojo, K.; Miyanami, K. Measurement of local solids concentration in a suspension by an optical method. *Powder Technol.* **1992**, *70*, 93–96.

26. Zhou, J.; Grace, J.R.; Qin, S.; Brereton, C.M.H.; Lim, C.J.; Zhu, J. Voidage profiles in a circulating fluidized bed of square cross-section. *Chem. Eng. Sci.* **1994**, *49*, 3217–3226.

27. Zhang, H.; Johnston, P.M.; Zhu, J.X.; de Lasa, H.I.; Bergougnou, M.A. A novel calibration procedure for a fiber optic solids concentration probe. *Powder Technol.* **1998**, *100*, 260–272.

28. San, J.M.J.; Olazar, M.; Alvarez, S.; Bilbao, J. Local bed voidage in conical spouted beds. *Ind. Eng. Chem. Res.* **1998**, *37*, 2553–2558.

29. Johnsson, H.; Johnsson, F. Measurements of local solids volume-fraction in fluidized bed boilers. *Powder Technol.* **2001**, *115*, 13–26.

30. Cui, H.; Mostoufi, N.; Chaouki, J. Comparison of Measurement Techniques of Local Particle Concentration for Gas-Solid Fluidization. In *Fluidization X*; Engineering Foundation: New York, NY, USA, 2001; pp. 779–786.

31. Liu, J.; Grace, J.R.; Bi, X. Novel multifunctional optical-fiber probe: II. High-density CFB measurements. *AIChE J.* **2003**, *49*, 1421–1432.

32. Issangya, A.S. Flow Dynamics in High Density Circulating Fluidized Beds. Ph.D. Thesis, The University of British Columbia, Vancouver, BC, Canada, 1998.

33. Magnusson, A.; Rundqvist, R.; Almstedt, A.E.; Johnsson, F. Dual fibre optical probe measurements of solids volume fraction in a circulating fluidized bed. *Powder Technol.* **2005**, *151*, 19–26.

34. Wei, W.; Qingliang, G.; Yuxin, W.; Hairui, Y.; Jiansheng, Z.; Junfu, L. Experimental study on the solid velocity in horizontal dilute phase pneumatic conveying of fine powders. *Powder Technol.* **2011**, *212*, 403–409.

35. Manyele, S.V.; Zhu, J.X.; Khayat, R.E.; Pärssinen, J.H. Analysis of the chaotic dynamics of a high-flux CFB riser using solids concentration measurements. *China Particuol.* **2006**, *4*, 136–146.

36. Rensner, D.; Werther, J. Estimation of the effective measuring volume of single-fibre reflection probes for solid volume concentration measurements. *Part. Part. Syst. Charact.* **1993**, *10*, 48–55.

37. Bos, A.S.; Heerens, J.J. Light backscattering as a technique to measure solids particle size and concentration in suspension. *Chem. Eng. Commun.* **1982**, *16*, 301–311.

38. Amos, G. Fluid Dynamics of Upward-Flowing Gas-Solids Suspensions. Ph.D. Thesis, University of Bradford, Bradford, UK, 1994.

Reprinted from *Sensors*. Cite as: Tsao, Y.-C.; Tsai, W.-H.; Shih, W.-C.; Wu, M.-S. An *In-situ* Real-Time Optical Fiber Sensor Based on Surface Plasmon Resonance for Monitoring the Growth of TiO2 Thin Films. *Sensors* **2013**, *13*, 9513–9521.

Letter

An *In-situ* Real-Time Optical Fiber Sensor Based on Surface Plasmon Resonance for Monitoring the Growth of TiO$_2$ Thin Films

Yu-Chia Tsao *, Woo-Hu Tsai, Wen-Ching Shih and Mu-Shiang Wu

Graduate Program in Electro-Optical Engineering, Tatung University, 40 Chongshan N. Rd., 3rd Sec. Taipei 104, Taiwan; E-Mails: whtsai@ttu.edu.tw (W.-H.T.); wcshih@ttu.edu.tw (W.-C.S.); mswu@ttu.edu.tw (M.-S.W.)

* Author to whom correspondence should be addressed; E-Mail: zaktaso@gmail.com; Tel.: +886-2-2182-2928 (ext. 6820); Fax: +886-2-2595-6393.

Received: 19 June 2013; in revised form: 12 July 2013 / Accepted: 18 July 2013 / Published: 23 July 2013

Abstract: An optical fiber sensor based on surface plasmon resonance (SPR) is proposed for monitoring the thickness of deposited nano-thin films. A side-polished multimode SPR optical fiber sensor with an 850 nm-LD is used as the transducing element for real-time monitoring of the deposited TiO$_2$ thin films. The SPR optical fiber sensor was installed in the TiO$_2$ sputtering system in order to measure the thickness of the deposited sample during TiO$_2$ deposition. The SPR response declined in real-time in relation to the growth of the thickness of the TiO$_2$ thin film. Our results show the same trend of the SPR response in real-time and in spectra taken before and after deposition. The SPR transmitted intensity changes by approximately 18.76% corresponding to 50 nm of deposited TiO$_2$ thin film. We have shown that optical fiber sensors utilizing SPR have the potential for real-time monitoring of the SPR technology of nanometer film thickness. The compact size of the SPR fiber sensor enables it to be positioned inside the deposition chamber, and it could thus measure the film thickness directly in real-time. This technology also has potential application for monitoring the deposition of other materials. Moreover, *in-situ* real-time SPR optical fiber sensor technology is in inexpensive, disposable technique that has anti-interference properties, and the potential to enable on-line monitoring and monitoring of organic coatings.

Keywords: surface plasmon resonance; optical fiber sensor

1. Introduction

Over the past decade, surface plasmon resonance (SPR) has been used in a wide range of chemical and biological sensing applications. SPR is a charge density oscillation that may exist at the interface of two media with dielectric constants of opposite sign, for instance, a metal and a dielectric. The excitation of a surface plasmon wave leads to the appearance of a dip in the measured intensity of reflected light, which must be considered in determining the sensitivity of SPR sensing. The first use of SPR in prism coupling was proposed by Kretschmann in 1968 [1]. In the early years after the basic theory of the SPR phenomenon was presented by Kretschmann, some attempts were made to develop an SPR sensor based on the optical interaction effect. Since then, surface plasmons have been studied extensively, and their major properties have been assessed [2]. The sensitivity of an SPR sensor is defined as the derivative of the monitored SPR parameter, which can detect changes of the refractive index. Several detection schemes have been demonstrated for SPR prism-based sensors; they include angular interrogation [3], wavelength interrogation [4], and intensity measurement [5].

Since Jorgenson and Yee proposed using optical fibers for SPR sensing in 1993 [6], many types of optical fiber sensor have been proposed, including single-mode dip fibers [7], single-mode D-type fibers [8], and D-shaped fibers [9]. In recent years, studies of SPR sensing systems have been focused on the attenuated total reflection geometry obtained by use of prism-coupling optics. However, those systems of optical fiber sensors need bulky structures as well as complicated signal processing to improve their high sensitivity. In contrast, the side-polished multimode fiber sensor provides a simple structure and system for chemical and biological sensing with high sensitivity in wavelength interrogation [10,11], and also has a high detection limit for biomolecules [11,12]. Time-dependent measurements of SPR fiber sensors are important for applications in biosensing and in environmental monitoring, and the sensitivity of intensity measurements with side-polished fiber sensors has been demonstrated [13,14].

Most of today's available techniques are restricted to certain type of films and many have difficulties with performing measurement *in-situ*. Various thin film measurement methods, such as surface profilometry and resistivity measurement are very difficult to carry out *in-situ*, and analysis of film growth rate is usually performed after the deposition run. Currently, ellipsometry is preferred as an *in-situ* monitoring tool [15–17]. However, this method lacks the ability to measure opaque films. Quartz crystal monitors are also available as an *in-situ* monitoring tool, but they only offer an indirect measurement of the thickness of a film grown on a surface [18,19]. To overcome these limitations, we have proposed a new *in-situ* monitoring technique utilizing a side-polished multimode optical fiber sensor based on surface plasmon resonance to measure the thickness of deposited nano-thin films.

2. Experiment

2.1. The Side-Polished SPR Optical Fiber Sensor

The SPR optical fiber sensor consists of a gold thin film and a side-polished structure. This sensor outputs the SPR response spectrum and provides a high detection limit for biomolecules [11–13]. The optical fiber sensor based on surface plasmon resonance with side-polished structure is shown in Figure 1a. The graded-index multi-mode fiber with 62.5 μm core diameter and 62.5 μm cladding thickness, fabricated by Prime Optical Fiber Corporation (POFC), was side-polished to make an SPR optical fiber sensor. For high yield rate polishing processes, a silicon V-groove must be fabricated to hold the bare fibers. Thus, we grew the oxides layer on a 4-inch silicon (1 0 0) wafer and used photolithography to etch a SiO₂ channel with 25% HF. The V-groove channel was etched by 45.3% KOH; the channel length and width were 5 mm and 125 μm, respectively. The multimode fiber was mounted on the V-groove holder with photoresist and monitored by optical microscopy. After the photoresist became hard, we polished the fibers that were embedded in the wafer using polishing diamond films with grain sizes of 6, 1 and 0.1 μm in turn. To increase the sensitivity of the SPR measurements, the length of the polished surface who set to 5 mm and the depth to 62.5 μm for the fundamental mode region. The dimensions of the polished surface, length and breadth, were 5 mm and 62.5 μm, and the depth of 62.5 μm was confirmed by optical microscopy (Olympus, BH, Tokyo, Japan).

Figure 1. (a) Schematic diagram of the SPR optical fiber sensor with sided-polished structure. **(b)** Schematic diagram of the PVD sputtering system and SPR optical fiber sensing system.

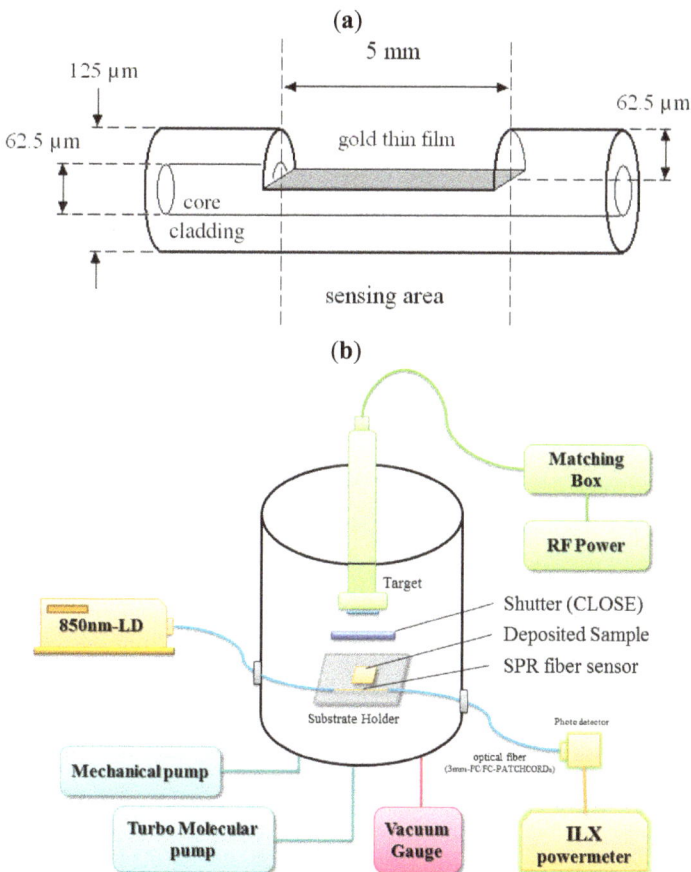

144

The gold thin film was deposited on the polished surface by a DC sputtering system (ULVAC Co., Kanagawa, Japan). During the deposition process, the vacuum chamber was evacuated to 4×10^{-2} torr, and the background pressure was 2×10^{-5} torr. The gold thin film was approximately 40 nm thick, and was observed using an SEM (scanning electron microscope, JEOL 7000, JEOL, Tokyo, Japan). All of these optical parts were manufactured by New Product Div., Forward Electronics Co., New Taipei City, Taiwan.

2.2. TiO₂ Sputtering System and SPR Optical Fiber Sensing System

As shown in Figure 1b, the *in-situ* TiO₂ sensing system of the SPR optical fiber sensor consists of a light source (850 nm-LD), a photodetector, an ILX power meter (with personal computer), an SPR fiber sensor, and an RF planar magnetron sputtering system (ANELVA SPF-210HS, Kanagawa, Japan). This system was used to demonstrate in-situ real-time monitoring of the growth of TiO₂ thin film based on the variation of the transmitted intensity of the SPR optical fiber sensor. The components are connected with 3-mm FC/FC-PATCHCORDs, which have a small insertion loss (<0.05 dB) and back reflection55(<dB) and are manufactured by General Optics Corporation (Zhongli, Taiwan). Because the surface plasmon can be excited by 850 nm light [12] via a suitable optical fiber, a light source (850 nm VCSEL MM Module, Appointech. Inc., Hsinch, Taiwan) with temperature-control is used. The light source also has high coupled power and reliability with pigtail-connected fiber. The Si-photodetector (Electro-Optics Technology, Inc., Traverse City, MI, USA) can monitor the output of an externally modulated cw laser. An ILX powermeter (OMM-6810B, ILX Lightwave, Irvine, CA, USA) suitable for a wavelength range of 350–1,700 nm, was also used. The sensitivity of the optical power meter is approximately 0.01 pW, and a personal computer was used to acquire data from measurements of the transmitted power.

An RF planar magnetron sputtering system (ANELVA SPF-210HS) was used to prepare the TiO₂ thin film for depositing. A 99.9% pure TiO₂ target disc, 2 inches in diameter, was used as the sputtering target. The chamber was pumped down to 1×10^{-6} torr using a turbo molecular pump before the sputtering gas (Ar) was introduced into the chamber through the mass flow controllers and controlled by the main valve of the pumping system. In all experiments, the target was pre-sputtered for 30 min under 60 W RF power before the actual sputtering to remove any contamination on the target surface. We fixed the target-to-substrate spacing at 50 mm, the deposition time to 30 min, and deposition pressure to 4×10^{-4} torr for the in-situ real time SPR monitoring of the growth of TiO₂ thin films.

2.3. Configuration of the SPR Optical Fiber Sensor and Deposited Sample

The detection area, side-polished surface, of the optical fiber sensor and the sputtering direction must be in the same direction, particularly in the horizontal plane of the deposited sample. The manufacture of the SPR optical fiber sensor is described in Section 2.2. The multimode optical fiber is installed in a V-groove on a polished holder to avoid the breaking the optical fiber between the edges of V-groove during the polishing step. To determine the direction with the SPR optical fiber sensor, the deposited sample is set on the polished holder; as shown in Figure 2. The distance between the centers of the deposited sample and optical fiber sensor is 7.5 mm. Figure 3 shows that the polished holder (2.5 × 3.75 cm) contains the optical fiber sensing area and deposited sample all in the 2-inch radius TiO_2 deposition area.

Figure 2. Schematic diagram of the setup for the SPR optical fiber sensor and deposited sample.

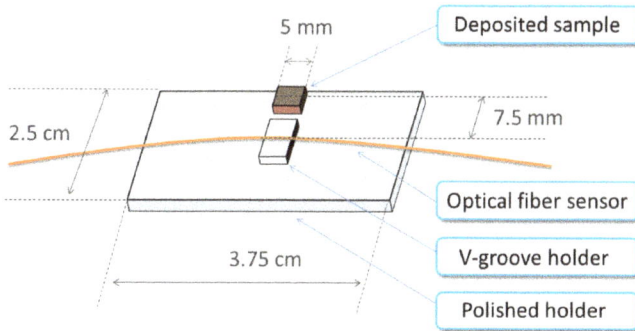

Figure 3. Deposition area and locations of the SPR optical fiber sensor and deposited sample.

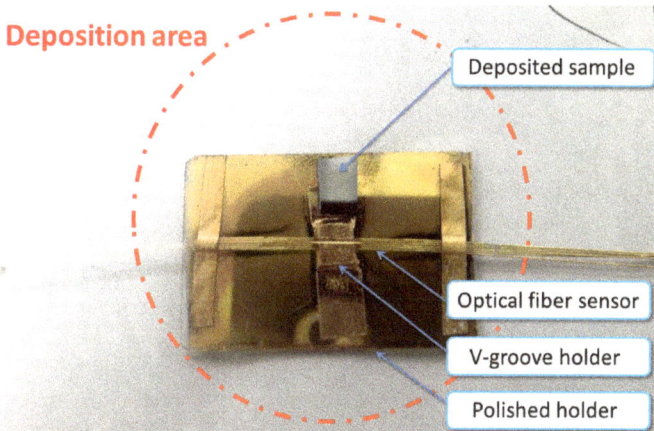

3. Results and Discussion

In this study, we compared of the deposited sample with the SPR response of an optical fiber to demonstrate the possibility of in-situ real-time monitoring. To monitor the growth of TiO_2 directly, the SPR optical fiber sensor was positioned near the deposited sample in the TiO_2 sputtering system, as shown in Figures 2 and 3. The thickness of TiO_2 thin film deposited in 30 min, measured using an FE-SEM (High Resolution Field-Emission Scanning Electron Microscope, JEOL JSM-6500F) is 50 nm, as shown on Figure 4.

The SPR response while monitoring the growth of TiO_2 thin film in real time is shown in Figure 5. The X-axis is time in minutes, and the Y-axis is the transmitted intensity. The transmitted intensity is the output power at the end of the optical fiber as measured by the ILX powermeter. There are two major steps in the sputtering process: pre-sputtering and sputtering. We use pre-sputtering to remove contaminant particles from the target surface; the transmitted intensity was not affected in this process. The location of shutter is between target and substrate. When the shutter is close, there

146

will be no anything deposited on the substrate. After pre-sputtering the shutter is opened for the deposition of TiO$_2$ thin film on the substrate. As shown in Figure 5, the sputtering time is 30 min.

Figure 4. The thickness of the 50 nm-TiO$_2$ thin film with 30 min. deposition time as observed by FE-SEM.

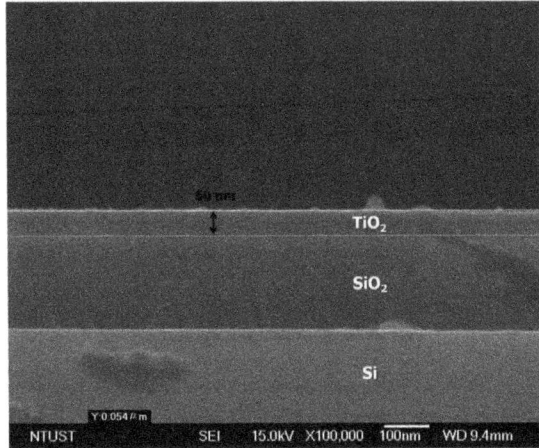

The SPR response measured in real-time during the 30 min pre-sputtering process (between points A and B in Figure 5) is approximately 163.85 µW (164.5 µW ~ 163.2 µW). The influence of the SPR optical fiber sensor cause less than 1.3 µW variation in the vacuum chamber, and does not significantly affect the signal of SPR optical fiber sensor in the vacuum chamber. After opening the shutter for TiO$_2$ sputtering, the SPR response monitored in real-time for 30-min of sputtering changes from 162.87 µW at point B to 132.33 µW at point D in Figure 5.

Figure 5. Dynamic experimental results for an SPR optical fiber sensor with 850 nm-LD monitoring the growth of deposited TiO$_2$ thin film in-situ in real time. The pre-sputtering process occurs between points A and B. The sputtering process occurs between points B and D.

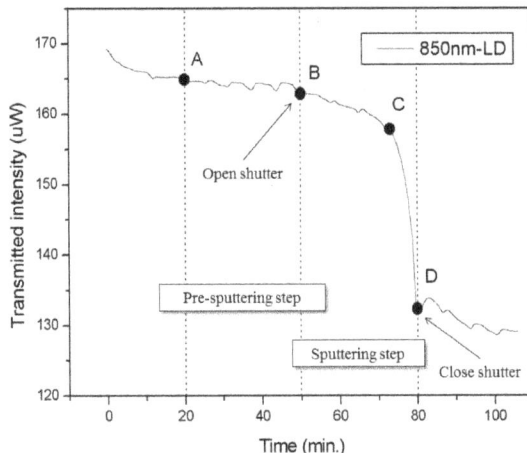

The SPR response shows that TiO_2 was deposited on the surface of the optical fiber. It also shows two distinct steps. It is well-known that the increased thickness of thin films is the factor that causes the SPR dip to shift to a longer-wavelength [11]. The most sensitive part of the SPR response occurs at the center of the SPR dip in wavelength. The first step of the SPR response, from point B to point C, shows minor variation and changes due to noise from 162.87 μW to 158.02 μW. The second step of the SPR response, from point C to point D, displays large changes from 158.02 μW to 132.33 μW. The TiO_2 sputtering finished after the shutter was closed at point D. This result shows that the deposition of TiO_2 thin films can be monitored *in-situ* in real-time by the SPR response of an optical fiber sensor.

The spectroscopic SPR response in spectra, measured by halogen light and an optical spectrum analyzer (OSA), is well-known [10,11,20]. The SPR spectra for layers of Au and TiO_2/Au are shown on Figure 6. The black and gray lines are the SPR spectra for a 40 nm-Au layer and a 50 nm-TiO_2/40 nm-Au layer, respectively, which were measured by optical spectrum analyzer (OSA, AQ6315A, Ando Electric Co., Ltd. Tokyo, Japan). The SPR spectral responses show that the transmitted intensity at 850 nm from E to F drops from 78.97 nW down to 63.77 nW. In other words, the transmitted intensity at 850 nm after 50 nm TiO_2 sputtering has dropped to 80.75% of the value before sputtering.

The real-time measurement of the SPR response shows a drop in transmitted intensity during sputtering to 81.24% of the pre-sputtering value, as seen in Figure 5 between points B and D. Thus, the same trend is seen in the real-time *in-situ* measurements as in the spectroscopic measurements. The SPR transmitted intensity response changes in real-time approximately 18.76% for a deposition of 50 nm thickness of TiO_2 thin film. We expect that thicknesses of coated TiO_2 thin film will correspond to a different SPR dip wavelengths. These measurements display the potential for real-time monitoring of nanometer film thickness with SPR optical fiber sensor technology. Further experiments and measurements investigating different coating film thicknesses and their corresponding dip wavelengths are under way.

Figure 6. The spectroscopic SPR response with a halogen light source for a 50 nm layer of TiO_2 deposited on an Au layer. The black and gray lines are the SPR spectra for a 40 nm-Au layer and a 50 nm-TiO_2/40 nm-Au layer, respectively, as measured by OSA. The dashed line is set to calculate the change of the intensity at 850 nm.

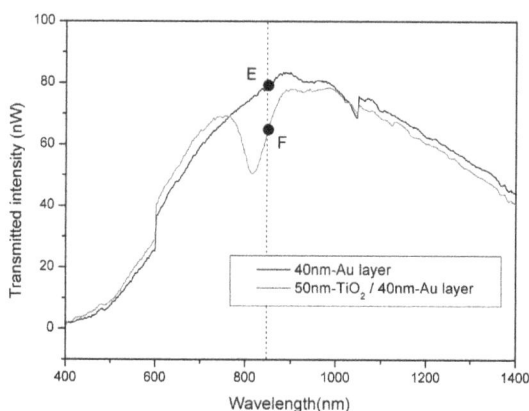

4. Conclusions

An *in-situ* thin film monitoring technique based on surface plasmon resonance and utilizing a side-polished multimode optical fiber sensor is proposed to measure the thickness of deposited nano-thin films. In this study, we positioned the SPR optical fiber sensor near the deposited sample in the TiO_2 sputtering system for real-time monitoring. The 850 nm-LD light source is suitable for exciting surface plasmon resonance in an optical fiber sensor. The sputtering process deposited TiO_2 on the material surface to change the refractive index, the effect of which was measured in real time by the SPR optical fiber sensor by monitoring the transmitted intensity of the SPR response. The decrease of the SPR response seen in real-time was related to the thickness of the TiO_2 thin film deposited and showed the same trend as the spectral SPR response. The SPR transmitted intensity measured in real-time changes by approximately 18.76% for a deposition of a 50 nm of TiO_2 thin film. We have shown the potential for real-time monitoring nanometer film thickness with the SPR response of optical fiber sensors. The compact size of the SPR fiber sensor allowed it to be set into the deposition chamber and thus measure directly in real-time. This technology also has the potential for application in other material deposition monitoring. Moreover, *in-situ* real-time SPR optical fiber sensor technology is inexpensive and disposable, has anti-interference properties, and has the potential to be used as an on-line monitor or organic coating monitor.

Conflicts of Interest

The authors declare no conflict of interest.

References

1. Raether, H. Surface plasmons on smooth and rough surfaces and on gratings. *Spring. Tract. Mod. Phys.* **1988**, *11*, 10–13.
2. Shen, S.; Lin, T.; Guo, J. Optical phase-shift detection of surface plasmon resonance. *Appl. Opt.* **1988**, *37*, 1747–1751.
3. Matsubara, K.; Kawata, S. Minami, S. Optical chemical sensor based on surface plasmon measurement. *Appl. Opt.* **1988**, *27*, 1160–1163.
4. Zhang, L.M.; Uttamchandani, D. Optical chemical sensing employing surface plasmon resonance. *Electron. Lett.* **1988**, *23*, 1469–1470.
5. Vidal, M.B.; Lopez, R.; Aleggret, S.; Alonso-Chamarro, J.; Garces, I.; Mateo, J. Determination of probable alcohol yield in musts by means of an SPR optical sensor. *Sens. Actuators B.* **1993**, *11*, 455–459.
6. Jorgenson, R.C.; Yee, S.S. A fiber-optic chemical sensor based on surface plasmon resonance. *Sens. Actuators B* **1993**, *12*, 213–220.
7. Díez, A.; Andrés, M.V.; Cruz, J.L. In-line fiber-optic sensors based on the excitation of surface plasma modes in metal-coated tapered fibers. *Sens. Actuators B* **2001**, *73*, 95–99.
8. Chin, M.H.; Wang, S.F.; Chang, R.S. D-type fiber biosensor based on surface-plasmon resonance technology and heterodyne interferometry. *Opt. Lett.* **2005**, *30*, 233–235.

9. Sorin, W.V.; Yu, M.H. Single-mode-fiber ring dye laser. *Opt. Lett.* **1985**, *10*, 550–552.

10. Tsai, W.H.; Tsao, Y.C.; Lin, H.Y.; Sheu, B.C. Cross-point analysis for multimode fiber sensor based on surface plasmon resonance. *Opt. Lett.* **2005**, *30*, 2209–2211.

11. Tsai, W.H.; Tsao, Y.C.; Lin, H.Y.; Sheu, B.C. Side-polished multimode fiber sensor based on surface plasmon resonance with halogen light. *Appl. Opt.* **2007**, *46*, 800–806.

12. Lin, H.Y.; Tsao, Y.C.; Tsai, W.H.; Yang, Y.W.; Yan, T.R.; Sheu, B.C. Development and application of side-polished fiber immunosensor based on surface plasmon resonance for the detection of Legionella pneumophila with halogens light and 850 nm-LED. *Sens. Actuators A* **2007**, *138*, 299–305.

13. Lin, Y.C.; Tsai, W.H.; Tsao, Y.C.; Tai, J.K. An enhance optical multimode fiber sensor based on surface plasmon resonance with cascaded structure. *IEEE Photonic. Technol. Lett.* **2008**, *20*, 1287–1289.

14. Tsai, W.H.; Lin Y.C.; Tai, J.K.; Tsao Y.C. Multi-step structure of side-polished fiber sensor to enhance SPR effect. *Opt. Laser Technol.* **2010**, *42*, 453–456.

15. Yang, Y.; Jayaraman, S.; Sperling, B.; Kim, D.Y.; Girolami, G.S.; Abelson, J.R. *In situ* spectroscopic ellipsometry analyses of hafnium diboride thin films deposited by single-source chemical vapor deposition. *J. Vac. Sci. Technol. A* **2007**, *25*, 200–206.

16. Yamada, Y.; Bao, S.; Tajima, K.; Okada, M.; Tazawa, M.; Roos, A.; Yoshimura, K. *In situ* spectroscopic ellipsometry study of the hydrogenation process of switchable mirrors based on magnesium-nickel alloy thin films. *J. Appl. Phys.* **2010**, *107*, 043517–043525.

17. Stanford, J.L. Determination of surface-film thickness from shift of optically excited surface plasma resonance. *J. Opt. Soc. Am.* **1970**, *60* 1, 49–53.

18. Kawashima, H.; Sunaga, K. A Sensor for Thin Film Thickness Monitor Using Torsional Quartz Crystal Resonators of Stepped, Free-Free Bar-Type: In *Proceedings of the 1998 IEEE International Frequency Control Symposium*, Pasadena, CA, USA, 27–29 May 1998; pp. 691–694.

19. Chao, W.K.; Wong, H.K. Reusable quartz crystals for thickness monitoring in thin film deposition. *J. Vac. Sci. Technol. A* **1990**, *8*, 150–151.

20. Lin, Y.C.; Tsao, Y.C.; Tsai, W.H.; Hung, T.S.; Chen, K.S.; Liao, S.C. The enhancement method of optical fiber biosensor based on surface plasmon resonance with cold plasma modification. *Sens. Actuators B* **2008**, *133*, 370–373.

Reprinted from *Sensors*. Cite as: Ahmad, H.; Zulkifli, M.Z.; Muhammad, F.D.; Samangun, J.M.; Abdul-Rashid, H.A.; Harun, S.W. Temperature-Insensitive Bend Sensor Using Entirely Centered Erbium Doping in the Fiber Core. *Sensors* **2013**, *13*, 9536–9546.

Article

Temperature-Insensitive Bend Sensor Using Entirely Centered Erbium Doping in the Fiber Core

Harith Ahmad [1,]*****, **Mohd Zamani Zulkifli** [1], **Farah Diana Muhammad** [1], **Julian Md Samangun** [1], **Hairul Azhar Abdul-Rashid** [2] **and Sulaiman Wadi Harun** [1]

[1] Photonics Research Centre, University of Malaya, 50603 Kuala Lumpur, Malaysia;
    E-Mails: mohdzamani82@yahoo.com (M.Z.Z.); faradibah90@yahoo.com (F.D.M.);
    juliansamangun@gmail.com (J.M.S.); swharun@um.edu.my (S.W.H.)
[2] Faculty of Engineering, Multimedia University, 63100 Cyberjaya, Selangor, Malaysia;
    E-Mail: hairul@mmu.edu.my

* Author to whom correspondence should be addressed; E-Mail: harith@um.edu.my;
    Tel.: +603-7697-6770; Fax: +603-7967-4146.

Received: 3 June 2013; in revised form: 15 July 2013 / Accepted: 17 July 2013 / Published: 23 July 2013

Abstract: A fiber based bend sensor using a uniquely designed Bend-Sensitive Erbium Doped Fiber (BSEDF) is proposed and demonstrated. The BSEDF has two core regions, namely an undoped outer region with a diameter of about 9.38 µm encompassing a doped, inner core region with a diameter of 4.00 µm. The doped core region has about 400 ppm of an Er_2O_3 dopant. Pumping the BSEDF with a conventional 980 nm laser diode gives an Amplified Spontaneous Emission (ASE) spectrum spanning from 1,510 nm to over 1,560 nm at the output power level of about −58 dBm. The ASE spectrum has a peak power of −52 dBm at a central wavelength of 1,533 nm when not spooled. Spooling the BSEDF with diameters of 10 cm to 2 cm yields decreasing peak powers from −57.0 dBm to −61.8 dBm, while the central wavelength remains unchanged. The output is highly stable over time, with a low temperature sensitivity of around ~0.005 dBm/°C, thus allowing for the development of a highly stable sensor system based in the change of the peak power alone.

Keywords: bending sensor; special fiber; temperature-insensitivity; erbium-doped fiber

1. Introduction

Fiber optic sensors have various industrial applications such as in monitoring of structural health [1–3], due to its advantages over electronic sensors such as being non-conducting, high sensitivity, resistivity to electromagnetic disturbance and robustness in erosive, conductive or explosive environments [4]. While most traditional applications for these sensors are focused on parameters such as temperature and pressure, other applications for these sensors also abound. Among these applications are bending sensors, which have been demonstrated in various types of fibers such as multimode, single-mode and curvature fibers [5–8]. These sensors have attracted significant attention due to its ability to measure many physical parameters such as pressures, forces, frequency, vibration and many other acoustic parameters, which can find many useful applications in sensing systems such as micro-displacement and acceleration to name but a few [9]. There have been several reports on fiber bending sensing techniques using fiber Bragg gratings (FBGs) [10,11] and long period fiber gratings (LPFGs) [12,13]. Nevertheless, the disadvantage of using fiber gratings as bending sensors is that they are also sensitive to temperature fluctuations and thus have the high additional cost of wavelength demodulation processes [1]. Several studies have therefore been made to investigate how to minimize the effect of the fiber sensitivity to the simultaneous temperature changes [14]. In addition, besides the drawback of simultaneous temperature sensitivity, most fiber bend sensor designs also use a separated gain medium and fiber sensing head, which will eventually increase the complexity of the sensor design. Thus, an alternative fiber bending sensor method with a simpler setup is becoming very necessary and would provide some advantages in terms of reducing the production and manufacturing costs.

Of late, there has been work done on the use of depressed-cladding erbium doped fibers (DC-EDF) [15,16] as a bend sensor. However, the drawback of this approach is that the peak wavelength of the amplified spontaneous emission (ASE) spectrum shifts as the bending radius decreases. It is an advantage and of great interest to have a fixed peak wavelength of the ASE spectrum with respect to the change of the bending radius as this will allow a simple detection system to be built, where a filter can be inserted to coincide with the peak wavelength, which then allows for the measurement of the output power against the bending radius.

This work describes a new type of bending sensor based on a bend sensitive erbium doped fiber (BSEDF), whereby the erbium dopants are concentrated within the inner ring of the core region, while leaving the outer region of the core undoped, but still having a higher refractive index as compared to the cladding. The erbium ions act as the source for the ASE spectrum and also as the medium for the sensing head. This design is different from the S-band DC-EDF, with a W-type refractive index configuration, having a three-layer structure—the core, depressed cladding and outer cladding—with typical dimensions of 3.5, 14.0 and 120.0 μm and refractive indices of 1.472, 1.452 and 1.457 [17]. The special fiber in this proposed work is not only temperature insensitive, but also easy to splice to single-mode fiber (SMF). Due to the location of the erbium dopant which is at the centre of the fiber core, a small bend on the fiber will induce a high transmission loss, thus making this bend sensor very sensitive towards bending as well as being temperature insensitive, with a slope efficiency of only ~0.005 dBm/°C. To the best of authors' knowledge, this is the first design of a fiber bend sensor where both the gain medium and the sensing head are created from a similar fiber, which at the time possesses temperature insensitivity. This will be described in detail in the next section.

2. Experimental Section

Figure 1a shows the doping profile of the BSEDF with the center of the core doped with erbium ions as obtained from an electron probe microanalyzer (EPMA). The core of the fiber comprises of two annular regions. The outer region consists of SiO_2, and co-doped with GeO_2, with an external diameter of about 9.38 μm and an inner diameter of about 4.00 μm, with a refractive index of about 1.4665 and externally surrounded by a cladding layer with a dimension of 123.21 μm and a refractive index of about 1.4624, with Δ~0.28%, which is shown in Figure 1b for the case of the fabricated preform.

Figure 1. (**a**) Elemental distribution of different dopants into doped and un-doped regions of the BSEDF obtained using EPMA, (**b**) refractive index profile of fabricated BSEDF preform measured by a commercially available Preform Analyser, with a core of roughly between −0.8 to 0.8 mm and the cladding between −4.0 to 4.0 mm, with the rest being oil, and (**c**) microscopic cross section of the BSEDF.

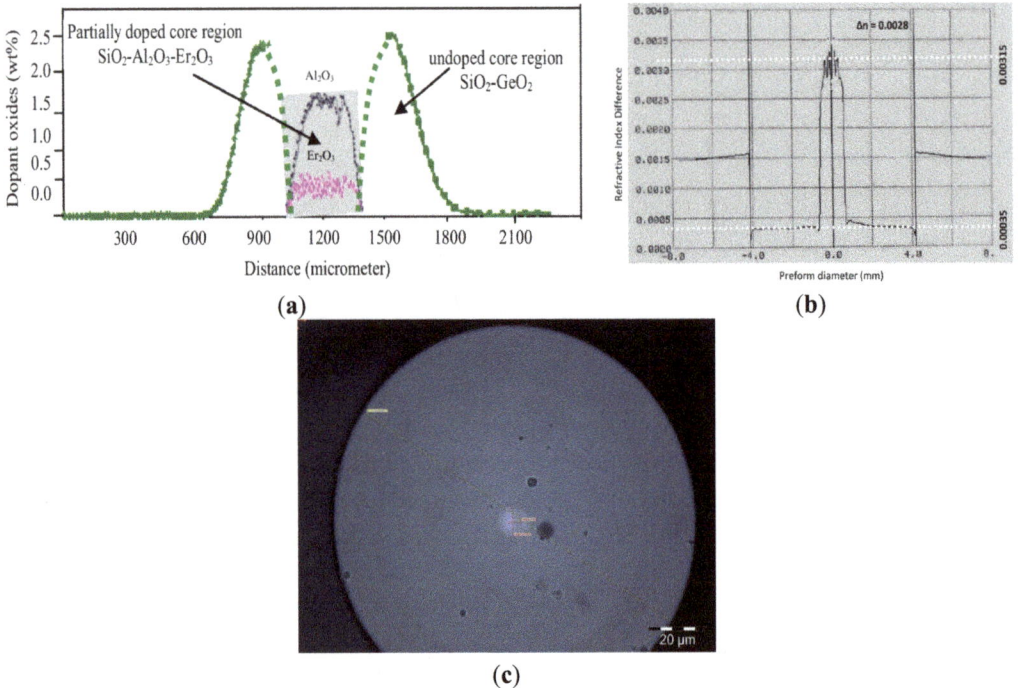

(a)

(b)

(c)

The core dopant concentrations are obtained using a preform analyzer, and it is determined that the central core portion is being doped with Er_2O_3 of around 0.425 wt%, which is equivalent to 0.20 mole% and co-doped with Al_2O_3. In terms of Er ion, this is equal to 400 ppm, as shown in Figure 1a (not to scale). The central diameter region is about 4.00 μm, giving a ratio of about 0.43 between the erbium doped diameter against the core diameter. Figure 1c shows the microscopic cross section of the BSEDF. The advantage of this inner core dopant of erbium ions within the core region enhances the sensitivity of the fiber towards bending as demonstrated in this work. Figure 2a shows the experimental setup of the proposed bend sensor, which uses the BSEDF as the gain medium.

Figure 2. (a) The experimental setup of the proposed bend sensor, **(b)** The schematic diagram of the spooled BSEDF with the Gaussian beam from the LD pump and the guided rays.

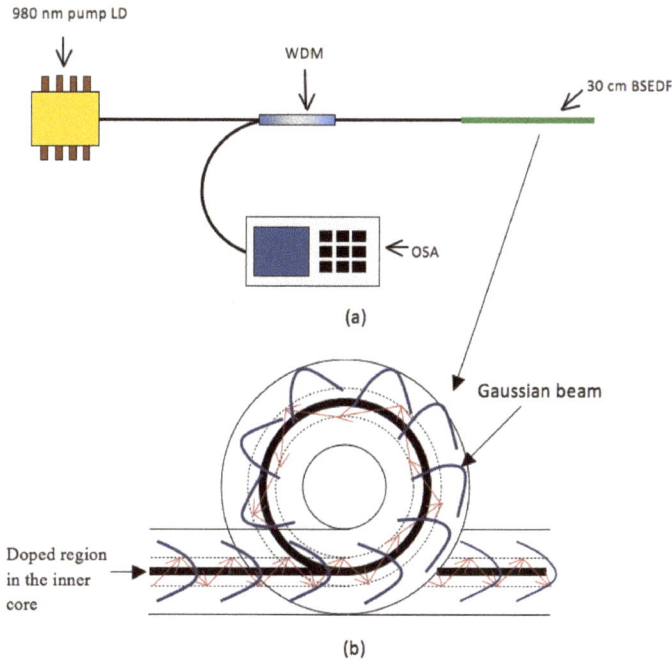

The BSEDF as described in the section above is pumped by a 980 nm laser diode operating at 50 mW and is connected to the 980 nm port of a 980/1,550 nm wavelength division multiplexer (WDM). The common-port of the WDM is then spliced to the BSEDF, which of a length of about 30 cm. The 1,550 port of the WDM is connected to a Yokogawa AQ6317 optical spectrum analyser (OSA) with a capability of giving a resolution of 0.02 nm, for spectral analysis. The BSEDF is then used as the bending sensor. As the beam from the laser diode propagates through the EDF, erbium ions are excited, hence producing the ASE that propagates in both forward and backward directions. The OSA would then detect the backward ASE spectrum that travels back through the WDM. The rationale for taking the backward ASE, instead of the forward ASE is so that there will be negligible contribution from the pump laser in the signal detection. The bend sensor mechanism is realized by spooling the EDF at different diameters from 10 cm to 2 cm and the output ASE spectrum as well as the average output power is measured against the different spooling diameters.

Figure 2b shows the schematic diagram of the spooled BSEDF with the Gaussian beam from the laser diode travelling around it. As the LD pump is injected through the fiber, Erbium ions are excited, hence producing the ASE. When the EDF is in a straight position, the propagating

Gaussian beam is at its maximum intensity at the core, thus giving the excitation of erbium ions within the inner core diameter of 4.00 μm. This will give the maximum ASE emission. In the case when the EDF is bent, the ASE output with a peak wavelength of 1,533 nm will experience attenuation at even small bending radii at longer wavelengths, and augers well with the findings of [18,19]. This results in a drop in the ASE output power level. With this unique profile of the BSEDF, the power loss during the bending of the fiber can be monitored easily and changes very fast with bending. This will provide a very sensitive bend sensor. As reported in [20,21], a linear relationship between the central displacement of the beam travelling within the fiber during bending and the decrease in transmitted light intensity could be easily measured, and this is one of the parameters discussed in this paper.

3. Results and Discussion

Figure 3 shows the ASE spectrum level with respect to different spooling diameters as taken from the OSA. As can be seen from the figure, the peak power level of the ASE spectrum also reduces with the decrease of the EDF spooling diameter, from −57.0 dBm to −61.8 dBm at the spooling diameter of 10 cm and 2 cm respectively. As the spooling diameter is reduced from 10 cm to 6 cm, the peak power decreases from 57.0 to −60.2 dBm (a total loss of 3.2 dB), whereas as the spooling diameter is further reduced from 6 to 2 cm, the peak power decreases from 60.2 to −61.8 dBm (a total loss of 1.6 dB), and this occurs at about the same peak wavelength of 1,533 nm. Thus, it can be deduced that at larger spooling diameters, which range from 6 cm to 10 cm, the loss of the peak power with respect to the decrease of the spooling diameter is higher compared to the peak power loss at smaller spooling diameters, which is within the range of 2 cm to 6 cm. This is shown in Figure 4, which shows a nearly linear slope. For the case of Depressed-Cladding Erbium Doped Fibers (DC-EDF), the peak wavelength of the ASE spectrum shifts as the bending radius decreases, as demonstrated by Rosolem et al. [16]. This will allow a simple detection system to be built, where a filter can be inserted to coincide the peak wavelength, which then allows for the measurement of the output power against the bending radius.

Figure 3. The ASE spectrum level of the BSEDF for different spooling diameter.

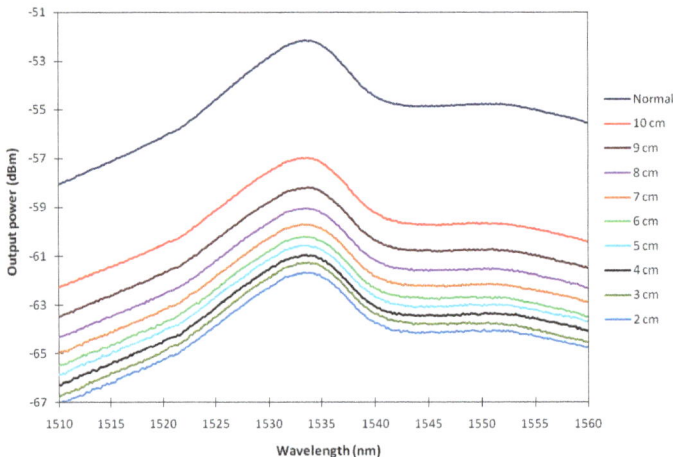

Figure 4. The peak power of the output ASE with respect to the bending diameter.

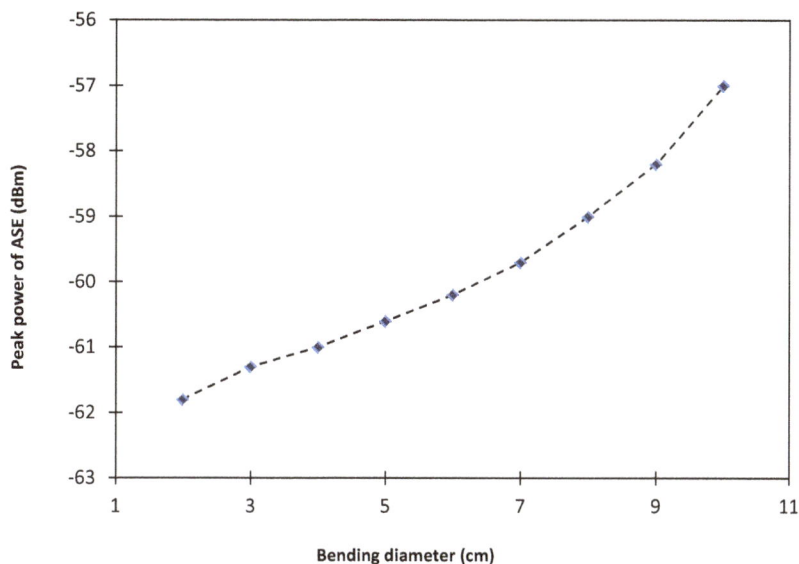

The relationship between the bending diameter and the average output power measured from the power meter is shown in Figure 5. As can be seen from the figure, the average output power increases exponentially with the increase of the bending diameter. As the bending diameter is increased from 2 to 6 cm, the average output power increases gradually from −40.9 to −40.0 dBm, giving a slope of 0.23 dBm/cm.

Figure 5. The average output power against the bending diameter.

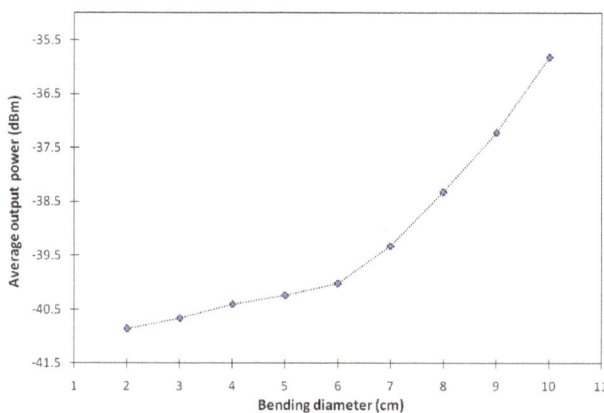

On the other hand, at larger bending diameters above 6 to 10 cm, the average output power increases abruptly from −40.0 to −35.8 dBm as the bending diameter is increased, with a slope of 1.1 dBm/cm. From this point, the fiber shows higher sensitivity to bending effect as the slope gets steeper, compared with the slope for bending diameters smaller than 6 cm. Based on this characteristic, this working mechanism can be developed into a reliable and simple bending sensor requiring only a simple filter at the peak wavelength, or measuring the total output power using a simple energy detector, for a very sensitive method for measuring bends or curvature. The present measurement of the diameter is not restricted to these values, a smaller diameter can be measured, and the ASE output will still be detectable as compared to DC-EDFs, where at very small diameters the ASE output becomes negligible.

In order to test the stability performance of the proposed system, a stability measurement of the ASE spectrum from the BSEDF is carried out within 60 min of observation time with an interval of 5 min between the measured outputs. This measurement is taken with the BSEDF in its unspooled position and the result is shown in Figure 6. With an interval of 5 min, the output spectrum of the ASE is observed to be constant for the 60 min observation time. No significant variation is observed in the ASE spectrum in terms of the output power and the output wavelength, and similar results are obtained for spooled cases. This proves and confirms the high stability of this proposed system.

Figure 6. The stability measurement of the ASE spectrum within 60 min of observation time.

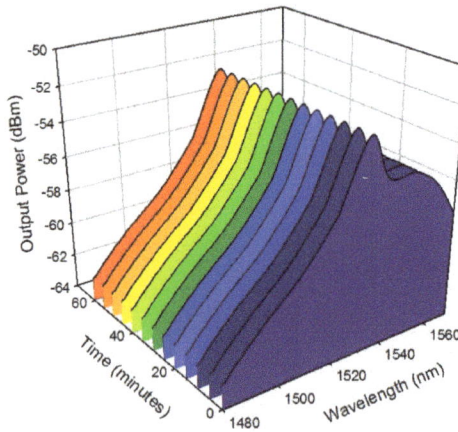

Figure 7. The setup for the sensitivity measurement of the BSEDF towards the temperature.

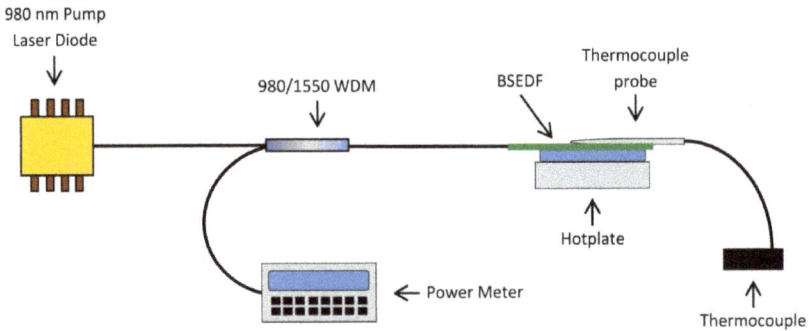

Another important parameter measurement of interest is the effect of temperature on the BSEDF or the sensitivity of the BSEDF towards the temperature change. Another set-up to test the sensitivity of the fiber to different temperatures is shown in Figure 7 and the result of the test is shown in Figure 8. As seen in Figure 7, the BSEDF is laid straight on a hot plate with a variable temperature. An aluminum tape is used to make the BSEDF stick to the surface of the hot plate. Aluminum is a good thermal conductor and helps keep the temperature constant along the fiber. A power meter is used to measure the average output power of the ASE from the BSEDF.

Figure 8 shows how the average output power of the ASE from the BSEDF varies with temperature. The average output power shift against the increase of the temperature is so small that it is hardly discriminated within the temperature range from 30 to 130 °C. The relationship between the average output power and temperature indicates that this sensing system is temperature insensitive, with a slope efficiency of only ~0.005 dBm/°C. Thus, this fiber can be said to have shown very minimum sensitivity towards temperature. Although there is a slight increment in the power meter's readings as the temperature increases, the value is too low to infer the fiber as temperature sensitive. This slight change in the reading is probably due to the air around the fiber's loose end being heated up resulting in a decrease in its reflective index. The hot air around the fiber's loose end reflects more ASE, hence giving a slight change in the power meter reading.

Figure 8. The average output power of the ASE against temperature.

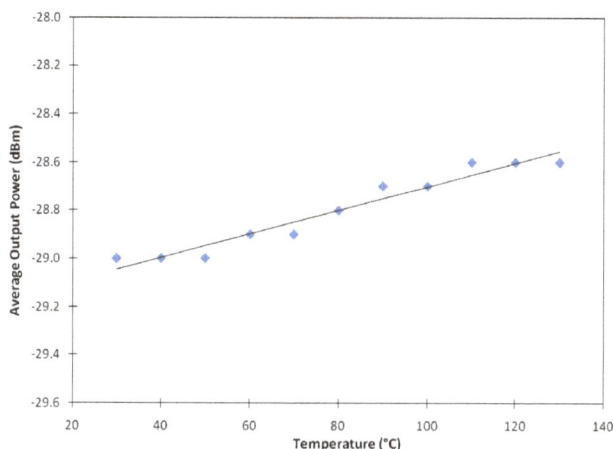

The same experiment as in Figure 7 is repeated using the BSEDF in a bent condition, with different spooling diameters, replacing the power meter with the OSA as to observe the ASE output. Figure 9 shows the ASE spectrum level of the BSEDF at different temperature for spooling diameters of 3, 4, 5 and 6 cm. It is observed that there is no significant variation of the ASE spectrum level detected by changing the temperature from 30 °C to 130 °C at each different spooling diameter. For instance, in the case of 3 cm spool diameter the ASE curves at different temperatures superimpose as a single line, which indicates clearly that the bend sensor is insensitive to temperature changes.

Besides this, it would also be of interest also to observe the output power at the peak wavelength, 1,533 nm of the ASE spectrum when exposed to different temperatures. This is done for different spooling diameters, and this is shown in Figure 10. As before, no observable changes can be seen in the 1,533 nm power at different temperatures, thereby validating the fact that the sensor is insensitive to temperature, especially at the 1,533 nm region.

Figure 9. The ASE spectrum level of the BSEDF at different temperature for different spooling diameter.

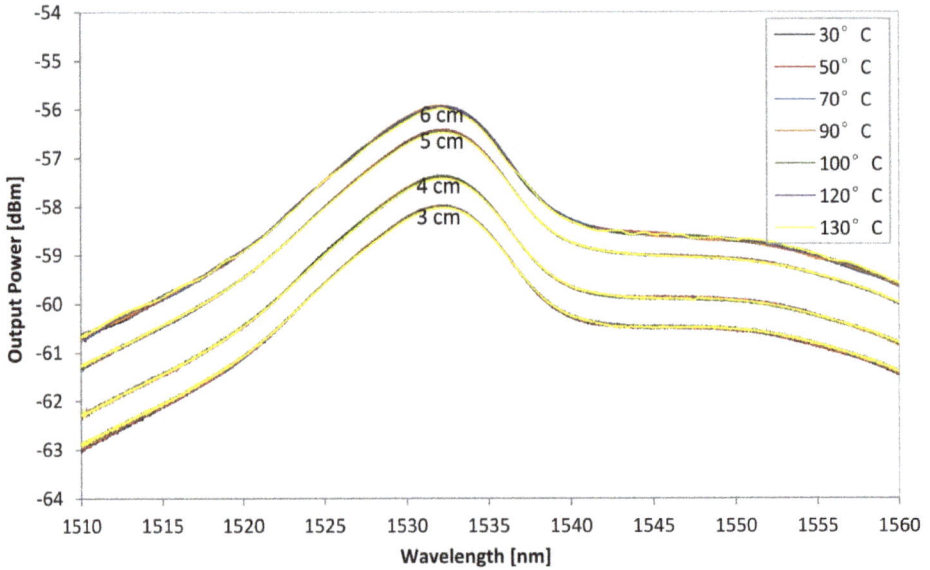

Figure 10. The output power at the peak wavelength of the ASE of 1,533 nm against temperature for different spooling diameter.

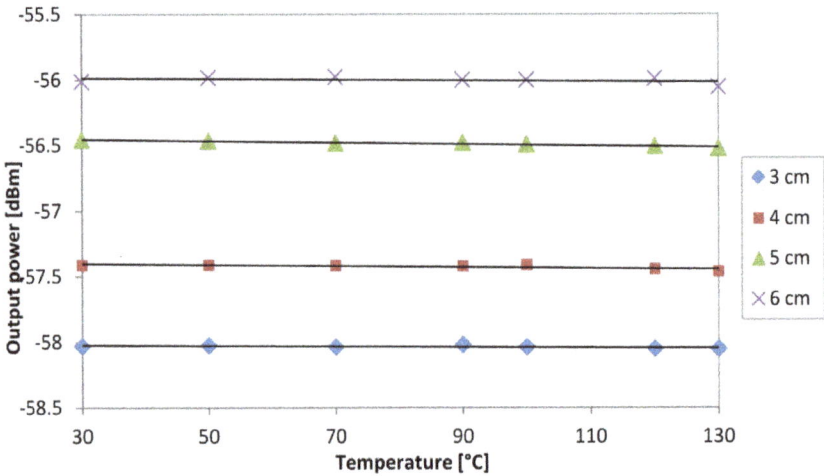

In comparison to other types of bend sensors, for instance in the DC-EDF, bend-sensor with FBG and normal single-mode fiber, there tend to be a sizeable dependence on the temperature, thereby giving results that need to be compensated. The ability of the proposed sensor to cancel out cross sensitivity to other parameters is characteristic of a good sensor.

4. Conclusions/Outlook

In this work, a bend sensor using a uniquely developed BSEDF is proposed and demonstrated. The BSEDF has a core with an un-doped outer region of about 9.38 μm and inner core region of about 4.00 μm with an Er_2O_3 dopant concentration of 400 ppm. The refractive index difference between the core and cladding is about 0.28%. The BSEDF is pumped by a 980 nm laser diode at 50 mW, and when unspooled emits an ASE spectrum spanning from 1,510 to over 1,560 nm at the output power level of −58 dBm, with a central wavelength of 1,533 nm and a peak power of −52 dBm. Spooling the BSEDF decreases the peak power, from an average of −57.0 dBm to −61.8 dBm at spooling diameters of 10 cm to 2 cm. However, the central wavelength remains unchanged, thus allowing for a simple sensor to be built based on the variation of the peak power alone. The output of the ASE is also highly stable, with no observable variation in the power output over a measurement period of one hour. The BSEDF is also temperature insensitive, with only a minor variation of about ~0.005 dBm/°C measured, thus reducing the effects of cross-sensitivity.

Acknowledgments

We would like to thank the University of Malaya for providing the funding for this research under the grants UM.C/625/1/HIR/MOHE/SCI/29, PV031/2012A and RP019-2012A.

Conflicts of Interest

The authors declare no conflict of interest.

References

1. Liu, W.; Guo, T.; Wong, A.C.; Tam, H.Y.; He, S. Highly sensitive bending sensor based on Er3+-doped DBR fiber laser. *Opt. Express* **2010**, *18*, 17834–17840.
2. Culshaw, B.; Kersey, A. Fiber-optic sensing: A historical perspective. *J. Lightwave Technol.* **2008**, *26*, 1064–1078.
3. Li, H.N.; Li, D.S.; Song, G.B. Recent applications of fiber optic sensors to health monitoring in civil engineering. *Eng. Struct.* **2004**, *26*, 1647–1657.
4. Xu, B.; Li, Y.; Sun, M.; Zhang, Z.W.; Dong, X.Y.; Zhang, Z.X.; Jin, S.Z. Acoustic vibration sensor based on non adiabatic tapered fibers. *Opt. Lett.* **2012**, *37*, 4768–4769.
5. Fan, Y.; Wu, G.; Wei, W.; Yuan, Y.; Lin, F.; Wu, X. Fiber-optic bend sensor using LP21 mode operation. *Opt. Express* **2012**, *20*, 26127–26134.
6. Lagakos, N.; Cole, J.H.; Bucaro, J.A. Microbend fiber-optic sensor. *Appl. Opt.* **1987**, *26*, 2171–2180.
7. Nguyen, N.Q.; Guptaa, N. Power modulation based fiber-optic loop-sensor having a dual measurement range. *J. Appl. Phys.* **2009**, *106*, 033502:1–033502:5.
8. Wang, Q.; Farrell, G.; Freir, T. Theoretical and experimental investigations of macro-bend losses for standard single mode fibers. *Opt. Express* **2005**, *13*, 4476–4484.

9. Pang, F.; Liang, W.; Xiang, W.; Chen, N.; Zeng, X.; Chen, Z.; Wang, T. Temperature-insensitivity bending sensor based on cladding mode resonance of special optical fiber. *IEEE Photonics Technol. Lett.* **2009**, *21*, 76–78.

10. Flockhart, G.M.H.; MacPherson, W.N.; Barton, J.S.; Jones, J.D.C.; Zhang, L.; Bennion, I. Two-axis bend measurement with Bragg gratings in multicore optical fiber. *Opt. Lett.* **2003**, *28*, 387–389.

11. Dong, X.; Meng, H.; Liu, Z.; Kai, G.; Dong, X. Bend measurement with chirp of fiber Bragg grating. *Smart Mater. Struct.* **2001**, *10*, 1111–1113.

12. Shao, L.Y.; Laronche, A.; Smietana, M.; Mikulic, P.; Bock, W.J.; Albert, J. Highly sensitive bend sensor with hybrid long-period and tilted fiber Bragg grating. *Opt. Commun.* **2010**, *283*, 2690–2694.

13. Liu, Y.; Williams, J.A.R.; Bennion, I. Long-period Fiber Grating Bend Sensor based on Measurement of Resonance Mode Splitting. In Proceedings of the Conference on Lasers and Electro-Optics (CLEO 2000), Nice, France, 7–12 May 2000; pp. 306–307.

14. Han, Y.G.; Lee, J.; Lee, S. Discrimination of bending and temperature sensitivities with phase-shifted long-period fiber gratings depending on initial coupling strength. *Opt. Express* **2004**, *12*, 3204–3208.

15. Rosolem, J.B.; Elias, M.B.; Bezerra, E.W.; Suzuki, C.K. Bending sensor based on S-band depressed cladding erbium-doped fiber. *IEEE Photonics Technol. Lett.* **2010**, *22*, 1060–1062.

16. Rosolem, J.B.; Elias, M.B.; Ribeiro, L.A.; Suzuki, C.K. Optical sensing systems based on depressed cladding Erbium doped fiber. *J. Lightwave Technol.* **2012**, *30*, 1190–1195.

17. Lei, D.; Yuan, J.Y.; Bo, X.J.; Zhen, Z.; Jun, S.J.; Cheng, L.K. A two-stage S-band Erbium-doped fiber amplifier based on W-type Erbium-doped fiber. *Chin. Phys. Lett.* **2010**, *27*, 094204.

18. Morgan, R.; Barton, J.S.; Harper, P.G.; Jones, J.D. Wavelength dependence of bending loss in monomode optical fibers: effect of the fiber buffer coating. *Opt. Lett.* **1990**, *15*, 947–949.

19. Faustini, L.; Martini, G. Bend loss in single-mode fibers. *J. Lightwave Technol.* **1997**, *15*, 671–679.

20. Gavalis, R.M.; Wong, P.Y.; Eisenstein, J.A.; Lilge, L.; Cao, C.G.L. Localized active-cladding optical fiber bend sensor. *Opt. Eng.* **2009**, *49*, 064401:1–064401:8.

21. Kuang, K.S.C.; Cantwell, W.J.; Scully, P.J. An evaluation of a novel plastic optical fiber sensor for axial strain and bend measurements. *Meas. Sci. Technol.* **2002**, *13*, 1523–1534.

Reprinted from *Sensors*. Cite as: Coelho, J.M.P.; Nespereira, M.; Abreu, M.; Rebordão, J. 3D Finite Element Model for Writing Long-Period Fiber Gratings by CO2 Laser Radiation. *Sensors* **2013**, *13*, 10333–10347.

Article

3D Finite Element Model for Writing Long-Period Fiber Gratings by CO_2 Laser Radiation

João M. P. Coelho [1,2,*], **Marta Nespereira** [1], **Manuel Abreu** [1] and **José Rebordão** [1]

[1] Laboratory of Optics, Lasers and Systems, Faculty of Sciences, University of Lisbon, Campus do Lumiar, Estrada do Paço do Lumiar, 22, Building D, 1649-038 Lisboa, Portugal; E-Mails: mcnespereira@fc.ul.pt (M.N.); maabreu@fc.ul.pt (M.A.); jmrebordao@fc.ul.pt (J.R.)

[2] Institute of Biophysics and Biomedical Engineering, Faculty of Sciences, University of Lisbon, Faculdade de Ciências da Universidade de Lisboa, Campo Grande, Lisboa 1749-016, Portugal

* Author to whom correspondence should be addressed; E-Mail: joao.coelho@fc.ul.pt; Tel.: +351-217-500-759; Fax: +351-217-163-048.

Received: 1 July 2013; in revised form: 6 August 2013 / Accepted: 8 August 2013 / Published: 12 August 2013

Abstract: In the last years, mid-infrared radiation emitted by CO_2 lasers has become increasing popular as a tool in the development of long-period fiber gratings. However, although the development and characterization of the resulting sensing devices have progressed quickly, further research is still necessary to consolidate functional models, especially regarding the interaction between laser radiation and the fiber's material. In this paper, a 3D finite element model is presented to simulate the interaction between laser radiation and an optical fiber and to determine the resulting refractive index change. Dependence with temperature of the main parameters of the optical fiber materials (with special focus on the absorption of incident laser radiation) is considered, as well as convection and radiation losses. Thermal and residual stress analyses are made for a standard single mode fiber, and experimental results are presented.

Keywords: laser processing; long-period fiber gratings; finite element modeling; fiber-based sensors; refractive-index modulation; thermo-mechanical processes

1. Introduction

Long-period fiber grating, or LPFG, play an important role in the development of fiber-based sensors in several areas of engineering. In the field of sensing systems, they are well suited to measure mechanical quantities: they can be applied as structural bend sensors, temperature sensors, axial strain sensors, refractive index sensors and biochemical optical sensors [1–7].

A LPFG can be considered a particular type of Fiber Bragg Grating, or FBG, in which the period of the index modulation, Λ, satisfies a phase matching condition between the fundamental core mode and a forward propagating cladding mode. This condition relates the resonant wavelength of the light into a particular cladding mode m, λ^m_{res}, the effective refractive index of the core, $n_{eff,co}$, and the effective refractive indexes of the mth-cladding mode, $n^m_{eff,cl}$

$$\lambda^m_{res} = \left(n_{eff,co} - n^m_{eff,cl} \right) \Lambda \tag{1}$$

These grating-based devices are produced by periodically creating a perturbation of the refractive indexes of the core and/or cladding along the length of the fiber. Typically the length of a FBG ranges from a few millimeters to about one centimeter, with modulation periods of a few dozens of microns. The length of a LPFG is of the order of a few centimeters with periods of hundreds of micrometers. LPFGs require therefore simpler fabrication processes and have lower costs, and show lower retro-reflection, higher sensitivity and robustness in sensing applications when compared with FBG [8].

A LPFG can be produced mechanically [9], chemically (etching) [10], by photonic processes (ultra-violet irradiation) [11] or thermally, either by applying an electric arc discharge [12] or a mid-infrared radiation (MIR) laser source [13,14]. Within the family of thermal techniques, MIR from CO_2 lasers has guaranteed better predictability and repeatability [15].

Since the first reports by Davis *et al.* [13] and Akiyama *et al.* [14] on the use of a 10.6 μm wavelength laser beam emitted by a CO_2 laser, different experimental methodologies have been used to write LPFGs [15]. The most common is, probably, using a static asymmetrical irradiation with a CW CO_2 laser and a cylindrical lens focusing a laser line on the fiber. This method has the advantage of requiring a simpler setup and, although the irradiation occurs on just one of the sides of the fiber, the line-shaped beam reduces the writing asymmetric effect when compared with focused circular laser beams.

In parallel to the development of writing techniques, research on the physical mechanisms responsible by refractive index changes has progressed. Based on several experimental analysis, most of the existing work has considered that the main mechanism responsible for inducing a refractive-index change is the relaxation of internal stresses by the exposure to the laser radiation [16–21]. Taking this in consideration, several analytical models were developed, based mainly on solutions of the general heat conduction equation [22,23]. These models, although giving proven data on the necessary engineering parameters, are based on expressions that do not consider all the physical phenomena (e.g., convection losses) and assume several simplifications (e.g., they ignore the temperature dependence of glass parameters).

In the following sections, the different thermo-mechanical effects of the interaction between a MIR laser beam and a standard silica-based single mode optical fiber will be analyzed, both theoretically and experimentally, and a 3D finite element method (FEM) model, implemented in the COMSOL Multiphysics program, is described, considering temperature dependence of the main parameters, in particular the absorption of laser radiation by the fiber's material.

2. Thermo-Mechanical Model

Considering a silica-based single mode optical fiber under tension irradiated by an elliptical 10.6 μm wavelength CO_2 laser beam (Figure 1), two main phenomena must be considered: the thermal heating due to the interaction between the photons and the glass molecular structure and the stress due to the differences between a relatively low-viscosity doped silica core and a relatively high-viscosity pure silica cladding [16].

Figure 1. Schematics of (**a**) coordinates used in this work and (**b**) main physical interfaces. The origin of the reference system is in the middle of the laser line.

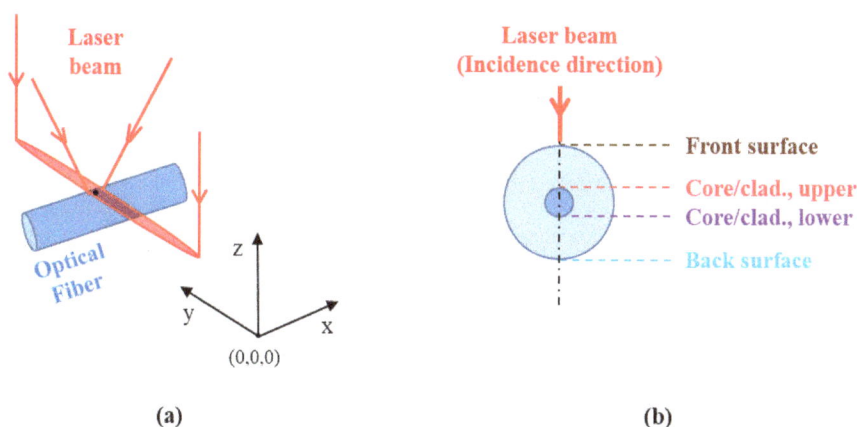

(a)   (b)

Differences between core and cladding thermal expansion coefficients and viscosity lead to residual thermal stresses and draw-induced residual stresses. In the case under study, these localized effects periodically induced along the fiber's length, can be responsible for the creation of the gratings. This effect is due to the refractive index change resulting from frozen-in viscoelasticity [17].

2.1. Theory

The refractive index change in a silica-based optical fiber can be approximated by the relation [17]:

$$\Delta n(T) \approx -6.35 \times 10^{-6} \sigma(T) \tag{2}$$

where $\sigma(T)$ represents the overall (both thermal, $\sigma_T(T)$, and drawn-induced, σ_x) residual stresses (in MPa) in the fiber's axial direction. According to Yablon [16], stresses in the other directions can be neglected.

The fiber is composed of a low viscosity, high thermal expansion coefficient core and a high viscosity, low thermal expansion coefficient cladding, the drawn-induced axial residual stresses act in opposition to the residual thermal stress. Either one can prevail or both can compensate each other.

The temperature distribution around a heat source can be obtained by solving the heat conduction problem. In what concerns the temperature variation with time, t, (transient regime) due to the action of a heat source $Q(x,y,z,t)$, the resulting energy balance leads to the heat conduction equation:

$$\left\{ \frac{\partial \rho}{\partial t} + \nabla \cdot (\rho \vec{v}) \right\} \int Cp \, dT + \rho Cp \left(\frac{\partial T}{\partial t} + \vec{v} \cdot \nabla T \right) = \nabla \cdot K \nabla T + q(T) + Q(x,y,z,t) \tag{3}$$

being $\vec{v}$ the velocity vector, ρ the density, Cp the specific heat, and K the thermal conductivity, these are the main parameters of the heated material. The factor $q(T)$ quantifies the convective and radiative heat flux [22]:

$$q(T) = h(T_{inf} - T) + \varepsilon \sigma_B \left(T_{amb}^4 - T^4 \right) \tag{4}$$

being T_{inf} the external temperature, T_{amb} the environment temperature, h the heat transfer coefficient, σ_B the Stefan-Boltzmann constant and ε the surface emissivity.

When considering the condition of mass conservation, an isotropic material with $K = K(T)$, and introducing the thermal diffusivity k, given by $K/(\rho \, Cp)$, Equation (3) can be simplified to:

$$\frac{\partial T}{\partial t} + \vec{v} \cdot \nabla T - k(\nabla \cdot \nabla T) + q(T) = \frac{Q(x,y,z,t)}{\rho C_p} \tag{5}$$

The power generation per unit volume of the material is given by $Q(x,y,z,,t)$ and depends on the characteristics of the heat source. For a laser beam incident on a surface and propagating in the z direction:

$$Q(x,y,z,t) = a_T \cdot (1 - R) I(x,y,z) g(t) \tag{6}$$

where a_T is the attenuation coefficient of the material, R its reflectance and $I(r,t)$ the irradiance. For continuous wave emission with a duration τ:

$$g(t) = \begin{cases} 0, \text{if } t \leq 0 \vee t > \tau \\ I(x,y,z), \text{if } 0 < t \leq \tau \end{cases} \tag{7}$$

If the laser beam has an elliptical Gaussian distribution (at the surface being irradiated), then [24]:

$$I(x,y,z) = \frac{8 a_T P}{\pi d_x d_y} \exp\left[-2\left(\frac{x^2}{d_x^2} + \frac{y^2}{d_y^2} \right) \right] \cdot \exp(-a_T z) \tag{8}$$

where d_x and d_y are the dimensions of the ellipse's axis.

Solving Equation (5), the thermally-induced residual stresses, $\sigma(T)$, can be obtained considering the constitutive equations for a linear isotropic thermoelastic material and the stress tensor obtained [25].

The residual axial elastic stresses in the cladding and core, σ_{cl} and σ_{co}, respectively, resulting from a draw tension F, over the equivalent cross-sectional areas A_{cl} and A_{co} can be obtained from [26]:

$$\sigma_{x,cl} = \frac{F}{A_{cl}} \left(\frac{A_{co} E_{co}}{A_{co} E_{co} + A_{cl} E_{cl}} \right) \tag{9}$$

and:

$$\sigma_{x,co} = F\left(\frac{E_{co}}{A_{co} E_{co} + A_{cl} E_{cl}}\right)$$

(10)

Besides stress-related refractive index change, localized heating can induce micro-deformation of the fiber and changes in the glass structure. The latter is likely to occur in the core for which the fictive temperature (glass structure doesn't change below the fictive temperature) is lower [18,27]. As reported, (e.g., [18]) for a Ge-doped core the fictive temperature ranges from 1,150 K to 1,500 K.

2.2. Physical Parameters

The optical fiber is a Corning SMF-28 fiber (125 μm diameter pure fused silica cladding and 8.2 μm diameter core of 3.5 mol% Ge-doped SiO_2) [28]. Figures 2–4 show the change on the different parameters with temperature. For some parameters (shown in Figure 2), the temperature dependence was modelled using native COMSOL functions for a Corning fused silica glass (7940). The doping effect on most of the parameters was disregarded mainly because the Ge concentration in the fiber's core is very low [29]. However, for the Young's modulus and Poisson's ratio (Figure 3), we extrapolated the function behavior [30]. Furthermore, the heat transfer coefficient was considered $h = 418.68 \text{ W·m}^{-2}\text{·K}^{-1}$ [22] and $R = 0.15$ [31].

Figure 2. Variation of the main parameters used of Corning 7940 fused silica glass as defined in the COMSOL's materials library.

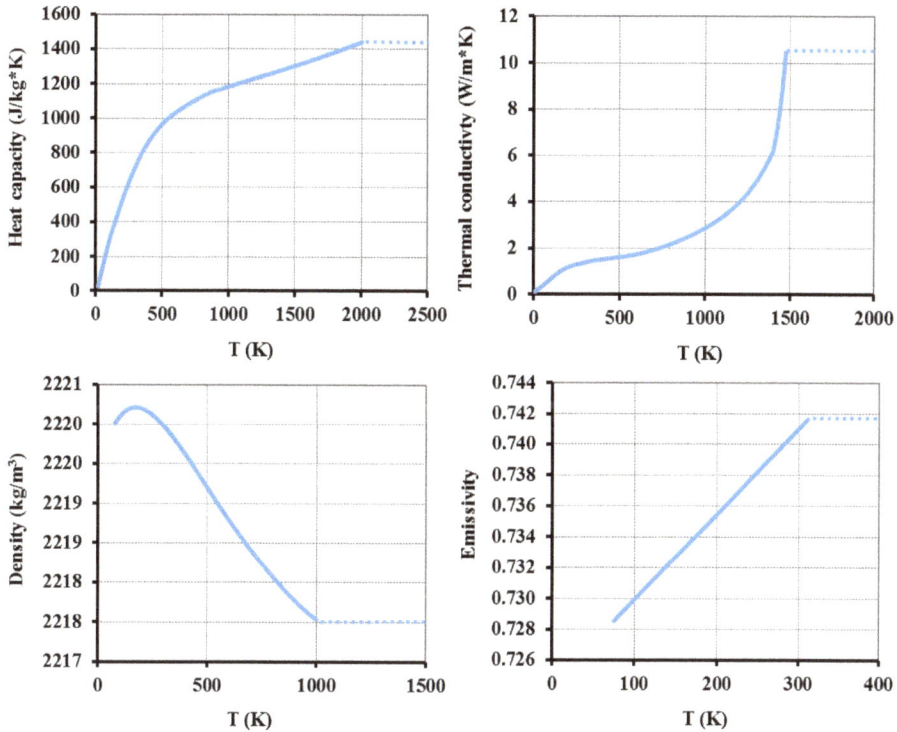

Figure 3. Variation of Young's module and Poisson's ratio for both fused silica (from COMSOL materials library) cladding and Ge-doped fused silica (extrapolated) core glasses.

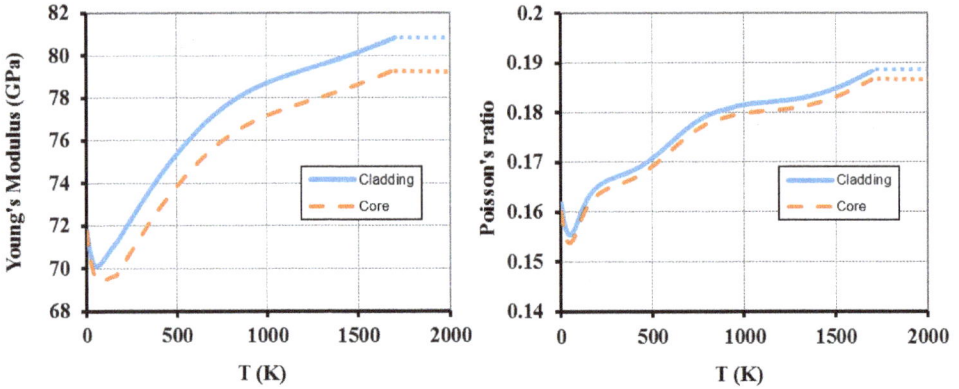

Figure 4. Absorption coefficient variation with temperature for fused silica glass accordingly with Equation (11).

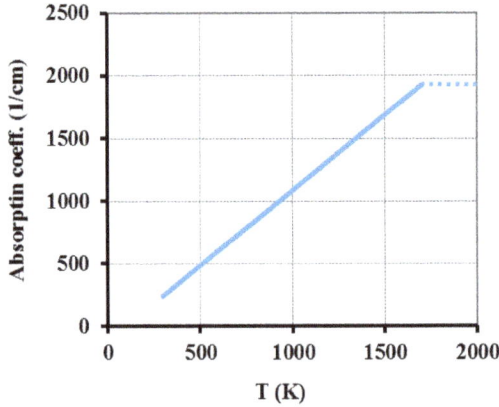

Another factor to take in consideration is the material's absorption coefficient for the wavelength of interest. Accordingly with Tian [32] the absorption coefficient of fused silica, a_T, for 10.59 μm for CO_2 laser wavelength (λ_l), within 298 K–2,073 K temperature range can be obtained by:

$$a_T\left(T\right) = \frac{4\pi}{\lambda_1}\left[1.82\times 10^{-2} - 10.1\times 10^{-5}\times\left(T - 273\right)\right] \tag{11}$$

This equation was used and the corresponding curve is presented in Figure 4. In the plots presented in Figures 2–4, the dotted lines represent the constant value that is assumed for higher temperature ranges.

2.3. Implementation

The physical problem was mathematically solved using the FEM model implemented using the COMSOL Multiphysics 3.5 program to create the transient heat conduction and (mechanical) stress-strain models under the conditions of this study. In order to introduce some of the complexity of stress-related issues regarding the processing of the optical fibers, the residual axial elastic stresses were implemented considering Equations (9) and (10) and the total resulting stress was obtained adding the thermally-induced residual stresses obtained with the program.

The implemented geometry consisted of a set of (concentric) cylinders with radius of curvatures accordingly with the characteristics of the core and cladding of the optical fiber described in the previous section. To avoid the influence of the external boundaries on the irradiated and analyzed zones, the overall length for the geometry was set as 11 mm. However, to reduce the computational load and loosen the mesh dimensions in zones not affected by the irradiation, the cylinders were implemented as three separate sets; the central one, where the laser incidence will be simulated, has a 700 µm length. Table 1 presents the 3D geometry data and the mesh statistics. Both outer boundary surfaces are defined as thermally isolated, being one of them fixed. The ambient temperature was considered to be 295 K and equal to the external temperature, T_{inf} in Equation (4).

Table 1. Geometry data and mesh statistics.

	Central Geometry		**External Geometries**	
	Cladding	**Core**	**Cladding**	**Core**
Geometry				
Length (mm)	0.7	0.7	5.15 × 2	5.15 × 2
Radius (µm)	62.5	4.1	62.5	4.1
Mesh (tetrahedral)				
# elements	24,122	1,231	27,917	1,704
min. quality	0.0474	0.1357	0.2347	0.2017
volume ratio	9.46×10^{-4}	0.0934	0.0019	0.4914

3. Experimental Methodologies

The experimental procedure was based on a simple point-by-point laser writing on the fiber characterized previously. The fiber was periodically moved (500 µm ± 1 µm grating period) along its axial direction with a linear translation stage (Thorlabs NRT100) and periodically irradiated (0.6 s ± 5 µs emission duration) by the beam emitted by a CW MIR CO_2 laser source (Synrad 48-2); the maximum available power was 25 W.

A schematic and a photograph of the set-up are shown in Figure 5. A 50 mm focal length cylindrical lens focuses the initial 1.75 mm laser beam into a 0.15 mm × 1.75 mm ($r_x \times r_y$) elliptical beam on the optical fiber. The dimensions of the beam on focus were measured through the knife-edge method, with an expected measurement error of ±5 µm [33].

In order to induce a constant strain to the fiber, a small weight (usually, several tenths of gram) is attached on one of the sides of the fiber, suspended. A broad band light source (Thorlabs S5FC1005S) and an optical spectrum analyzer (OSA, Agilent 86140B) allows monitoring the LPFG fabrication, while a fast camera (PCO SensiCAM), perpendicular to the irradiation axis, was used to optically visualize the process. The irradiated zones were analyzed through an optical microscope (Zeiss AxioScope A1) with maximum amplification of 1,000×.

Figure 5. (a) Schematic and **(b)** photograph of the setup used for CO_2 laser writing a LPFG on a Corning SMF-28 fiber.

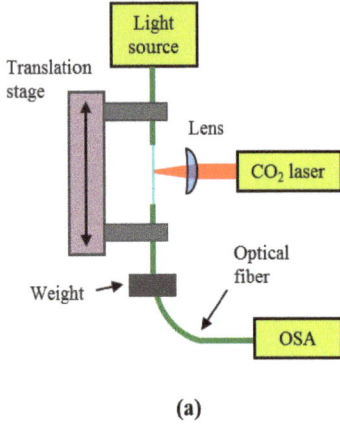

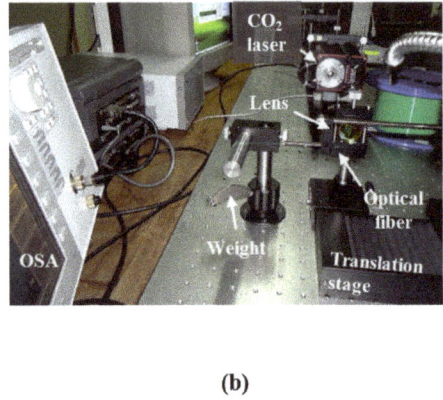

(a)   (b)

4. Results and Analysis

An example of the temperature distribution is shown in Figure 6, including a zoom view of the irradiated zone, for 6 W (±0.5 W) laser power, duration of 0.6 s (±1 ms) and 47 g (±0.5 g) weight ($F = 0.461$ N), the base parameters for this study, taking in consideration the experimental results.

Figure 6. Temperature distribution in the implemented 3D geometry for the laser irradiation of an optical fiber ($P = 6$ W; $\tau = 0.6$ s; $F = 0.461$ N). Color bar values are in K.

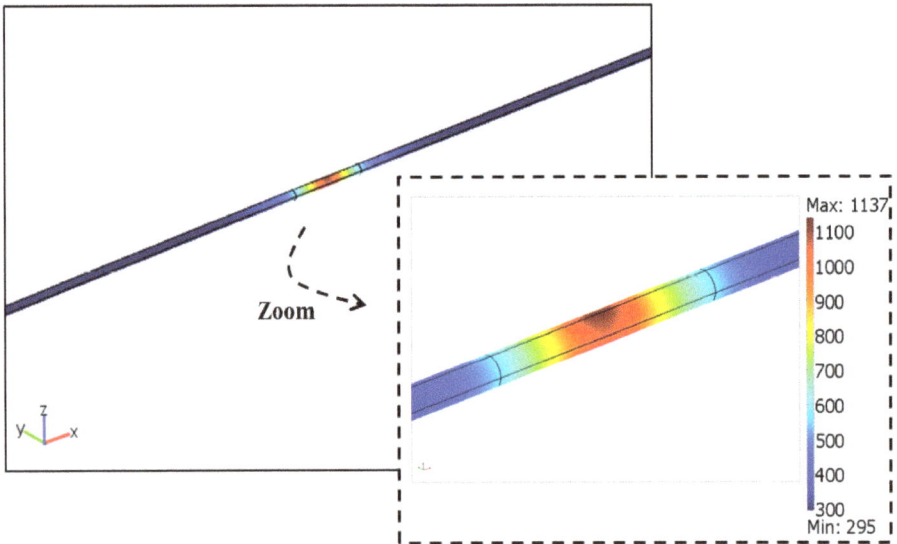

Figure 7 shows the plots resulting from the simulation using Equation (5) under the conditions mentioned before for the irradiated front surface, core/cladding interfaces (upper and lower) and the back surface of the fiber (Figure 1), and $x = y = 0$ m. In Figure 7a, the duration of the simulation was made larger than the laser emission duration in order to visualize the cooling process. The plot in Figure 7b, representing the variation of temperature along the fiber's axial direction, shows that the temperature was slightly larger than 1,050 K at the interfaces between the core and cladding, and one could assume that the core could be considered as being at the same temperature. This assumption cannot hold for the cladding since its temperature varied about 100 K along its thickness.

Figure 7. Plots of the temperature (**a**) evolution during laser irradiation and cooling and (**b**) distribution at the fiber's axial direction simulated for $t = 0.6$ s. ($P = 6$ W; $F = 0.461$ N).

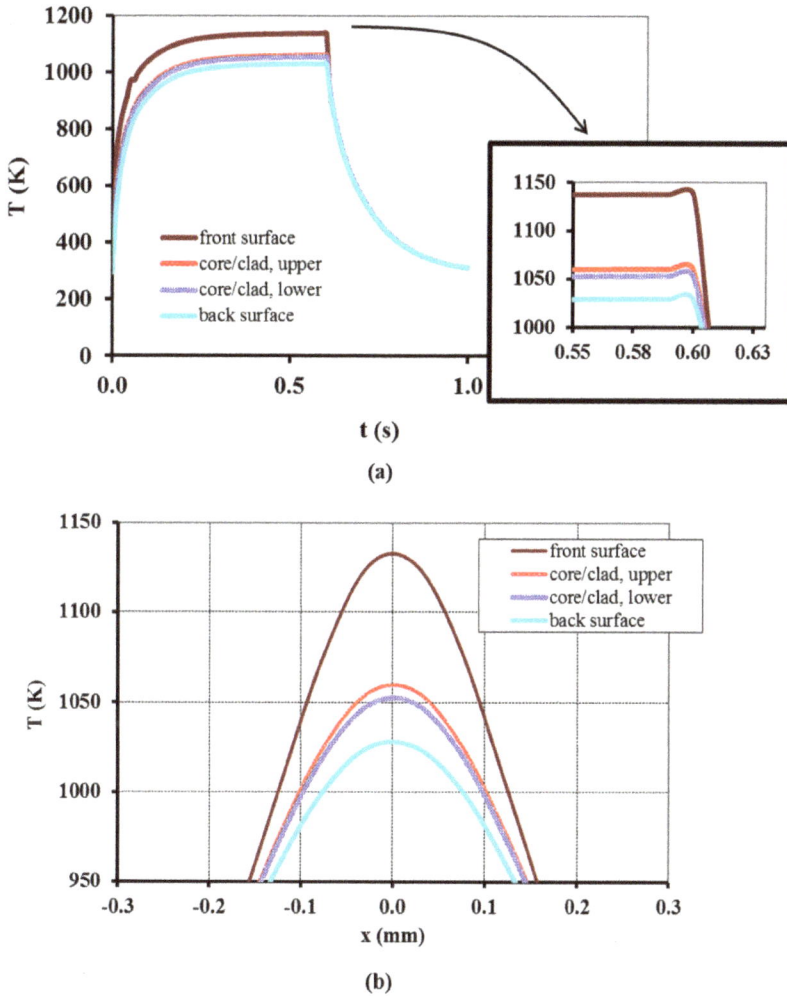

As the laser power increases, the temperature increases. Figure 8 shows the variation of temperature at the upper surface (at $x = y = 0$ m) of the optical fiber with laser power as obtained from the model.

For the considered example, at $t = 0.6$ s, axial residual thermal stresses values along z-axis were determined as having a maximum of about -0.8 MPa. Axial elastic stresses act in the opposite direction and were calculated as being $\sigma_{x,co} = 35$ MPa and $\sigma_{x,cl} = 0.153$ MPa. The resulting refractive index change (the difference between final and initial values) is calculated to be in the order of -2×10^{-4} for the core and 4×10^{-6} for the cladding. The refractive index distribution in the fiber (along the z-axis), before and after the laser irradiation is showed in Figure 9a and the maximum refractive index change (core and cladding) for different draw forces is plotted in Figure 9b. Figure 10 shows the calculated (maximum) refractive index change at the core and cladding for different laser power. Under the considered conditions, even if the refractive index of the core shows minor changes by increasing the laser power, it is possible to observe the beginning of the contribution of thermal stresses around 5 W. This is clearly observed regarding the refractive index change for the cladding, where a well-defined step occurs between 4.5 W and 5 W.

Figure 8. Plot of temperature variation with incident laser power simulated at $x = y = z = 0$ m. ($\tau = 0.6$ s; $F = 0.461$ N).

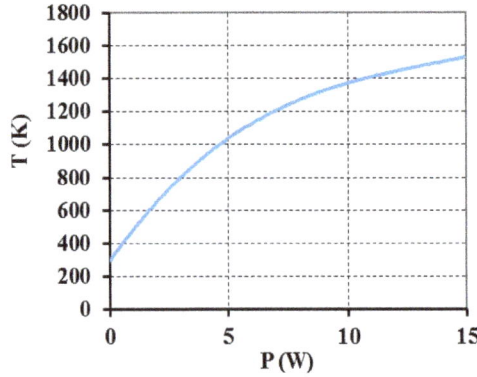

Figure 9. Calculated (**a**) refractive index profiles of the fiber, before and after laser irradiation ($F = 0.461$ N); and (**b**) refractive index change (maximum change for core and cladding) for different applied draw tensions. ($P = 6$ W; $\tau = 0.6$ s).

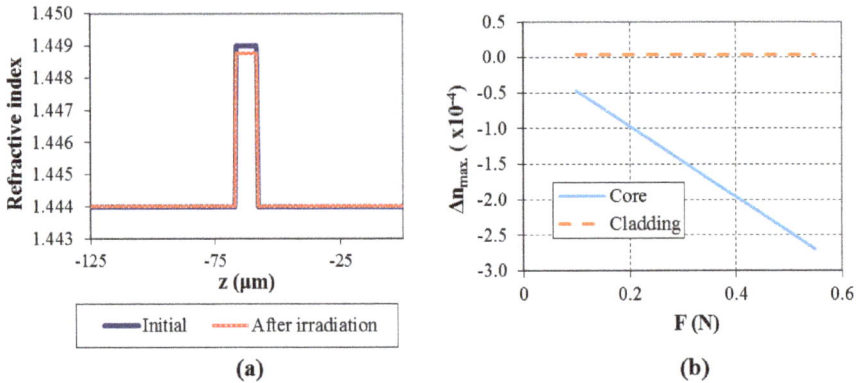

(a) (b)

Figure 10. Calculated (maximum) refractive index change at the (**a**) core and (**b**) cladding for different applied laser powers. (τ = 0.6 s; F = 0.461 N).

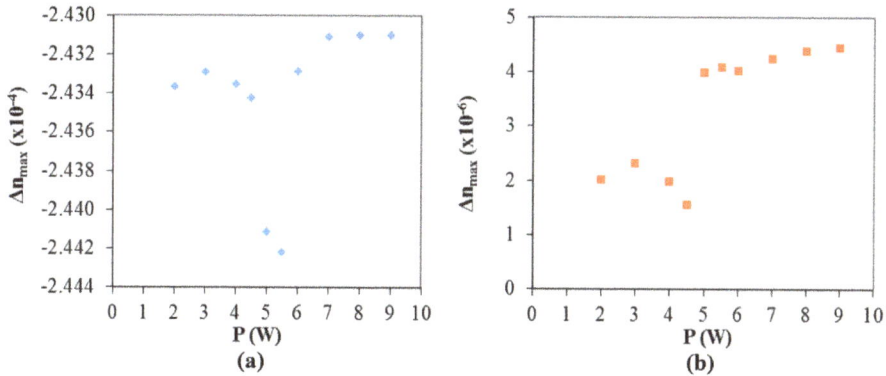

(a)   (b)

Figure 11a shows a microscope photograph of an irradiated fiber under the conditions considered in this work. The imaged zone was part of a 25-mm grating with a period of 500 μm, and the visible affected area along the fiber's axis was about 130 μm (supported on several measurements along the grating). Also visible was a (small) micrometric deformation of the fiber. Figure 11b shows the spectral transmission of the resultant LPFG, comparing the experimental data with the simulated spectrum. The latter was obtained using the refractive index changes obtained by the FEM model (and mentioned before) and using a simulation tool developed by Baptista [34] based on the three layer model developed by Erdogran [35,36].

Figure 11. (**a**) Picture showing an irradiated zone from a 25 mm LPFG with 500 μm period and (**b**) experimentally obtained and simulated relative normalized spectral transmission. (P = 6 W; τ = 0.6 s; F = 0.461 N).

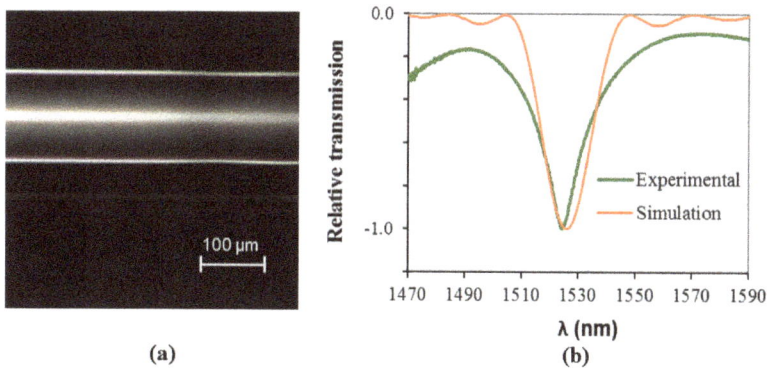

(a)   (b)

Analyzing both theoretical and experimental data, besides the relative spectral transmission data agreement, one can consider that it is necessary to reach temperatures higher than 1,000 K to accomplish an LPFG under these conditions. This assumption relies on the cross-analysis between the visible affected length of the fiber (around 130 μm) and the fiber's back surface temperature for $x = -65$ μm or $x = 65$ μm. This is in accordance with experimental evidences that, even for higher power conditions for which tapering occurs, the laser focal dimension influence on the affected volume prevail, at least at the surface [37].

Also, although we simulated the impact of different laser powers and weights, experimentally, using lower laser powers (typically <5 W) no LPFGs were obtained. For higher laser powers (typically >8 W) or higher weights (typically >60 g, $F > 0.588$ N) tapering occurs, a phenomena not included in the developed model. These experimental observations can also be inferred from results obtained in the FEM model, based on the analysis of the plots shown in Figure 10, in particular regarding Figure 10b. In the latter case, and based on the analysis previously made, it is easily identified the required minimum applied power (5W) for writing LPFG under the considered conditions.

The values obtained by the model are also in agreement with those estimated by other authors for the refractive index modulations necessary for achieving a fiber optic grating. Temperatures obtained by the model are similar to those obtained by other authors for arc-induced LPFG (e.g., in the range 1,100 K–1,400 K according to [27]) and the refractive index changes obtained are also within the overall range mentioned (in the order of magnitude of 10^{-4} for the core and 10^{-6} for the cladding) in other works [17,19–21,27]. The obtained behavior of the refractive index change as the applied drawing force increases complies with the most recent experimental indications that the refractive index of the core decreases while the opposite occurs in the cladding, and that this change occurs primarily in the core [20,21].

Nevertheless, due to the complexity of the physical phenomena involved, refractive index change dependence on stress also requires further research. Future work should focus on experimental measurements of temperature, stresses and refractive index changes induced by laser radiation. Although published works can contribute in assessing the validity of the results, the influence of specific characteristics of the fibers is a well-recognized issue. In particular, the effect of pre-existing stresses (e.g., from the fiber manufacture), differences in the materials, or other unaccounted phenomena can influence the performance of the FEM model when compared with real data. Similarly, the influence of the experimental data uncertainties on the model must be analyzed in detail, as well as the impact of the several approximations considered (e.g., transverse stresses are neglected), unaccounted phenomena like eventual changes on the glass polarizability and using standard material data.

5. Conclusions

The FEM model presented in this work demonstrated its potential to simulate the thermo-mechanical processes involved in writing LPFGs using CO_2 laser radiation. It takes in account the influence of temperature on the most relevant thermal and mechanical parameters of the fiber material, as well as convective and radiative effects. The model is 3D, considers a focused laser line irradiating a single-mode fiber for a given duration, and generates temperature and residual stresses distributions and the required refractive index change. An example of irradiating a single-mode optical fiber was presented and both theoretical (simulated) and experimental data are analyzed. Although additional work should be performed to further validate the analysis done (mainly regarding stresses acting in the optical fiber), the FEM results are in accordance with literature and experimental data.

Acknowledgments

This work was partially supported by FEDER funding through the Programa Operacional Factores de Competitividade–COMPETE and by national funding by the FCT–Portuguese Fundação para a Ciência e Tecnologia through the project PTDC/FIS/119027/2010. The authors gratefully acknowledge José Luis Santos, Orlando Frazão and Pedro Jorge from INESC-Porto and Catarina Silva and D. Castro Alves for their advices and contributions. A special thanks to Fernando Monteiro and António Oliveira for their technical support to the activities described in this paper.

Conflicts of Interest

The authors declare no conflict of interest.

References

1. Patrick, H.; Chang, C.; Vohra, S. Long period fiber gratings for structural bending sensing. *Electron. Lett.* **1998**, *34*, 1773–1775.
2. Bhatia, V. Applications of long-period gratings to single and multi-parameter sensing. *Opt. Express* **1999**, *4*, 457–466.
3. Falciai, R.; Mignani, A.; Vannini, A. Long period gratings as solution concentration sensors. *Sens. Actuators B Chem.* **2001**, *74*, 74–77.
4. James, S.; Tatam, R. Optical fibre long-period grating sensors: Characteristics and application. *Meas. Sci. Technol.* **2003**, *14*, R49–R61.
5. Chen, G.; Xiao, H.; Huang, Y.; Zhou, Z.; Zhang, Y.A. Novel long-period fiber grating optical sensor for large strain measurement. *Proc. SPIE* **2009**, *7292*, 729212.
6. Wang, J.; Tang, J.-L. Feasibility of fiber bragg grating and long-period fiber grating sensors under different environmental conditions. *Sensors* **2010**, *10*, 10105–10127.
7. Silva, C.; Coelho, J.M.P.; Caldas, P.; Jorge, P. Fiber Sensing System Based on Long-Period Gratings for Monitoring Aqueous Environments. In *Fiber Optic Sensors*; Yasin, M., Harun, S., Arof, H., Eds.; InTech: Ridjeka, Croatia, 2012; pp. 317–341.
8. Kersey, A.D.; Davis, M.A.; Heather, J.P.; LeBlanc, M.; Koo, K.P.; Askins, C.G.; Putnam, M.A.; Friebele, E.J. Fiber grating sensors. *J. Lightwave Technol.* **1997**, *15*, 1442–14463.
9. Savin, S.; Digonnet, M.J.F.; Kino, G.S.; Shaw, H.J. Tunable mechanically induced long-period fiber gratings. *Opt. Lett.* **2000**, *25*, 710–712.
10. Vaziri, M.; Che, C.L. An etched two-mode fiber modal coupling element. *J. Lightwave Technol.* **1997**, *15*, 474–481.
11. Vengsarkar, A.M.; Lemaire, P.J.; Judkins, J.B.; Bhatia, V.; Erdogan, T.; Sipe, J.E. Long-period fiber gratings as band-rejection filters. *J. Lightwave Technol.* **1996**, *14*, 58–65.
12. Estudillo-Ayala, J.; Mata-Chavez, R.; Hernandez-Garcia, J.; Rojas-Laguna, R. Long Period Fiber Grating Produced by Arc Discharges. In *Fiber Optic Sensors*; Yasin, M., Harun, S., Arof, H., Eds.; InTech: Ridjeka, Croatia, 2012; pp. 295–316.

13. Davis, D.D.; Gaylord, T.K.; Glytis, E.N.; Kosinski, S.G.; Mettler, S.C.; Vengsarkar, A.M. Long period fiber grating fabrication with focused CO_2 laser pulses. *Electron. Lett.* **1998**, *34*, 302–303.

14. Akiyama, M.; Nishide, K.; Shima, K.; Wada, A.; Yamauchi, R. A Novel Long-Period Fiber Grating Using Periodically Releases Residual Stress of Pure-Silica Core Fiber. In *Proceedings of the Optical Fiber Communication Conference (OFC)*, San José, CA, USA, 22–27 February 2008; pp. 276–277.

15. Coelho, J.M.P.; Nespereira, M.; Silva, C.; Pereira, D.; Rebordão, J.M. Advances in Optical Fiber Laser Micromachining for Sensors Development. In *Current Developments in Optical Fiber Technology*; Harun, W., Arof, H., Eds.; InTech: Rijeka, Croatia, 2013; pp. 375–401.

16. Yablon, A.D. Optical and mechanical effects of frozen-in stresses and strains in optical fibers. *IEEE J. Sel. Topics Quantum Electron.* **2004**, *10*, 300–311.

17. Yablon, A.D.; Yan, M.F.; Wisk, P.; DiMarcello, F.V.; Fleming, J.W.; Reed, W.A.; Monberg, E.M.; DiGiovanni, D.J.; Jasapara, J. Refractive index perturbations in optical fibers resulting from frozen-in viscoelasticity. *Appl. Phys. Lett.* **2004**, *84*, 19–21.

18. Lancry, M.; Réginier, E.; Poumellec, B. Fictive temperature in silica-based glasses and its application to optical fiber manufacturing. *Progr. Mater. Sci.* **2012**, *57*, 63–94.

19. Kim, B.H.; Ahn, T.-J.; Kim, D.Y.; Lee, B.H.; Chung, Y.; Paek, U.-C.; Han, W.-T. Effect of CO_2 laser irradiation on the refractive-index change in optical fibers. *Appl. Opt.* **2002**, *41*, 3809–3815.

20. Li, Y.; Wei, T.; Montoya, J.A.; Saini, S.V.; Lan, X.; Tang, X.; Dong, J.; Xiao, H. Measurement of CO_2-laser-irradiation-induced refractive index modulation in single-mode fiber toward long-period fiber grating design and fabrication. *Appl. Opt.* **2008**, *47*, 5296–5304.

21. Hutsel, M.R.; Gaylord, T.K. Residual-stress relaxation and densification in CO_2-laser-induced long-period fiber gratings. *Appl. Opt.* **2012**, *51*, 6179–6187.

22. Grellier, A.J.C.; Zayer, N.K.; Pannell, C.V. Heat transfer modelling in CO_2 laser processing of optical fibres. *Opt. Commun.* **1998**, *152*, 324–328.

23. Coelho, J.M.P.; Nespereira, M.C; Abreu, M.; Rebordão, J.M. Modeling refractive index change in writing long-period fiber gratings using mid-infrared laser radiation. *Photonic Sens.* **2012**, *2*, 1–7.

24. Coelho, J.M.P.; Abreu, M.A.; Carvalho-Rodrigues, F. Modelling the spot shape influence on high-speed transmission lap welding of thermoplastics films. *J. Opt. Lasers Eng.* **2008**, *46*, 55–61.

25. Kong, F.; Kovacevic, R. 3D finite element modeling of the thermally induced residual stress in the hybrid laser/arc welding of lap joint. *J. Mater. Process. Technol.* **2010**, *210*, 941–950.

26. Timoshenko, S.P.; Goodier, J.N. *Theory of Elasticity*, 2nd ed.; McGraw-Hill: New York, NY, USA, 1951; pp. 409–410.

27. Rego, G.M. *Arc-Induced Long-Period Fibre Gratings. Fabrication and Their Application in Communications and Sensing*. Ph.D. Thesis, Department Electrical Computer Engineering, University of Porto, Porto, Portugal, 17 July 2006.

28. Corning Inc. Corning® SMF-28 Optical Fiber Product Information, 2002. Available online: http://www.corning.com/WorkArea/showcontent.aspx?id=41261 (accessed on 16 June 2013).

29. André, P.; Rocha, A.; Domingues, F.; Facão, M. Thermal Effects in Optical Fibers. In *Developments in Heat Transfer*; Bernardes, M.A.S., Ed.; InTech: Ridjeka, Croatia, 2011; pp. 1–20.
30. Clowes, J.; Syngellakis, S.; Zervas, M. Pressure sensitivity of side-hole optical fiber sensors. *IEEE Photon. Technol. Lett.* **2009**, *10*, 857–859.
31. Yang, S.; Matthews, M.; Elhadj, S.; Draggoo, V.; Bisson, S. Thermal transport in CO_2 laser irradiated fused silica: *In situ* measurements and analysis. *J. Appl. Phys.* **2009**, *106*, doi:10.1063/1.3259419.
32. McLachlan, A.; Meyer, F. Temperature dependence of the extinction coefficient of fused silica for CO_2 laser wavelengths. *Appl. Opt.* **1987**, *26*, 1728–1731.
33. Siegman, A.E.; Sasnett, M.W.; Johnston, T.F. Choice of clip level for beam width measurements using knife-edge techniques. *IEEE J. Quantum Electron.* **1991**, *27*, 1098–1104.
34. Baptista, F.D.V. Simulação do Comportamento Espectral de Redes de Período Longo em Fibra Óptica. M.Sc. Thesis, Centro de Ciências Exactas e da Engenharia, University of Madeira: Funchal, Portugal, November 2009.
35. Erdogan, T. Cladding-mode resonances in short- and long-period fiber grating filters. *J. Opt. Soc. Am. A* **1997**, *14*, 1760–1773.
36. Erdogan, T. Cladding-mode resonances in short- and long-period fiber grating filters: Errata. *J. Opt. Soc. Am. A* **2000**, *17*, doi:10.1364/JOSAA.17.002113.
37. Nespereira, M.; Coelho, J.M.P.; Monteiro, F.; Abreu, M.; Rebordão, J.M. Optical fiber tapers produced by near-infrared laser radiation. *Proc. SPIE* **2013**, *8785*.

Reprinted from *Sensors*. Cite as: Reyes, M.; Monzón-Hernández, D.; Martínez-Ríos, A.; Silvestre, E.; Díez, A.; Cruz, J.L.; Andrés, M.V. A Refractive Index Sensor Based on the Resonant Coupling to Cladding Modes in a Fiber Loop. *Sensors* **2013**, *13*, 11260–11270.

Article

A Refractive Index Sensor Based on the Resonant Coupling to Cladding Modes in a Fiber Loop

Mauricio Reyes [1], David Monzón-Hernández [2,*], Alejandro Martínez-Ríos [2], Enrique Silvestre [3], Antonio Díez [1], José Luis Cruz [1] and Miguel V. Andrés [1]

[1] Departamento de Física Aplicada, Instituto de Ciencia de los Materiales, Universidad de Valencia, Burjassot 46100, Spain; E-Mails: Mauricio.Reyes@uv.es (M.R.); antonio.diez@uv.es (A.D.); jose.l.cruz@uv.es (J.L.C.); miguel.andres@uv.es (M.V.A.)

[2] Grupo de Sensores Ópticos y Microdispositivos, Centro de Investigaciones en Óptica A.C., Loma del Bosque 115, col. Lomas del Campestre, Leon, Guanajuato 37150, Mexico; E-Mail: amr6@cio.mx

[3] Departamento de Óptica, Universidad de Valencia, Burjassot 46100, Spain; E-Mail: enrique.silvestre@uv.es

* Author to whom correspondence should be addressed; E-Mail: dmonzon@cio.mx; Tel./Fax: +52-477-441-4200.

Received: 23 July 2013; in revised form: 2 August 2013 / Accepted: 16 August 2013 / Published: 23 August 2013

Abstract: We report an easy-to-build, compact, and low-cost optical fiber refractive index sensor. It consists of a single fiber loop whose transmission spectra exhibit a series of notches produced by the resonant coupling between the fundamental mode and the cladding modes in a uniformly bent fiber. The wavelength of the notches, distributed in a wavelength span from 1,400 to 1,700 nm, can be tuned by adjusting the diameter of the fiber loop and are sensitive to refractive index changes of the external medium. Sensitivities of 170 and 800 nm per refractive index unit for water solutions and for the refractive index interval 1.40–1.442, respectively, are demonstrated. We estimate a long range resolution of 3×10^{-4} and a short range resolution of 2×10^{-5} for water solutions.

Keywords: mode coupling; optical fiber devices; refractive index sensor

1. Introduction

Refractometric techniques have experienced a remarkable evolution in the last decade, motivated by the fact that refractive index (RI) measurement gives important information in medical, chemical, industrial and environmental applications. In most modern applications, there is a demand for miniaturization, real time and *in-situ* sensing. Fiber-based technology provides feasible, simple, low cost, and highly sensitive alternatives. Product of this interest, several RI sensors, based on mature fiber technologies, such as highly sensitive long period fiber gratings [1], core-exposed fiber Bragg gratings [2,3], modal all-fiber interferometers [4], or fiber tapers [5,6], have been recently proposed. Furthermore, attractive approaches have been reported taking the new possibilities opened by the advent of the holey fibers [7]. Most fiber-based RI sensors exploit the interaction of the evanescent field with the external medium. Therefore, either polishing, etching, tapering or drilling a section of fiber will be required to prepare a sensor head [1–7]. These post-processing techniques are technically demanding, and make the fiber fragile. In order to overcome this problem, standard long period gratings (LPG) [8], LPG-based interferometers [9] and short multimode fiber interferometers [10], which preserve the integrity of the fiber, can be used.

Here, we propose an easy-to-build and robust optical fiber refractive index sensor, based on the resonant coupling between the fundamental mode of a standard single-mode fiber and the cladding modes, *i.e.*, the high order modes of the fiber itself, in a circular fiber loop. Thus, the fiber integrity is preserved, requiring no Bragg or long period grating, but a simple circular loop of a relatively small radius. However, the polymer coating has to be removed in order to allow the interaction of the cladding modes with the external medium, which reduces in part the robustness of the fiber. The coupling between core and cladding modes is produced by the refractive index profile perturbation that the curvature of the fiber generates. The resonant transfer of power takes place at the wavelengths where the core mode satisfies the phase matching condition with a cladding mode. Therefore the dependence of cladding modes modal index with the external refractive index makes the wavelength of a given resonance to shift. Thus, measuring the wavelength shift of the transmission notches it is possible to determine small refractive index changes of the external medium. Our approach avoids the chemical etching process reported in [11], where a single loop of standard single-mode fiber with a reduced cladding diameter is used. Moreover, we provide an improved theoretical analysis, we estimate the temperature effects and we implement a wavelength codified operation of the sensor, in addition to the simple power loss characterization. Wavelength codified sensors are more reliable than amplitude codified sensors, so this approach is in principle preferable.

2. Mode Coupling in Curved Fiber

The strain generated along the fiber, when it is bent, produces anisotropic and asymmetric changes of the dielectric constant due to the photo-elastic effect [12]. In addition, the geometrical effect associated to the curvature of a waveguide can be modeled by an effective refractive index perturbation in an equivalent straight waveguide [13]. The combination of these two effects can be described as a perturbed straight fiber with an effective relative dielectric constant profile:

$$\varepsilon_x(r,\phi) = \varepsilon_y(r,\phi) = \frac{\varepsilon(r)(1+2\kappa r \cos\phi)}{(1-\varepsilon(r)\alpha\kappa r \cos\phi)} \quad , \quad r < a$$

$$\varepsilon_z(r,\phi) = \frac{\varepsilon(r)(1+2\kappa r \cos\phi)}{(1-\varepsilon(r)\beta\kappa r \cos\phi)} \quad , \quad r < a \tag{1}$$

$$\varepsilon_x(r,\phi) = \varepsilon_y(r,\phi) = \varepsilon_z(r,\phi) = \varepsilon_{ext}(1+2\kappa r \cos\phi) \quad , \quad r < a$$

where ε_x, ε_y and ε_z are the effective relative dielectric constants of the perturbed medium in the transverse x and y directions and the longitudinal z direction, respectively; ε is the dielectric constant profile of the fiber; κ is the curvature of the fiber ($\kappa = 1/R$, R: radius of curvature); r and ϕ are the cylindrical coordinates ($x = r \cos\phi$), as it is depicted in Figure 1; $\alpha = \sigma p_{11} + (\sigma-1) p_{12}$ and $\beta = 2 \sigma p_{12} - p_{11}$, being σ the Poisson's ratio ($\sigma = 0.16$ for silica) and p_{ij} the strain-optic tensor ($p_{11} = 0.12$, $p_{12} = 0.27$, for silica); ε_{ext} is the external dielectric constant; and a is the cladding radius of the fiber. In the external medium, being either air or a liquid, no strain effect is assumed. A first order approximation of Equation (1) was used in [14] to model bending effects on long period fiber gratings. Other previous theoretical analysis of bending losses in single mode fibers, as those reported in [11,15–17], develop a perturbation method based on a scalar approximation for the electromagnetic fields, assuming a plane interface, and considering no photo-elastic effect, i.e., only the geometrical effect were taken into account. In comparison, here we develop a truly vector modal analysis of the bent fibers, which includes both geometrical and the photo-elastic effects, where the oscillations of reported bend-loss curves are explained in terms of the resonant couplings between the fundamental mode and the cladding modes of the bent fiber.

Figure 1 includes a plot of the effective index perturbation in the x direction, $\delta n_e = n_e(r, \phi) - n(r)$ when $r < a$ and $\delta n_e = n_e(r, \phi) - n_{ext}$ when $r > a$, along the x axis, where $n_e = (\varepsilon_x)^{1/2}$ and $n = (\varepsilon)^{1/2}$, computed for a standard SMF28 single-mode fiber (core radius: 4.1 µm, cladding radius: 62.5 µm, numerical aperture: 0.13) with a bending radius of 6.35 mm. In all our simulations we have taken into account the chromatic dispersion of silica (cladding index: 1.444024 at 1,550 nm).

As a result of bending, the modal index of higher order modes increases strongly –while the modal index of the fundamental mode has only a small increase–, and, eventually, when it matches the fundamental mode index a resonant coupling is produced, and efficient power transfer between the fundamental mode and the corresponding cladding mode is enabled. This will be observed as an attenuation dip in the transmission spectrum of the fundamental mode. In order to implement a refractive index sensor, we can take advantage of the natural dependence of the cladding modes modal index with the external refractive index, which will make the spectral position of the transmittance notches to shift as a function of the external refractive index.

In order to calculate the propagation factors of the modes guided by the bent fibers, we followed the method described in [18]. This numerical technique allows fast and accurate modal analysis of special fibers by using an iterative Fourier method. This approach permits dealing with arbitrary spatial refractive-index distributions. The spatial refractive index distribution used for the simulations of the bent fibers corresponds to the effective relative dielectric constant of Equation (1).

In Figure 1c, we can see the first resonant coupling by plotting the modal index deviation, δn_m, with respect the fundamental mode index at $\kappa = 0$, $\lambda = 1,550$ nm, as a function of curvature, when the fiber loop is surrounded by air. This coupling corresponds to the original modes LP_{01} and LP_{02}.

The fields of mode LP$_{02}$ are already strongly distorted by the perturbation. One can follow easily the dispersion curve of the fundamental mode and the higher order cladding mode around the anticrossing point.

Figure 1. (a) Geometry of bent fiber. **(b)** Refractive index perturbation for a bending radius of 6.35 mm; the position of the core is indicated with two dotted lines. **(c)** First resonant coupling predicted by theory as a function of curvature; the insets show the mode intensity patterns of the two coupled modes in a box of 133 × 133 μm.

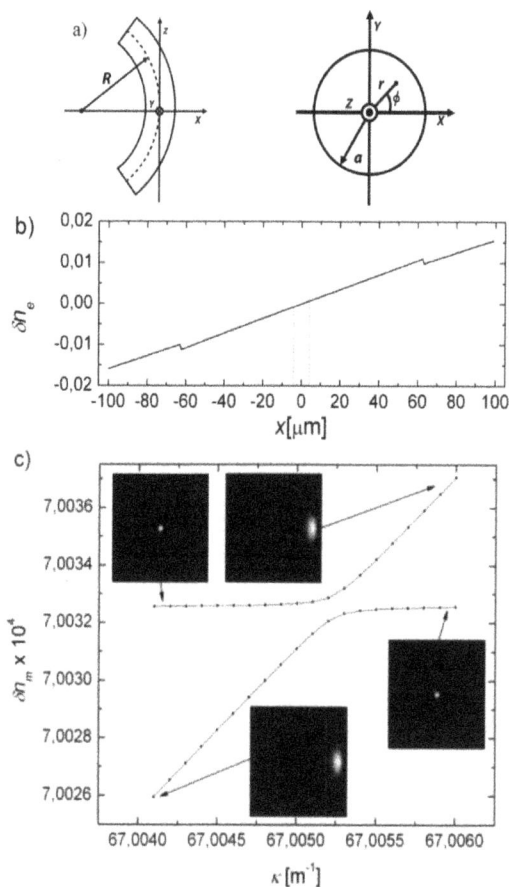

As it is shown in Figure 1, the first resonant coupling takes place when the curvature radius is ~15 mm. This first coupling has a coupling coefficient $k = 1.6 \times 10^{-2}$ m^{-1}, which can be derived from the minimum value of the separation, $\delta n_{m,min}$, between the dispersion curves: $k = \pi \, \delta n_{m,min}/\lambda$ [19]. Such a small coupling coefficient would require a relatively large length of interaction to produce a significant power transfer: about 48 m for 50% power transfer. With the objective to make a practical proposal, we focused our experiments on smaller radius with higher coupling coefficients that may give rise to compact sensor heads. For higher curvatures the cladding modes involved in the experiments will be relatively high order modes. In addition, the curvature enhances the evanescent fields of the cladding modes since the fields are moved towards the external part of the loop (see inset in Figure 1c and, consequently, we expect to achieve higher external refractive index sensitivities by reducing the curvature radius of the fiber.

The modal analysis that we have carried out demonstrates that only at the wavelengths where a phase matching condition is satisfied, as that depicted in detail in Figure 1c a notch in the transmittance is produced, as a result of the distributed transfer of power between the fundamental mode and a perturbed cladding mode along the bent fiber. This type of transfer of power between modes, is what we call resonant coupling. Alternatively, an equivalent modal interferometer description can be used, in which at the point where the curvature starts one should calculate the projection of the fundamental mode onto the perturbed modes of the bent fiber, and after propagation along the length of one loop the modes should be projected back onto the fundamental output mode. We prefer the description in terms of resonant coupling to the modal interferometric approach, due to the modal perspective that we adopt throughout our analysis and because both theory and experiments show that the transfer of power is produced in relatively narrow wavelength bands, while one uses typically a modal interferometer perspective when the transmittance of the system presents a sinusoidal dependence with the wavelength.

3. Refractive Index Sensor

The experimental set-up used to test the proposed RI sensor is shown in Figure 2. The sensor head consists of a small loop of bare optical fiber (SMF-28). The fiber is coiled around a metallic rod with a diameter d, which determines the radius of curvature of the fiber, $R = d/2$. This metallic rod is a support for the fiber loop, to ensure a constant radius of curvature, but has no influence in the optical response of the device since the fields of the guided modes are pushed towards the outer part of the fiber because of the curvature. All our experiments were carried out using rods with diameters around 12 mm. The polymer coating of the fiber was removed by sinking the fiber during few minutes in a methyl chloride solution, avoiding the use of a mechanical stripper in order to ensure no damage of the fiber surface. The transmission spectrum of the fiber previous to bending was used as a reference signal to normalize the transmission spectra obtained when the fiber loop is formed and immersed in different refractive index solutions.

Figure 2. Schematic representation of the experimental set-up. WLS and OSA stand for white light source and optical spectrum analyzer, respectively.

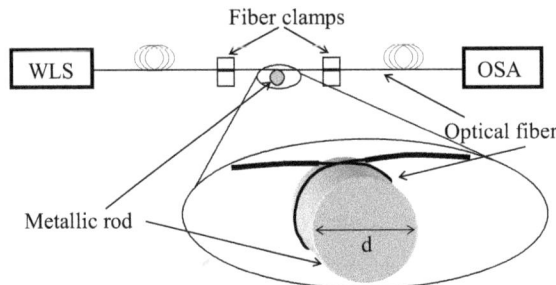

Figure 3a gives the calculated high order mode index difference, with respect the fundamental mode index at zero curvature, *versus* wavelength. These values have been computed assuming a radius of curvature of 6.35 mm and air as external medium. The dispersion curves of the modes that exhibit resonant couplings, and give rise to anti-crossing points, are plotted with solid lines, while some other high order modes exhibit no coupling because of the asymmetry of the fields and are plotted with dashed lines. We can observe, in the spectral range 1,375–1,700 nm, four theoretical resonant couplings at 1,390, 1,460, 1,597 and 1,735 nm, with $\delta n_{m,min}$ of about 10^{-6}, 10^{-7}, 10^{-6} and 10^{-6}. If we follow the dispersion curves of the modes by reducing the curvature of the fiber down to $\kappa = 0$, then we can identify the modes of the unperturbed fiber that evolve to the intensity distributions depicted in Figure 3a and give rise to the resonant coupling: LP_{24}, LP_{81}, LP_{62} and LP_{33}, from lower resonant wavelength to higher.

Figure 3. (a) High order mode index difference with respect the fundamental mode index with $\kappa = 0$: coupled modes with anti-crossing curves (solid lines), and non coupled modes (dashed lines). The insets show the mode intensity patterns of the coupled modes in a box of 133 × 133 µm. (b) Theoretical transmission spectrum. (c) Experimental transmission spectrum. The fiber was wound around a cylinder of 12.70 mm diameter (6.35 mm radius).

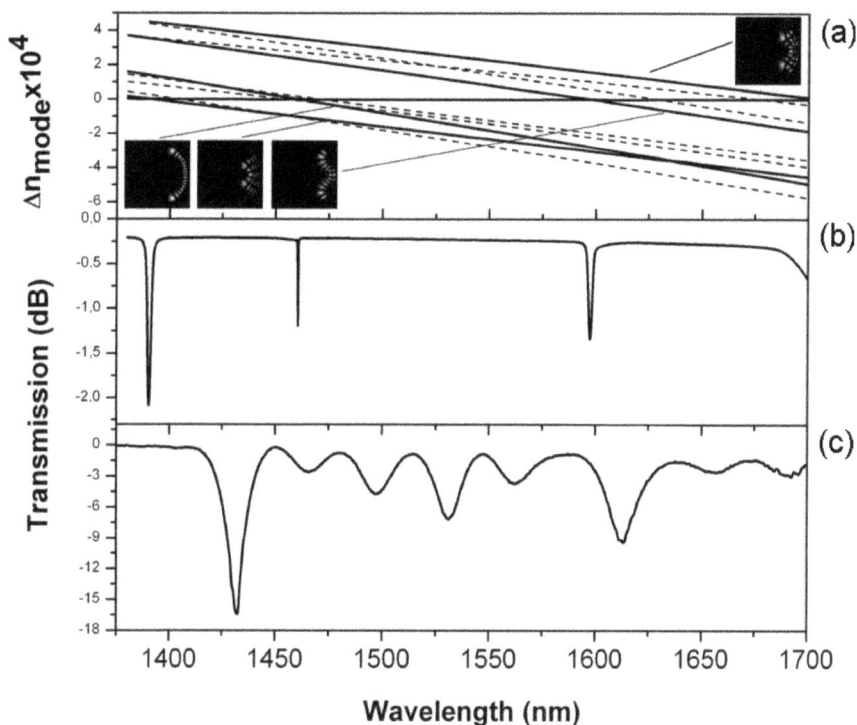

The theoretical transmittance of the fiber loop is shown in Figure 3b. This simulation has been computed projecting the input fundamental mode onto the perturbed modes of the bent section of fiber, allowing the perturbed modes to propagate along the fiber loop with their own propagation factors and projecting the fields onto the output fundamental mode of the fiber at the end of the loop. The spectral position of the notches are determined by the anticrossing points of the dispersion curves, while the minimum transmittance of each dip depends on the accumulated phase difference and the amplitude of excitation of the modes when the projection is computed.

The experimental transmission spectrum of the fiber wound around a cylinder of 12.70 mm diameter is shown in Figure 3c. It is possible to distinguish several peaks, but those with higher losses are located at 1,432 and 1,613 nm. A relative good correspondence between the theoretical analyses and the experimental results can be established if we assume that the experimental wavelengths of the resonant couplings are shifted about 40 nm towards longer wavelengths with respect the theoretical values. Such a shift could be produced by the mechanical tension required to hold the fiber around the metallic rod and the approximations involved in the theoretical simulations. In addition, we have to point out that no artificial adjustment of the nominal values of the fiber has been carried out and that the parameters considered for the simulations are the nominal values depicted in Section 2. Any simulation involving cladding modes, as for example in the case of long period gratings, is highly sensitive to small changes of fiber and material parameters. In our case, no attempt to force a match between theory and experiment has been carried out, since we are more interested in the discussion of the physical mechanism exploited by the sensor proposal.

The difference in amplitude between the theoretical and the experimental notches are probably due to the transient from $\kappa = 0$ to the curvature determined by the rod, which is not considered in the simulations and that may have a significant effect on the projection of the input field onto the fields of the bent fiber. The shallow dips that the experimental transmission spectrum exhibits between 1,450 and 1,575 nm are produced by smaller couplings to asymmetric higher order modes. Even asymmetric modes that, in principle, should be uncoupled, can give rise to small resonant couplings because of any break of the symmetry in the experiment. Experimentally, a reduction of the radius of curvature of 0.175 mm produces a shift of the transmission dips of 25 nm towards longer wavelengths, showing a strong dependence on the radius of curvature. However, the transmission spectra are highly repeatable; and we have not observed changes in the spectrum every time the fiber was wound around the same cylinder.

According to our theoretical results, the transmission spectrum depicted in Figure 3c should not be interpreted as interference fringes between two modes which phase difference changes as a function of wavelength. In the wavelength range of the experiment we have eight notches (the three stronger notches at 1,432, 1,526 and 1,613 nm and five shallow dips at 1,463, 1,500, 1,562, 1,658 and 1,692 nm) that correspond to the eight cladding modes shown in the theoretical dispersion curves of Figure 3a (three symmetric cladding modes that give rise to the stronger couplings and five asymmetric modes that exhibit weaker couplings). Each cladding mode produces separately a notch in the transmission spectrum, although the relatively large number of couplings generates some overlapping between notches.

In order to analyze the possibility to use the small fiber loop as a refractive index sensor, we measured experimentally the wavelength shift of the resonant couplings as a function of the external refractive index. These measurements were carried out, first, using commercial Cargille liquids with calibrated RI from 1.360 to 1.442 (nominal values at 519 nm and 25 °C), and later using standard solvents as water, acetone and isopropyl alcohol.

Between consecutive measurements, the fiber loop was cleanedwith acetone and then dried with air. As an example, Figure 4a gives the transmission spectra whenwater, acetone and isopropyl alcohol were applied, while in Figure 4b we show the transmission spectra for four different RI liquids. The wavelength ofeach resonant coupling shift towards longer values as the refractive index of the external medium augments. A more detailed calibration is given in Figure 5 for the strongest experimental coupling, the one located at 1,432 nm when the fiber is surrounded by air.

Figure 4. (**a**)Transmission spectra of the sensor for different liquids, when the diameter of the loop was 12.70 mm. (**b**) Transmission spectra of the sensor for different RI liquids. The diameter of the loop was 12.70 mm.

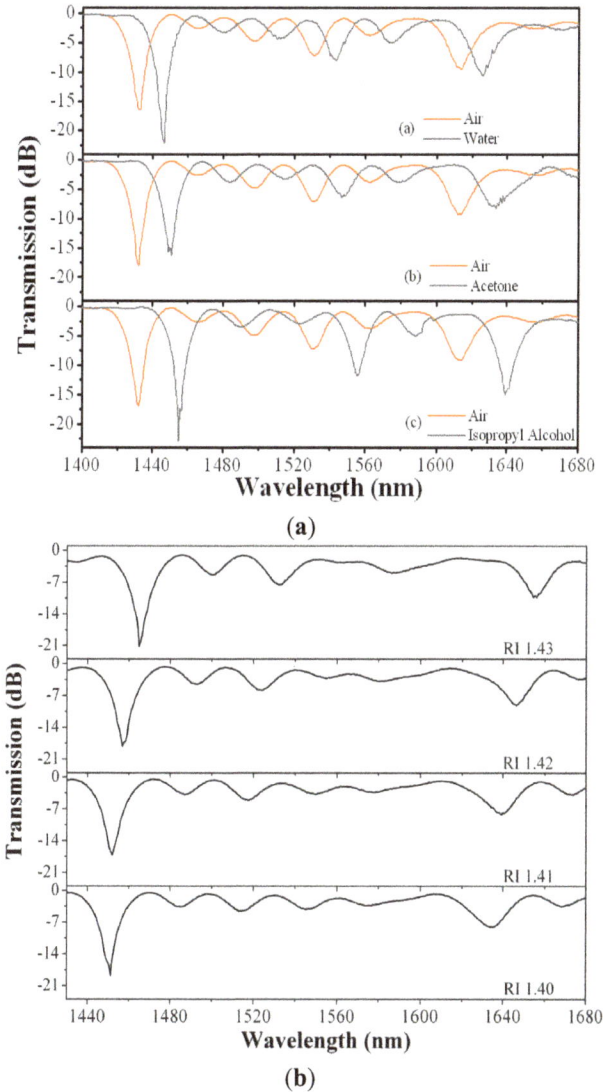

(a)

(b)

184

Figure 5. Wavelength shift of a resonant coupling versus external refractive index: (solid line) theoretical shift of the resonance, (solid circles) experimental values obtained with Cargille liquids, (open circles) experimental values obtained with standard liquids (see Figure 4), (dashed line) eye guide.

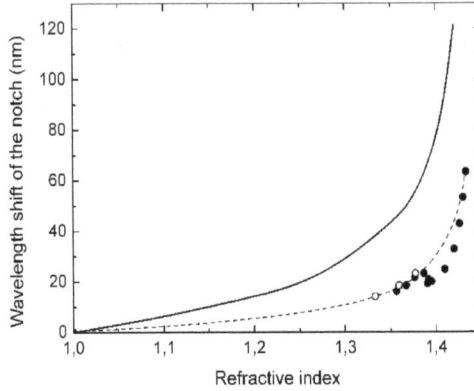

The theory predicts correctly the tendency, although it fails to give the exact values for the wavelength shifts. In addition, the experimental points show an anomalous behavior around 1.39. That anomaly is likely produced by a resonant coupling between three modes. The simultaneous phase matching of three modes has been observed in the numerical simulations around that point, between two coupled modes and, in principle, one uncoupled mode. However, as we have mentioned already, the experiment shows that it exists a break of the symmetry and most of the high order modes are slightly coupled. Thus, around 1.39 we have a three modes coupling that will give rise to a complex anticrossing point in which the gap between the dispersion curves of the modes has a non trivial dependence on the wavelength. We think that this is the reason for the somewhat anomalous experimental behavior. Nevertheless, around the refractive index of water solutions (~1.33)——where most practical applications are——the response is monotonic and has a sensitivity of 170 nm per refractive index unit (RIU). This sensitivity is higher than the values reported for standard LPG which are typically around or below 50 nm/RIU for water solutions [20], as well as higher than the values reported in the references discussed in the introduction. Using chemically etched LPGs, the sensitivity can be improved and a sensitivity of 262 nm/RIU has been reported [21]. Thus, we can estimate a detection limit of 3×10^{-4} for water solutions, assuming an OSA resolution of 50 pm. However, this estimation might be somewhat unrealistic since the line width of the dips is about 10 nm. Thus if we assume that we can detect 1/20 of the line width, *i.e.*, a shift of 500 pm, then the estimation would be a detection limit of 3×10^{-3}. At higher refractive index values, around 1.41, the sensitivity increases to 800 nm/RIU, which gives a detection limit between 6×10^{-4} and 6×10^{-5}. Although these values are relatively low, they are better than that of some commercial bulk refractometers, and similar to that obtained in [11] by etching the fiber diameter to 81 μm. However, our proposal is a wavelength codified sensor which is intrinsically more reliable than amplitude codified sensors as that reported in [11].

Temperature effects are an important issue for any refractive index sensor. A detailed simulation has been carried out in which the thermal expansion of both the core and cladding radii have been taken into account ($\alpha_T = 0.55 \times 1^{-6}$ K^{-1}), as well as the thermo-optic coefficients of the Ge-doped core ($\partial n_{core}/\partial T = 1.15 \times 10^{-5}$ K^{-1}) and the pure silica cladding ($\partial n_{clad}/\partial T = 1.06 \times 10^{-5}$ K^{-1}). The theoretical estimation is a drift of 0.1 nm/K, towards shorter wavelengths, for the resonant couplings shown in Figure 3. Such a small drift is comparable to the resolution, so it can be considered negligible provided the measurements are carried out maintaining a constant temperature within ±1 K.

Moreover, if we decided to operate our sensor as an amplitude codified sensor by measuring transmittance changes *versus* external refractive index sensor, then a detection limit of 2×10^{-5} is estimated, since the slope at the optimum point of the transmission spectrum for water is 2.9 dB/nm and a resolution of 0.01 dB is usual in commercial optical power meters. Similarly, a detection limit of 4×10^{-6} is obtained for an external refractive index of about 1.41. Thus, combining wavelength and power measurements our proposal exhibits large refractive index range with moderate resolution and short ranges around a given refractive index value with high resolution.

4. Conclusions

We have demonstrated an easy-to-build optical fiber refractometer that consists of a single mode fiber wound around a cylinder to form a single loop. The theoretical principle of operation has been discussed and, according to the experimental results, a long range resolution of 3×10^{-4} for water solutions is demonstrated, by measuring wavelength changes of the spectral transmission notches. Amplitude measurements, *i.e.*, optical power transmission measurements, would give a resolution of 2×10^{-5}. At higher refractive index values, the detection limits can be as low as 4×10^{-6}.

Acknowledgments

This work was supported in part by the Ministerio de Economía y Competitividad and the Generalitat Valenciana of Spain (projects TEC2008-05490 and PROMETEO/2009/077, respectively).

Conflicts of Interest

The authors declare no conflict of interest.

References

1. Zhu, T.; Rao, Y.-J.; Wang, J.-L.; Song, Y. A highly sensitive fiber-optic refractive index sensor based on an edge-written long-period fiber grating. *IEEE Photon. Technol. Lett.* **2007**, *19*, 1946–1948.
2. Iadiccico, A.; Campopiano, S.; Cutolo, A.; Giordono, M.; Cusano, A. Nonuniform thinned fiber Bragg gratings for simultaneous refractive index and temperature measurements. *IEEE Photon. Technol. Lett.* **2005**, *17*, 1495–1497.
3. Liang, W.; Huang, Y.; Xu, Y.; Lee, R.K.; Yariv, A. Highly sensitive fiber Bragg grating refractive index sensors. *Appl. Phys. Lett.* **2005**, *86*, 151122:1−151122:3.

4. Salceda-Delgado, G.; Monzón-Hernández, D.; Martínez-Ríos, A.; Cárdenas-Sevilla, G.A.; Villatoro, J. Optical microfiber mode interferometer for temperature-independent refractometric sensing. *Opt. Lett.* **2012**, *37*, 1974–1976.

5. Monzón-Hernández, D.; Villatoro, J. High-resolution refractive index sensing by means of a multiple-peak surface plasmon resonance optical fiber sensor. *Sens. Actuators B Chem.* **2005**, *115*, 227–231.

6. Xu, F.; Pruneri, V.; Finazzi, V.; Brambilla, G. An embedded optical nanowire loop resonator refractometric sensor. *Opt. Express.* **2008**, *16*, 1062–1067.

7. Huy, M.C.P.; Laffont, G.; Dewynter, V.; Ferdinand, P.; Roy, P.; Auguste, J.-L.; Pagnoux, D.; Blanc, W.; Dussardier, B. Three-hole microstructured optical fiber for efficient fiber Bragg grating refractometer. *Opt. Lett.* **2007**, *32*, 2390–2392.

8. Bhatia, V.; Vengsarkar, A.M. Optical fiber long-period gratings sensors. *Opt. Lett.* **1996**, *21*, 692–694.

9. Mosquera, L.; Sáez-Rodríguez, D.; Cruz, J.L.; Andrés, M.V. In-fiber Fabry-Perot refractometer assisted by a long-period grating. *Opt. Lett.* **2010**, *35*, 613–615.

10. Wu, Q.; Semenova, Y.; Yan, B.; Ma, Y.; Wang, P.; Yu, C.; Farrell, G. Fiber refractometer based on a fiber Bragg grating and single-mode-multimode-single-mode fiber structure. *Opt. Lett.* **2011**, *36*, 2197–2199.

11. Wang, P.; Semenova, Y.; Wu, Q.; Farrel, G.; Ti, Y.; Zheng, J. Macrobending single-mode fiber-based refractometer. *Appl. Opt.* **2009**, *48*, 6044–6049.

12. Smith, A.M. Birefringence induced by bends and twists in single mode optical fiber. *Appl. Opt.* **1980**, *19*, 2606–2611.

13. Marcuse, A. Influence of curvature on the losses of doubly clad fibers. *Appl. Opt.* **1982**, *21*, 4208–4213.

14. Block, U.L.; Dangui, V.; Digonnet, M.F.; Fejer, M.M. Origin of apparent resonance mode splitting in bent long-period fiber gratings. *J. Lightw. Technol.* **2006**, *24*, 1027–1034.

15. Wang, Q.; Farrell, G.; Freir, T. Theoretical and experimental investigations of maxcro-bend losses for standard single mode fibers. *Opt. Express* **2005**, *13*, 4476–4484.

16. Renner, H. Bending losses of coated single-mode fibers: A simple approach. *J. Lightw. Technol.* **1992**, *10*, 544–551.

17. Fasutini, L.; Martini, G. Bend loss in single-mode fibers. *J. Lightw. Technol.* **1997**, *15*, 671–679.

18. Silvestre, E.; Pinheiro-Ortega, T.; Andrés, P.; Miret, J.J.; Ortigosa-Blanch, A. Analytical evaluation of chromatic dispersion in photonic crystal fibers. *Opt. Lett.* **2005**, *30*, 453–455.

19. Yariv, A. Propagation and Coupling of Modes in Optical Dielectric Waveguides-Periodic Waveguides. In *Optical Electronics in Modern Communications*, 5th ed.; Oxford University Press: New York, NY, USA, 1997; pp. 491–539.

20. Chamorro Enríquez, D.A.; da Cruz, A.R.; Rocco Giraldi, M.T.M. Hybrid FBG–LPG sensor for surrounding refractive index and temperature simultaneous discrimination. *Opt. Las. Technol.* **2012**, *44*, 981–986.

21. Yan, J.; Zhang, A.P.; Shao, L.-Y.; Ding, J.-F.; He, S. Simultaneous measurement of refractive index and temperature by using dual long-period gratings with an etching process. *IEEE Sens. J.* **2007**, *7*, 1360–1361.

Reprinted from *Sensors*. Cite as: Barrera, D.; Sales, S. A High-Temperature Fiber Sensor Using a Low Cost Interrogation Scheme. *Sensors* **2013**, *13*, 11653–11659.

Article

A High-Temperature Fiber Sensor Using a Low Cost Interrogation Scheme

David Barrera* and Salvador Sales

Optical and Quantum Communications Group, Institute of Telecommunications and Multimedia Applications (iTEAM), Universitat Politècnica de València, Valencia, 46022, Spain; E-Mail: ssales@dcom.upv.es

* Author to whom correspondence should be addressed; E-Mail: dabarvi@iteam.upv.es; Tel.: +34-963-877-007; Fax: +34-963-879-583.

Received: 16 June 2013; in revised form: 24 August 2013 / Accepted: 29 August 2013 / Published: 4 September 2013

Abstract: Regenerated Fibre Bragg Gratings have the potential for high-temperature monitoring. In this paper, the inscription of Fibre Bragg Gratings (FBGs) and the later regeneration process to obtain Regenerated Fiber Bragg Gratings (RFBGs) in high-birefringence optical fiber is reported. The obtained RFBGs show two Bragg resonances corresponding to the slow and fast axis that are characterized in temperature terms. As the temperature increases the separation between the two Bragg resonances is reduced, which can be used for low cost interrogation. The proposed interrogation setup is based in the use of optical filters in order to convert the wavelength shift of each of the Bragg resonances into optical power changes. The design of the optical filters is also studied in this article. In first place, the ideal filter is calculated using a recursive method and defining the boundary conditions. This ideal filter linearizes the output of the interrogation setup but is limited by the large wavelength shift of the RFBG with temperature and the maximum attenuation. The response of modal interferometers as optical filters is also analyzed. They can be easily tuned shifting the optical spectrum. The output of the proposed interrogation scheme is simulated in these conditions improving the sensitivity.

Keywords: optical fibre sensor; high-temperature; Regenerated Fibre Bragg Grating; interrogation; modal interferometer

1. Introduction

Optical fibre sensors offer a good performance under extreme conditions because they have small dimensions and low weight, they are immune to electromagnetic interference, chemically inert and spark free [1]. Among the various types of optical fibre sensors, Fibre Bragg Gratings (FBGs) offer greater multiplexing capabilities for multipoint measurement and the fluctuations of the received power do not affect the measurements. However, for high-temperature monitoring FBGs present a progressive decay, limiting the range of operation [2]. Several techniques have been proposed to extend the operating temperature range of the FBGs: modifications of the composition of the fibre, inscription of Type II gratings and high-temperature annealing [3–5]. Regenerated Fiber Bragg Gratings (RFBGs) are obtained from a seed FBG after a high-temperature annealing process. During the annealing process, the reflected optical power of the seed FBGs decays, followed by an increase. This is known as the regeneration process, and gives rise to a RFBG with improved temperature stability. In the last years, the regeneration process has been studied under different conditions and with several optical fibres [6–10].

In this paper the fabrication of RFBGs in high-birefringence optical fibres is shown. The optical spectrum shows two resonances corresponding to the fast and slow polarizations which show a different wavelength shift with temperature. This effect is going to be used to reduce the complexity and cost of the interrogation of the high-temperature sensor. The proposed interrogation setup is based on the use of optical filters in order to convert the wavelength shifts of each of the polarizations into changes in the optical power. The optical filters have a significant influence in the output of the interrogation setup and their design is also addressed. The paper is organized as follows: Section 2 details the fabrication process of the RFBG in high-birefringence optical fibre. Section 3 presents the proposed interrogation setup and the design of the ideal filter. In Section 4, the use of modal interferometers as the optical filters is discussed. Finally, concluding remarks are provided.

2. RFBG Regeneration

A Fibercore HB1500T high-birefringence optical fibre is used. The core of the optical fibre is flanked by areas of high-expansion, boron-doped glass that shrink-back more than the surrounding silica as the fibre is drawn and freeze the core in tension. This tension induces birefringence and when a FBG is inscribed originates two grating peaks at different wavelength for the slow and fast axis. The difference in the thermal expansion coefficient of the boron-doped areas makes that the wavelength separation of the two Bragg resonances changes with temperature.

In order to be able to measure high-temperatures a RFBG is needed. In first place the optical fibre is introduced in a hydrogen loading chamber. The seed FBGs is then inscribed in the optical fibre using an argon-ion frequency doubled laser at 244 nm and a phase mask technique. Finally, the FBG is introduced in a tubular oven and a high-temperature annealing is performed. The optical spectrum is continuously monitored using during the annealing process by a Micron Optics SM125 FBG interrogator.

Figure 1a shows the optical spectrum of one of the seed gratings and the RFBG obtained after the regeneration process described in Figure 1b. The regeneration process consist in a 10 °C/min temperature ramp up to 900 °C followed by a 5 °C/min temperature ramp up to 1,000 °C where the temperature is stabilized for 2 h. After this stabilization period, the oven is allowed to return to ambient temperature. It can be noticed the two resonances corresponding to the fast and slow polarizations. At the same temperature, the obtained RFBGs present a permanent wavelength shift and have a reduced reflectivity compared to the original FBG but the dynamic range is still around 20 dB.

Figure 1. (**a**) Spectra of the seed FBG and RFBG; (**b**) Maximum reflected optical power of the fast and slow axis during the regeneration process.

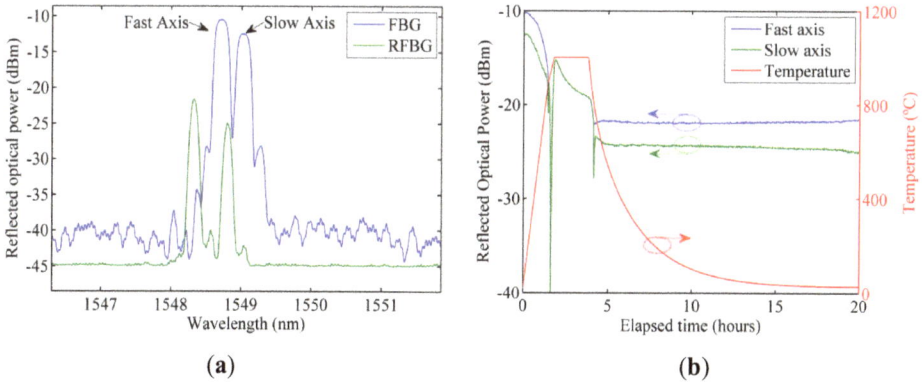

(**a**)   (**b**)

The wavelength shift of the two resonances in the RFBG is characterized using temperature cycles. These characterization cycles use three concatenated temperature cycles that follows the same heating scheme of the regeneration process previously described. Nonetheless, in contrast with the regeneration process, between two consecutive temperature cycles the temperature is left to drop only down to 300 °C avoiding the long period of time needed to cool down the oven to ambient temperature. For temperatures higher than 800 °C the birefringence is smaller and is no longer possible to distinguish the two resonances with the instrumentation used and only one resonance is observed. The results and the fitting with a fifth order polynomial are shown in Figure 2a. Figure 2b shows the separation between the slow and fast axis Bragg resonances and the error between the fitting and the experimental data.

Figure 2. (**a**) Wavelength shift with temperature of the slow and fast axis Bragg resonances; (**b**) Separation between the slow and fast axis Bragg resonances.

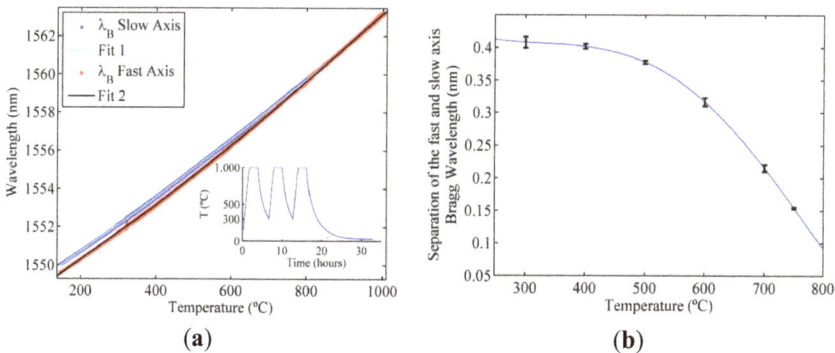

(**a**)   (**b**)

3. Interrogation Setup

In order to reduce the cost of the sensor interrogation, the scheme that is shown in Figure 3 is proposed. The light from a non-polarized optical light source illuminates the RFBG. The reflected signal is then divided into the two polarizations. A variable attenuator can be used to compensate possible differences in the optical power between the two polarizations. Each of the polarizations passes through an optical filter that converts the wavelength shift into power changes that is measured with a photodiode. Finally, the outputs of the two photodiodes are subtracted to determine the temperature variation. Optical power fluctuations are compensated due to the differential detection scheme on condition that optical variations affect the two polarizations at the same time. The use of the optical fibre filters results in a significant cost reduction compared with tunable lasers and tunable filters within the actual commercial interrogation units.

Figure 3. Proposed interrogation setup. ASE stands for Amplified Spontaneous Emission and PD for Photodiode.

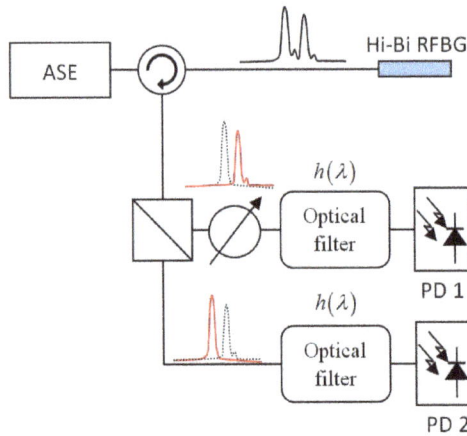

The response of the interrogation scheme can be tuned designing the optical spectrum of the optical filters. Standard optical edge filters can be used. In this case the output of the interrogation setup would show a non-lineal response, according to the temperature characterization, requiring a later electronic compensation. Alternately, signal processing can be performed optically designing and implementing the optical edge filter [11]. In an ideal situation, the two filters are the same and the output of the interrogation setup has a linear dependence with temperature. Since the optical filter is the same for the two polarizations the optical spectrum of the optical filter can be determined by a recursive relationship:

$$\Delta P = h(\lambda_1) - h(\lambda_2) = A \cdot T + B \tag{1}$$

$$h(\lambda_1) = h(\lambda_2) + A \cdot T + B \tag{2}$$

where ΔP is the optical power difference measured in the photodiodes, $h(\lambda)$ is the spectrum of the optical filter where λ_1 and λ_2 are the wavelengths of the Bragg resonances for the fast and slow axis. A and B are the constants that define the linear dependence with temperature. The value of A is limited by the maximum attenuation of the optical filter and B can be obtained from the temperature T_M where the two Bragg resonances match:

$$B = -A \cdot T_M \tag{3}$$

Figure 4a shows the optical spectrum of the ideal filter obtained using the recursive method and the temperature characterization of the RFBGs. Figure 4b represents the simulated output of the interrogation setup when the ideal filter is used. As can be noticed, the output has a linear dependence with temperature, as expected. Nonetheless, the slope of the ideal filter is limited by the high wavelength shift of the RFBG reducing the resolution of the interrogation setup. This limitation can be solved allowing the output of the interrogation setup be non-lineal.

Figure 4. (a) Optical spectrum of the ideal filter; **(b)** Simulated output of the proposed interrogation setup using the ideal filter.

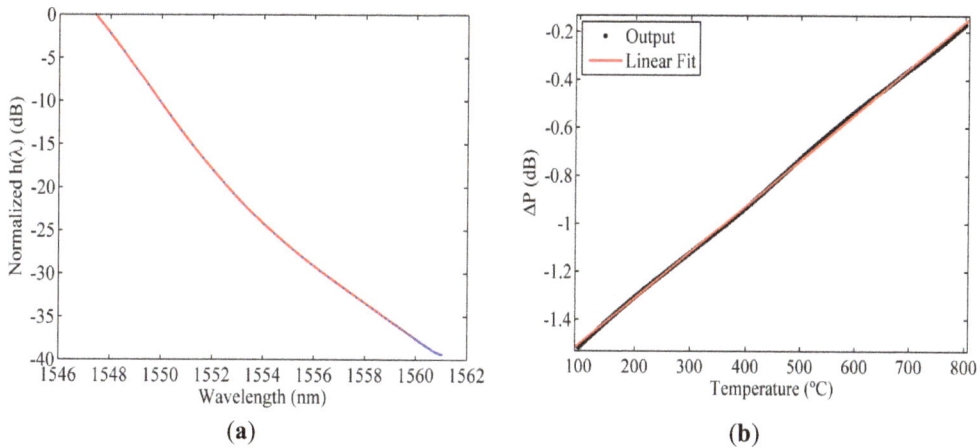

(a)   (b)

4. Modal Interferometers

The use of highly tunable filters permits the use of the same optical filter for both polarizations. Tuning one of the optical filters allows one to change the output of the interrogation setup improving the sensitivity of the interrogation setup. Using the interrogation setup proposed previously, the use of modal interferometers as optical filters is discussed. The spectra of modal interferometers show a periodic response that can be modelled as a sinusoidal response [12]. The period of the sinusoid can be easily selected during the fabrication of the modal interferometer and strain or temperature can be used to shift the spectral response. Figure 5a shows the spectrum of a modal interferometer with a period of 14.3 nm, where the period of the modal interferometer has been selected to be larger than the wavelength shift of the RFBG for high temperatures.

Figure 5. (a) Optical spectrum of a modal interferometer with a period of 14.3 nm; (b) Simulated output of the proposed interrogation setup using modal interferometers with relative phase differences.

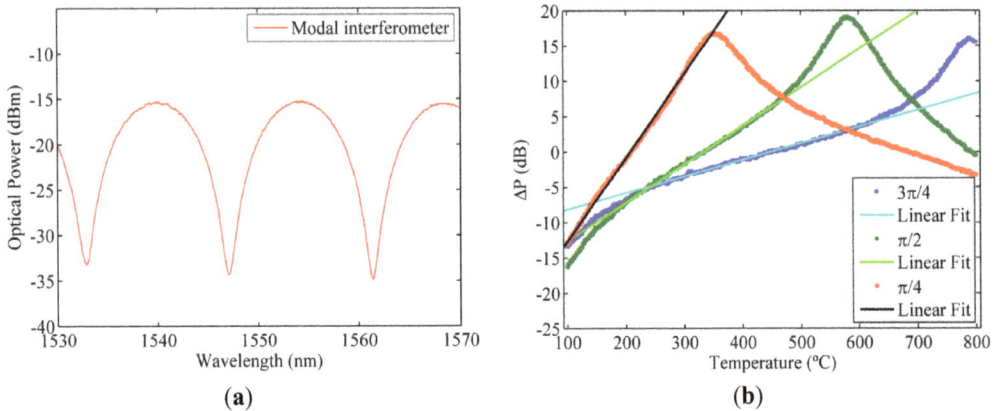

(a)   (b)

The output of the interrogation setup is simulated using the measured optical spectra of the modal interferometer and the RFBG. As a result of the small separation of the two Bragg resonances the performance of the interrogation setup can be improved introducing a wavelength shift between the two optical filters. This wavelength shift is equivalent to introduce a phase shift in the sinusoidal optical spectrum of the modal interferometers. It is worth to mention that, because of the periodic response of the modal interferometers, the unambiguous range is reduced at the same time. Figure 5b shows the effect of the relative phase difference induced between the two optical filters.

Using the modal interferometers the linear range obtained is limited by the period of the modal interferometer and the relative phase difference. The sensitivity, which is determined by the slope of the output of the interrogation setup, depends on the relative phase difference. The linear range can also be tuned applying the same wavelength shift to both optical filters.

5. Conclusions

The fabrication and characterization of regenerated fiber Bragg gratings in high-birefringence optical fiber for use in high-temperature applications has been shown. The optical spectrum shows two Bragg resonances for the slow and fast polarizations. The separation of the two Bragg resonances can be used to reduce the cost of an interrogation unit using optical filters to convert the wavelength shifts into optical power changes. The spectral response of the optical filter has been studied. The ideal filter which linearizes the output of the interrogation setup has been obtained but the sensitivity is limited by the large wavelength shift of the RFBG and the maximum attenuation of the filter. To avoid the limitations of the ideal optical filter the output of the interrogation setup is studied using modal interferometers. Modal interferometers are highly tunable devices which optical spectrum can be easily shifted by temperature or strain. Due to the periodic response of the modal interferometers the output of the interrogation setup is also periodic with a quasi linear response in a limited range. The range and slope of the interrogation setup can be tuned by modifying the phase difference between the modal interferometers.

Acknowledgments

The authors wish to acknowledge the financial support of the Infraestructura FEDER UPVOV08-3E-008, FEDER UPVOV10-3E-492, the Spanish MCINN through the project TEC2011-29120-C05-05 and the Valencia Government through the Ayuda Complementaria ACOMP/2013/146. The authors also acknowledge the collaboration of Alvarez from Fibercore for providing the high birefringence optical fiber.

Conflicts of Interest

The authors declare no conflict of interest.

References

1. Othonos, A.; Kalli, K. *Fiber Bragg Gratings: Fundamentals and Applications in Telecommunications and Sensing*; Artech House Optoelectronics Library: Boston, MA, USA, 1999.
2. Erdogan, T.; Mizrahi, V.; Lemaire, P.; Monroe, D. Decay of ultraviolet-induced fiber Bragg gratings. *J. Appl. Phys.* **1994**, *76*, 73–80.
3. Butov, O.V.; Dianov, E.M.; Golant, K.M. Nitrogen-doped silica-core fibres for Bragg grating sensors operating at elevated temperatures. *Meas. Sci. Technol.* **2006**, *17*, 975–979.
4. Grobnic, D.; Mihailov, S.J.; Smelser, C.W.; Ding, H. Sapphire fiber bragg grating sensor made using femtosecond laser radiation for ultrahigh temperature applications. *IEEE Photonics Technol. Lett.* **2004**, *16*, 2505–2507.
5. Canning, J.; Stevenson, M.; Bandyopadhyay, S.; Cook, K. Extreme silica optical fibre gratings. *Sensors* **2008**, *8*, 6448–6452.
6. Cook, K.; Shao, L.; Canning, J. Regeneration and helium: regenerating Bragg gratings in helium-loaded germanosilicate optical fibre. *Opt. Mater. Express* **2012**, *2*, 1733–1742.
7. Lindner, E.; Canning, J.; Chojetzki, C.; Brückner, S.; Becker, M.; Rothhardt, M.; Bartelt, H. Post-hydrogen-loaded draw tower fiber Bragg gratings and their thermal regeneration. *Appl. Opt.* **2011**, *50*, 2519–2522.
8. Trpkovski, S.; Kitcher, D.J.; Baxter, G.W.; Collins, S.F.; Wade, S.A. High-temperature-resistant chemical composition Bragg gratings in Er3+-doped optical fiber. *Opt. Lett.* **2005**, *30*, 607–609.
9. Wang, T.; Shao, L.; Canning, J.; Cook, K. Regeneration of fiber Bragg gratings under strain. *Appl. Opt.* **2013**, *52*, 2080–2085.
10. Barrera, D.; Finazzi, V.; Villatoro, J.; Sales, S.; Pruneri, V. Packaged optical sensors based on regenerated fiber bragg gratings for high temperature applications. *IEEE Sens. J.* **2012**, *12*, 107–112.
11. Fernandez-Ruiz, M.R.; Carballar, A.; Azana, J. Design of ultrafast all-optical signal processing devices based on fiber bragg gratings in transmission. *J. Light. Technol.* **2013**, *31*, 1593–1600.
12. Barrera, D.; Villatoro, J.; Finazzi, V.; Cárdenas-Sevilla, G.; Minkovich, V.; Sales, S.; Pruneri, V. Low-loss photonic crystal fiber interferometers for sensor networks. *J. Light. Technol.* **2010**, *28*, 3542–3547.

Reprinted from *Sensors*. Cite as: Grassi, A.P.; Tremmel, A.J.; Koch, A.W.; El-Khozondar, H.J. On-Line Thickness Measurement for Two-Layer Systems on Polymer Electronic Devices. *Sensors* **2013**, *13*, 15747–15757.

Article

On-Line Thickness Measurement for Two-Layer Systems on Polymer Electronic Devices

Ana Perez Grassi [1,*]**, Anton J. Tremmel** [1]**, Alexander W. Koch** [1] **and Hala J. El-Khozondar** [2]

[1] Institute for Measurement Systems and Sensor Technology, Technische Universität München, Theresienstr. 90/N5, Munich 80333, Germany; E-Mails: a.tremmel@tum.de (A.J.T.); a.w.koch@tum.de (A.W.K.)

[2] Electrical Engineering Department, Islamic University of Gaza, Gaza P.O.Box 108, Palestine; E-Mail: hkhozondar@iugaza.edu

* Author to whom correspondence should be addressed; E-Mail: a.perez@tum.de; Tel.: +49-89-2892-5107.

Received: 25 September 2013; in revised form: 24 October 2013 / Accepted: 11 November 2013 / Published: 18 November 2013

Abstract: During the manufacturing of printed electronic circuits, different layers of coatings are applied successively on a substrate. The correct thickness of such layers is essential for guaranteeing the electronic behavior of the final product and must therefore be controlled thoroughly. This paper presents a model for measuring two-layer systems through thin film reflectometry (TFR). The model considers irregular interfaces and distortions introduced by the setup and the vertical vibration movements caused by the production process. The results show that the introduction of these latter variables is indispensable to obtain correct thickness values. The proposed approach is applied to a typical configuration of polymer electronics on transparent and non-transparent substrates. We compare our results to those obtained using a profilometer. The high degree of agreement between both measurements validates the model and suggests that the proposed measurement method can be used in industrial applications requiring fast and non-contact inspection of two-layer systems. Moreover, this approach can be used for other kinds of materials with known optical parameters.

Keywords: optical films; TFI; thin film interferometry; two-layer model; TFR; thin film reflectance measurement; polymer electronics

1. Introduction

The use of electronic functional polymers in the production of integrated circuits has been increasing significantly in recent years. Polymer electronics require new production techniques different from those used for silicon. The production of polymer electronics is based on a printing process similar to that used on paper. In particular, the circuit layers are successively printed on a substrate, which moves on a conveyor belt at a high velocity. The correct thickness of such layers is essential for guaranteeing the electrical behavior of the final product. Therefore, the thickness and other parameters must be monitored carefully during the production through a fast and non-contact process.

The conveyor belt of the printing setup complicates the implementation of transmission-based methods [1,2] for monitoring the film thickness. For this reason, we focused on reflection-based approaches. Common methods for measuring thin film thickness based on reflection include thin film interferometry (TFI) [3,4], thin film reflectometry (TFR) [5] and spectral ellipsometry [6]. Spectral ellipsometers can achieve a higher accuracy in thickness measurements than thin film reflectometers [7]. However, they require a more complex setup and are potentially slower [7]. TFI is based on a moving repetitive scanning process, which makes it only appropriate for static measurements [8,9]. As a result, TFR is advantageous for applications, such as on-line thickness monitoring, where measuring time should be kept short and/or the high accuracy of spectral ellipsometers is not needed.

The reflected signal measured by TFR is a function of the involved film thickness [5]. Therefore, by fitting it with a valid model, the thickness values can be obtained. A reflectance model for a single-layer system of polycrystalline silicon was presented by Hauge [10]. Hauge considers ideal interfaces for his model. However, in practice, irregular interfaces affect the reflectance significantly. The interfaces can be evaluated through the effective media approximation (EMA) [11] or by altering the Fresnel coefficients through different interface models [12]. Additionally, Montecchi [13] presents a model based on the perturbative method to measure thickness by considering inhomogeneities, roughness and slanted interfaces. This model is limited to one layer and discards perturbations on the layer-substrate and on the air-layer interfaces. Swanepoel [14] presents an approach to consider irregular interfaces on transmission spectra.

All mentioned contributions present models for single-layer systems. The most popular approaches for resolving multilayer systems are based on matrix methods [15] and recursive algorithms [16]. Most contributions in the inspection of multilayer systems are made for transmittance measurements [17,18] and X-ray reflectometry [16]. On the contrary, for white reflectometry, the literature lacks concrete works. From the reflectance expression of a single-layer system, like that given in [10], a two-layer reflectance model can be derived directly. Although the expression for a two-layer system is well known, the literature lacks models that can be directly applied to real measurements in white light reflectometry. Interface inhomogeneities and distortions introduced by the measurement equipment significantly affect the signal captured by the sensor. If the latter is neglected, the fitting process will compensate for these distortions with the thickness parameters. As a result, the model will match the measured signal for incorrect values.

Our approach uses Stearns' method [12] to incorporate the interface irregularities to the model. Stearns proposes to model the interface profile by using an analytical function. This method is largely used and well known in X-ray reflectometry [19,20]. However, to the best of our knowledge, its application in white light reflectometry for multilayer systems has not been published until now.

In the same way, we analyze the influence of the chromatic effect [21,22] introduced by the setup on the captured signal. This effect should be considered to avoid distortions in the measured results. Moreover, this model can be applied to measure other materials with known optical parameters.

2. Fundamentals of Thin Film Reflection

The complex refractive index, $\mathbf{n}(\lambda)$, of a material can be denoted as: $\mathbf{n}(\lambda) = n(\lambda) - jk(\lambda)$, where $n(\lambda)$ is the index of refraction, $k(\lambda)$ is the absorption coefficient and λ is the wavelength of the light. For the sake of simplicity, the dependency on the wavelength, λ, will be suppressed in the notation throughout this paper. In the case of normal incidence of light, the Fresnel coefficient, $\mathbf{r}_{lm}$, for two successive films, $m = l + 1$, with refractive indices, $\mathbf{n}_l$ and $\mathbf{n}_m$, respectively, is defined as follows [23]:

$$\mathbf{r}_{lm} = \frac{\mathbf{n}_l - \mathbf{n}_m}{\mathbf{n}_l + \mathbf{n}_m} \tag{1}$$

The total reflection coefficient, $\mathbf{r}$, of a single-layer system composed of a thin film ($\mathbf{n}_1$) on an absorbing substrate ($\mathbf{n}_2$) surrounded by air ($\mathbf{n}_0$) is given by [10]:

$$\mathbf{r} = \frac{\mathbf{r}_{01} + \mathbf{r}_{12}e^{-j\mathbf{d}_1}}{1 + \mathbf{r}_{01}\mathbf{r}_{12}e^{-j\mathbf{d}_1}} \tag{2}$$

where $\mathbf{d}_1 = 4\pi \mathbf{n}_1 d_1/\lambda$ is representing the thickness of the film. As given in Equation (1), $\mathbf{r}_{01}$ is the Fresnel coefficient between air and film and $\mathbf{r}_{12}$ between film and substrate. Finally, the reflectance, $R(\lambda)$, is given by:

$$R(\lambda) = \mathbf{r} \cdot \mathbf{r}^* \tag{3}$$

where $\mathbf{r}^*$ indicates the complex conjugate of $\mathbf{r}$.

Figure 1 illustrates a two-layer system. In this case, two films ($\mathbf{n}_1$ and $\mathbf{n}_2$) are deposited successively on an absorbing substrate ($\mathbf{n}_3$). The whole system is surrounded by air ($\mathbf{n}_0$). To calculate the reflection coefficient for this system, Equation (2) can be extended as follows [24]:

$$\mathbf{r} = \frac{\mathbf{r}_{01} + \mathbf{r}_{12}e^{-j\mathbf{d}_1} + \mathbf{r}_{23}e^{-j\mathbf{d}} + \mathbf{r}_{01}\mathbf{r}_{12}\mathbf{r}_{23}e^{-j\mathbf{d}_2}}{1 + \mathbf{r}_{01}\mathbf{r}_{12}e^{-j\mathbf{d}_1} + \mathbf{r}_{12}\mathbf{r}_{23}e^{-j\mathbf{d}_2} + \mathbf{r}_{01}\mathbf{r}_{23}e^{-j\mathbf{d}}} \tag{4}$$

where $\mathbf{d}_l = 4\pi \mathbf{n}_l d_l/\lambda$ and $\mathbf{d} = \mathbf{d}_1 + \mathbf{d}_2$. Again, by using Equation (3), the reflectance, $R(\lambda)$, for a two-layer system with an absorbing substrate can be obtained.

3. Model Extension

Equation (4) describes a system with ideal interfaces between layers. However, in practice, irregular interfaces affect the reflectance and must be considered in the model. One approach to model interfaces is based on the effective media approximation (EMA) [11]. By EMA, the inhomogeneous interfaces between layers are replaced by fictitious homogeneous layers, which are incorporated as such in the model [25]. Another approach proposes to modify the Fresnel coefficients in order to reproduce the effect of the interfaces on the reflectance [12]. In this case, the Fresnel coefficients, $\mathbf{r}_{lm}$, are altered by multiplying them with a function, $f(g_l)$, where g_l assigns a thickness to the interface, s_l, proportional to its grade of inhomogeneity. The modified Fresnel coefficients are defined as follows:

$$\check{\mathbf{r}}_{lm} = \mathbf{r}_{lm} \cdot f(g_l) \tag{5}$$

Introducing Equation (5) in Equation (4), the modified reflection coefficient, $\check{\mathbf{r}}$, is obtained. This approach yields a simpler and faster solution than EMA, which makes it advantageous for our application. The form of $f(g_l)$ depends on the considered interface model. As explained in [12], $f(g_l)$ could be ideally defined if the exact three dimensional structure of the interface was known. In general, however, such detailed knowledge of the interface is unavailable, and it is more reasonable to model the interface profile using an analytical function. Four different interface functions are presented in [12].

The principal causes of interface inhomogeneities are: the roughness of a layer surface and the mix of materials originated when two layers came in contact. In the case of polymer electronics, the substrate surface is smooth and does not mix with the first applied layer. Therefore, we can consider the interface, s_2 (Figure 1), as ideal. On the contrary, we cannot discard the presence of roughness and material mix on the interface, s_1, between the first and second layer. In this case, $f(g_1)$ must model both kinds of inhomogeneities [16]. Finally, the reflection coefficient of the interface, s_0, between air and the first layer is modified only by the surface roughness. In this case, g_0 approximates the RMSvalue of the surface roughness [16]. For our measurements, we use roughness function $f(g_l) = \exp\left(-2(2\pi g_l/\lambda)^2\right)$, which is a modification factor for the Fresnel reflection coefficient. It was generated by modeling the interface profile by an error function [12,16]. As suggested by Stearns, this function describes an interface produced by the diffusion of two materials. In most of the cases, g_l gives only a qualitative description of the interface, but, as will be shown, its consideration is essential to measure the layer thickness correctly.

Figure 1. Two-layer system surrounded by air (n_0, k_0).

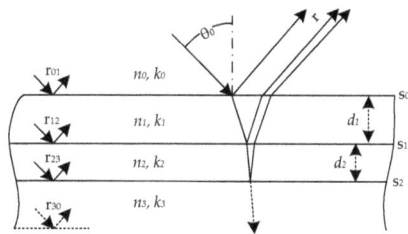

Equations (2) and (4) assume a completely absorbing or infinitely thick substrate. For thin transparent substrates, these assumptions are not valid. In this case, it might be necessary to consider the backside reflection coefficient, $\mathbf{r}_{30}$, at the interface between substrate and air (see Figure 1). This latter requires a recalculation of $\mathbf{r}$. However, if the coherence length of light is much smaller than the

198

axial dimensions of the substrate, the reflectance can be approximated as $R(\lambda) = (\check{\mathbf{r}} \cdot \check{\mathbf{r}}^*) + h(k_3, d_3)$, where $h(k_3, d_3)$ is defined as in [26]:

$$h(k_3, d_3) = \begin{cases} r_{30}^2, & k_3 \to 0 \\ 0, & k_3 \gg 0 \,\text{and/or}\, d_3 \to \infty \end{cases} \tag{6}$$

Figure 2a illustrates our measurement setup. A complete description of this measurement system is presented in [21,22]. In this setup, light from a spectrally broad source is coupled to a multimode fiber (Y-branch) and guided by two focusing lenses to the layer system at normal incidence. The reflected light intensity is then coupled back into the fiber and led to a spectrometer. The focusing lenses introduce a focal shift in wavelength known as chromatic aberration [23], which affects the signal captured by the spectrometer. This aberration depends on the axial position, z, and wavelength λ. This distortion is important for our application, because the moving sample is affected by vertical vibration movements. These movements alter the variable, z, during the measurement and, therefore, also the form of the captured signal. To avoid the compensation of this effect by invalid thickness values during the fitting process, chromatic aberration must be considered in the model. The chromatic effect can be described through a function, $c(z, \lambda)$, of the wavelength, λ, and the axial position, z, considering all system apertures. Figure 2b shows a set of chromatic functions for different values of z [22]. The chromatic effect on the reflectance can be described by multiplying both signals $R(\lambda) \cdot c(z, \lambda)$. The function, $c(z, \lambda)$, can be obtained by modeling the setup or by performing a set of off-line measurements. In this work, $c(z, \lambda)$ results from a set of measurements as described in [22]. By regenerating or recomputing $c(z, \lambda)$, the model can be adapted for a new measurement setup or for new measurement conditions.

Now, we combine all described extensions to obtain a reflectance model for two-layer systems. Equation (4) gives an analytical expression for the total reflection, $\mathbf{r}$, on absorbing substrates. If the Fresnel coefficients, $\mathbf{r}_{lm}$, in Equation (4) are replaced by those $\check{\mathbf{r}}_{lm}$ defined in Equation (5), we obtain an approximation of the total reflection denoted by $\check{\mathbf{r}}$ that considers the selected interface models. From $\check{\mathbf{r}}$, a reflectance signal can be calculated using Equation (3). To consider the possibility of transparent substrates, function $h(k_3, d_3)$, according to Equation (6), must be added. Finally, the complete result should be multiplied by $c(z, \lambda)$ to include the chromatic effect described before:

$$R(\lambda) = [\check{\mathbf{r}} \cdot \check{\mathbf{r}}^* + h(k_3, d_3)] \cdot c(z, \lambda) \tag{7}$$

Note that $R(\lambda)$ denotes the modeled reflectance spectra, while $R_0(\lambda)$ describes the measured reflectance. Both spectra will be compared in the next section in order to find the thickness of the inspected layers.

Figure 2. (a) Schematic representation of the measurement setup. The source light is coupled into a Y-branch fiber and guided to the measured object through the lenses, which introduced a chromatic effect. The multimode fiber has a numerical aperture of 0.22 and a diameter of 100 μm. Both lenses have a diameter of 12.5 mm and a focal length of 14 mm. The fill factor is 0.32; (b) Chromatic map, $c(z, \lambda)$, for $-250 \leq z \leq 250$ μm. The dashed lines show $c(100 \, \mu m, \lambda)$.

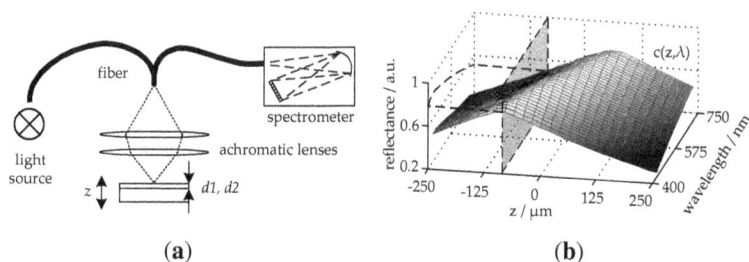

(a)   (b)

4. Measurements and Results

4.1. Measured Samples

The analyzed samples are used in polymer electronic applications. They consist of PMMA and P3HT layers applied successively by spinning on both transparent (PET) and non-transparent (Si) substrates. The optical parameters of all materials, $n(\lambda)$ and $k(\lambda)$, are well-known. Each sample presents a border, where each layer thickness can be measured by a profilometer. The device used is a Veeco, Dektak 6M Profilometer. The profilometer measurements provide the thickness reference values, d_1^0 and d_2^0, used to validate the proposed model. The layer application method generates a thickness variability of around $\pm 10\%$. This should be considered, as the profilometer measurements were performed at the samples' edges, whereas the TFR measurements are performed in their centers. Table 1 lists all analyzed samples and their reference values. The first three samples have a non-transparent substrate of Si, and the last three have a transparent substrate of PET.

Table 1. Best fitting results and reference values for two-layer systems on non-transparent (measurements 1–3) and transparent substrates (measurements 4–6). Values are in nanometers.

Meas.	d_1	d_2	g_0	g_1	d_1^0	d_2^0	z
1 (Figure 4a)	254	104	3	27	250	100	10×10^3
2 (Figure 4b)	257	183	1	20	250	180	6×10^3
3 (Figure 5a)	261	218	2	35	250	200	-46×10^3
4 (Figure 5b)	254	152	2	32	250	150	-38×10^3
5 (Figure 6a)	252	247	1	27	250	250	-85×10^3
6 (Figure 6b)	252	192	3	31	250	190	-200×10^3

4.2. Algorithm, Conditions and Computational Times

The measured reflectance, $R_0(\lambda)$, is compared with the modeled one, $R(\lambda)$, for different values of film thickness (d_1, d_2), interfaces (g_0, g_1) and axial displacement (z). By fitting both signals, the best parameters are selected in order to characterize the two inspected layers.

The starting points of the parameters and the searching intervals for the fitting algorithm must be correctly defined to achieve a valid result. This requires *a priori* knowledge of the minimum and maximum possible values for each parameter. This knowledge is based on, amongst others, the samples' materials, layer application method, measurement setup and experience.

The algorithm can be divided in two steps: the search for the initial points and the measurement fitting. The search for the initial points is performed with a recursive least squares algorithm. This is computationally intensive, but it allows larger searching intervals than other more efficient algorithms without falling in invalid minima. In this case, the limits of the searching intervals for the thickness values must not exceed $\pm 50\%$ of their real values. This means that the real values should be known *a priori* with a precision of $\pm 50\%$, which is true for almost all practical industrial cases. The initial points are then precise enough to run the measurement fitting algorithm using smaller searching intervals, allowing us to use a Levenberg–Marquardt algorithm to solve the system.

The signal capture can be considered instant and not affected by the movement of the sample. The average fitting time is about 8.9 s, which allows us to control the production every 10 m. In the case of detecting failures on the thickness values, this rate gives a loss of material between 10 and 20 m, which is economically acceptable for these kinds of material. Moreover, if the production line guaranties no abrupt variations of the layers' thickness, the searching of the initial points can be performed off-line as part of a calibration process. This latter one reduces the measurement time considerably, allowing for a higher control frequency.

4.3. Results

We use the first sample in Table 1 to show the influence of each model parameter on the measured values. The measured reflectance, $R_0(\lambda)$, for this sample is shown in Figure 3 through a solid line. In the first approach, we limit the model to $R(\lambda) = \mathbf{r} \cdot \mathbf{r}^*$, with $\mathbf{r}$ defined as in Equation (4), *i.e.*, without considering interface irregularities or chromatic effect. The resulting best fitting (curve 1 in Figure 3) gives thickness values for the first and second layer with deviations of 9.5% and 5% with respect to the reference values. In the second approach, we incorporate the interface function into the model. In this case, the Fresnel coefficients are modified as indicated in Equation (5), where $f(g_l)$ was defined as a roughness function [12,16] (the chromatic effect was not considered). Now, the best fitting (curve 2 in Figure 3) gives deviations of 1.6% and 8% for the first and second layer, respectively. Finally, we test the complete model described in Equation (7). The best fitting (curve 3 in Figure 3 and also in Figure 4a), which is shown in Table 1, gives deviations of less than 1.6% and 4% compared to the reference values.

Figures 4a through 6b show the measured $R_0(\lambda)$ and best fitting $R(\lambda)$ reflectance for all samples presented in Table 1. For all measurements, the obtained thickness values present a deviation of less than 9% with respect to those obtained using the profilometer. These results validate our model for two-layer systems on both transparent and non-transparent substrates.

Figure 3. Comparison of different definitions of $R(\lambda)$ for a two-layer system on the absorbing substrate. The solid line represents the measured reflectance, $R_0(\lambda)$. (**1**) $R(\lambda)$ without considering inhomogeneity or chromatic effect; (**2**) $R(\lambda)$ considering interface irregularities, but no chromatic function; (**3**) $R(\lambda)$ considering roughness and chromatic function.

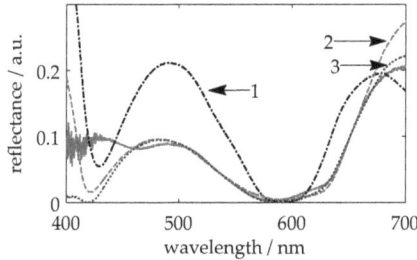

Figure 4. Measurement of two-layer systems on an absorbing substrate. The solid line represents the measured reflectance, and the dotted one is the best fitting result.

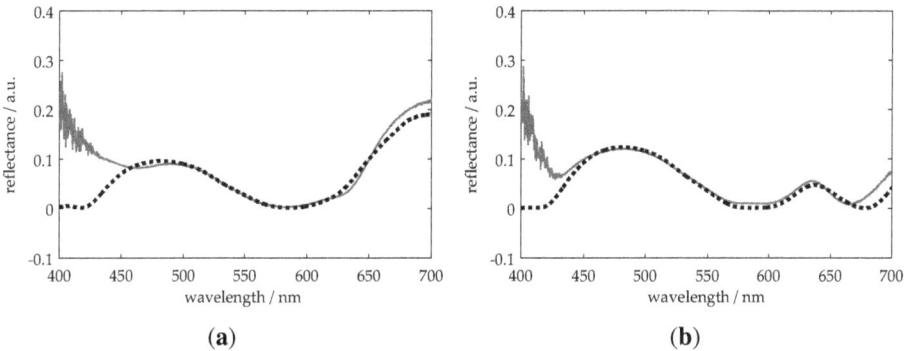

(**a**)   (**b**)

Additionally, although the parameters, g_0, g_1 and z, are considered in the model to approximate the values of d_1 and d_2 to the real ones, they also provide some extra information. Tactile measurements of the roughness in the first and second layer give mean peak to peak values of around 6.9 nm and 1.3 nm, respectively. The irregularities of the first interface, s_0 (air-first layer), can be completely described by the roughness of the first layer. Therefore, as expected, the tactile measurement and the values of g_0 are in the same order of magnitude. On the contrary, the tactile measurement for the second layer, which can only be performed by isolating it with respect to the first one, does not represent the interface, s_1. For this interface, g_1 describes a combination between roughness and interface inhomogeneity, which results in a significantly higher value than that given by the profilometer. Finally, the obtained values of z are inside the expected ones, and together with the obtained thickness, they can be used together to describe the setup configuration during the measurement.

202

Figure 5. Measurement of two-layer systems on absorbing (**a**) and transparent (**b**) substrates. The solid line represents the measured reflectance, and the dotted one is the best fitting result.

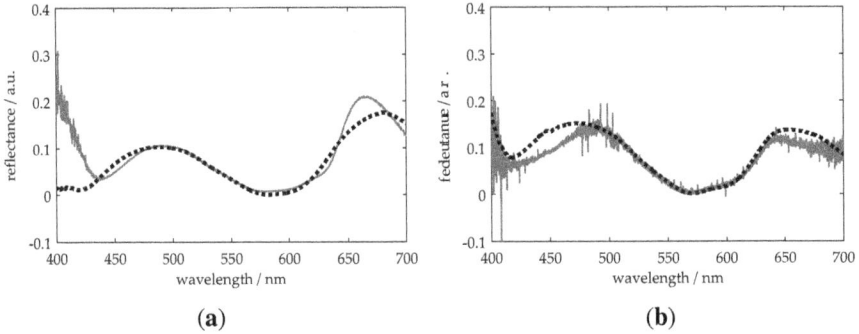

(a)

(b)

Figure 6. Measurement of two-layer systems on a transparent substrate. The solid line represents the measured reflectance, and the dotted one is the best fitting result.

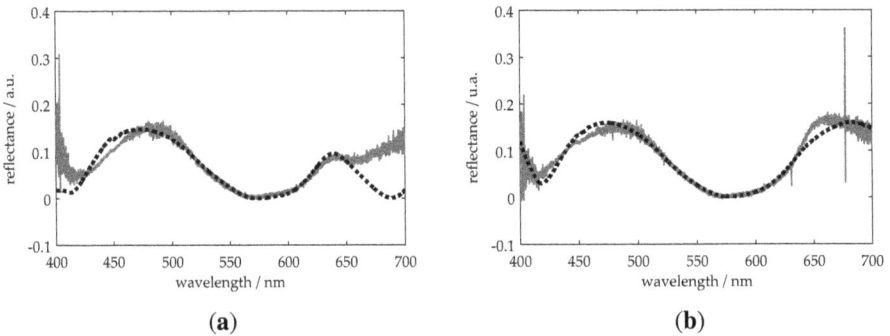

(a)

(b)

5. Conclusions and Outlook

A model was proposed to measure the thickness of two-layer systems. This model was tested on samples used on polymer electronic applications; however, it can be easily extended to other kinds of surfaces with known optical characteristics. The model follows from an analytical derivation of the reflectance signal and is extended to consider surface interfaces, measuring setup characteristics and vertical vibration movements for absorbing and transparent substrates. Our results show that considering interface irregularities and chromatic effects increases the accuracy of the thickness measurements for both layers. Our method requires one to know *a priori* the expected value of the inspected thickness with a precision of $\pm 50\%$, which is given for almost all real applications. The validity of the presented measurements was shown by comparing the obtained results with those of a profilometer device. The deviation between these two methods remains under the expected layer thickness deviation of 10%. The first interface parameter, g_0, shows a meaningful match with the reference measurements. In the case of the second interface, the obtained value describes not only the roughness, but also the inhomogeneity, of the interface between both layers and cannot be compared with the profilometer results. The accuracy of this value will be checked in a future work by using an effective media approach. The presented results suggest that the proposed approach can be successfully used to measure the thicknesses of two-layer systems. The proposed method is especially adequate for applications in which a fast and non-contact inspection is required. This is the case, e.g., for the quality control of electronic components. An extension of this model to N-layer systems is also planned as future work.

Acknowledgments

The authors want to thank the German Science Foundation (DFG) for supporting this work by funding the project "Hyperspectral chromatic reflectometer for measuring moving objects" (Hyperspektrales chromatisches Reflektometer zur Vermessung bewegter Objekte).

Conflicts of Interest

The authors declare no conflict of interest.

References

1. Poelman, D.; Smet, P. Methods for the determination of the optical constants of thin films from single transmission measurements: A critical review. *Appl. Phys. D* **2003**, *36*, 1850–1857.
2. Jung, C.; Rhee, B. Simultaneous determination of thickness and optical constants of polymer thin film by analyzing transmittance. *Appl. Opt.* **2002**, *41*, 3861–3865.
3. Ruprecht, M.; Koch, A. Interferometrische messung von schichtparametern. *Meß-und Automatisierungstechnik* **1997**, *2*, 129–139.
4. Koch, A.; Ruprecht, M.; Toedter, O.; Häusler, G. *Optische Messtechnik an Technischen Oberflächen*; Expert-Verlag: Renningen-Malmsheim, Germany, 1998.
5. Born, M.; Wolf, E.; Bhatia, A. *Principles of Optics*; Cambridge University Press: Cambridge, UK, 2000.
6. Azzam, R.; Bashara, N. *Ellipsometry and Polarized Light*; North-Holland: Amsterdam, The Netherlands, 1977.
7. Ye, S.; Kim, S.; Kwak, Y.; Cho, H.; Cho, Y.; Chegal, W. An ellipsometric data acquisition method for thin film thickness measurement in real time. *Meas. Sci. Technol.* **2008**, *19*, 047002.
8. Conroy, M. Advances in thick and thin film analysis using interferometry. *Wear* **2009**, *266*, 502–506.
9. Chen, W.; Saunders, J.E.; Barnes, J.A.; Yam, S.S.H.; Loock, H.P. Monitoring of vapor uptake by refractive index and thickness measurements in thin films. *Opt. Lett.* **2013**, *38*, 365–367.
10. Hauge, P. Polycrystalline silicon film thickness measurement from analysis of visible reflectance spectra. *J. Opt. Soc. Am.* **1979**, *69*, 1143–1152.
11. Niklasson, G.; Granqvist, C.; Hunderi, O. Effective medium models for the optical properties of inhomogeneous materials. *Appl. Opt.* **1981**, *20*, 26–30.

12. Stearns, D. The scattering of X-rays from nonideal multilayer structures. *J. Appl. Phys.* **1989**, *65*, 491–506.

13. Montecchi, M.; Montereali, R.M.; Nichelatti, E. Reflectance and transmittance of a slightly inhomogeneous thin film bounded by rough, unparallel interfaces. *Thin Solid Films* **2001**, *396*, 264–275.

14. Swanepoel, R. Determination of surface roughness and optical constants of inhomogeneous amorphous silicon films. *J. Phys. E: Sci. Instrum.* **1984**, *17*, 896–903.

15. Katsidis, C.; Siapkas, D. General transfer-matrix method for optical multilayer systems with coherent, partially coherent, and incoherent interference. *Appl. Opt.* **2002**, *41*, 3978–3987.

16. Windt, D. IMD Software for modeling the optical properties of multilayer films. *Comput. Phys.* **1998**, *12*, 360–370.

17. Ylilammi, M.; Ranta-aho, T. Optical determination of the film thicknesses in multilayer thin film structures. *Thin Solid Films* **1993**, *232*, 56–62.

18. Oraizi, H.; Afsahi, M. Analysis of planar dielectric multilayers as FSS by transmission line transfer matrix method (TLTMM). *Prog. Electromagn. Res.* **2007**, *74*, 217–240.

19. Modi, M.; Lodha, G.; Nayak, M.; Sinha, A.; Nandedkar, R. Determination of layer structure in Mo/Si multilayers using soft X-ray reflectivity. *Phys. B: Cond. Matter* **2003**, *325*, 272–280.

20. Durand, O.; Berger, V.; Bisaro, R.; Bouchier, A.; de rossi, A.; Marcadet, X.; Prévot, I. Determination of thicknesses and interface roughnesses of GaAs-based and InAs/AlSb-based heterostructures by X-ray reflectometry. *Mater. Sci. Semiconduct. Process.* **2001**, *4*, 327–330.

21. Hirth, F.; Buck, T.; Steinhausen, N.; Koch, A. Effect of Chromatic Aberrations on Resolution in Thin Film Reflectometry. In Proceedings of the Society of Photo-Optical Instrumentation-SPIE, Monterey, USA, 3 June 2010; Volume 7729, p.77291L.

22. Hirth, F.; Buck, T.; Pérez Grassi, A.; Koch, A. Depth-sensitive thin film reflectometer. *Meas. Sci. Technol.* **2010**, *21*, 125301.

23. Hecht, E. *Optics*; Addison Wesley Longman: Chicago, IL, USA, 2002.

24. Heavens, O.S. *Optical Properties of Thin Solid Films*; Courier Dover Publications: Mineola, NY, USA, 1955.

25. Aspnes, D.; Theeten, J.; Hottier, F. Investigation of effective-medium models of microscopic surface roughness by spectroscopic ellipsometry. *Phys. Rev. B* **1979**, *20*, 3292–3302.

26. Minkov, D. Calculation of the optical constants of a thin layer upon a transparent substrate from the reflection spectrum. *J. Phys. D: Appl. Phys.* **1989**, *22*, 1157, doi:10.1088/0022-3727/22/8/021.

Reprinted from *Sensors*. Cite as: Murr, P.J.; Schardt, M.; Koch, A.W. Static Hyperspectral Fluorescence Imaging of Viscous Materials Based on a Linear Variable Filter Spectrometer. *Sensors* **2013**, *13*, 12687–12697.

Article

Static Hyperspectral Fluorescence Imaging of Viscous Materials Based on a Linear Variable Filter Spectrometer

Patrik J. Murr *, Michael Schardt and Alexander W. Koch

Institute for Measurement Systems and Sensor Technology, Technische Universität München, Theresienstrasse 90/N5, Munich 80333, Germany; E-Mails: m.schardt@tum.de (M.S.); a.w.koch@tum.de (A.W.K.)

* Author to whom correspondence should be addressed; E-Mail: patrik.murr@tum.de; Tel.: +49-0-89-289-23352; Fax: +49-0-89-289-23348.

Received: 25 June 2013; in revised form: 29 August 2013 / Accepted: 13 September 2013 / Published: 23 September 2013

Abstract: This paper presents a low-cost hyperspectral measurement setup in a new application based on fluorescence detection in the visible (Vis) wavelength range. The aim of the setup is to take hyperspectral fluorescence images of viscous materials. Based on these images, fluorescent and non-fluorescent impurities in the viscous materials can be detected. For the illumination of the measurement object, a narrow-band high-power light-emitting diode (LED) with a center wavelength of 370 nm was used. The low-cost acquisition unit for the imaging consists of a linear variable filter (LVF) and a complementary metal oxide semiconductor (CMOS) 2D sensor array. The translucent wavelength range of the LVF is from 400 nm to 700 nm. For the confirmation of the concept, static measurements of fluorescent viscous materials with a non-fluorescent impurity have been performed and analyzed. With the presented setup, measurement surfaces in the micrometer range can be provided. The measureable minimum particle size of the impurities is in the nanometer range. The recording rate for the measurements depends on the exposure time of the used CMOS 2D sensor array and has been found to be in the microsecond range.

Keywords: fluorescence; hyperspectral imaging; linear variable filter; CMOS 2D sensor array

1. Introduction

Today, fluorescence measurements are established in various applications and have a broad spectrum of functionality. The main focus of current fluorescence setups can be found in the fields of biology, food technology and medicine. The quality control of fruits, for example, is realized with fluorescence measurements [1,2]. Most of these fluorescence measurement systems are based on image scanning with grayscale resolution without a separation of wavelengths. Damages and variances at the measurement surface can be detected in recorded images, due to the grayscale resolution. Other measurements with fluorescence in reflection are developed for the detection of skin cancer and are currently tested on animals, like mice or chickens [3,4]. The use of such measurement setups has some drawbacks. One of them is that at the moment, only solid and liquid materials can be investigated. Another drawback is the low scanning speed of the currently available measurement setups. A further problem is the integration of the measurement setups into a running operation. Actually, this is not possible with low effort. In addition, the current measurement setups are very expensive and massive. Thus, at the moment, there is no opportunity to get the information of a fast moving object or to integrate the setups into an established process. Yet, there are some approaches for small and cheap fluorescence spectrometers. These spectrometers, for example, consist of a combination of a linear variable filter (LVF) and a complementary metal oxide semiconductor (CMOS) 2D sensor array [5,6].

In this paper, a small and low-cost setup for fluorescence measuring of hyperspectral images with the option for moving measurements in a new application is presented. This concept is based on the method of hyperspectral imaging [7,8]. There are three well-known implementation possibilities: push-broom scan, optical bandpass filter and Fourier transform spectroscopy. For the presented setup, the push-broom method that analyzes one image per line was used. This method is currently established in aerospace applications and in scanning environmental pollution [9,10]. An advantage of this method is that spectral information corresponding to the location may be obtained. The evaluation of the measurement setup was made with viscous materials. So far, no approaches for fluorescence measurements with viscous materials have been published. At the moment, quality controls for viscous materials are realized by controlling the acoustic sounds or signals [11]. This method takes a lot of time and is very expensive, because in most of the cases, the evaluation is realized in laboratories and not directly in the production process or application site. In addition, this method does not enable continuous and contactless measurements. The consequence is that only random samples can be evaluated, and no statement for the complete product during a process can be made.

Hence, for this application, a small, low-cost and universally-usable measurement setup with illumination in the ultraviolet (UV) range was developed. The recorded images show hyperspectral fluorescence spectra in a desired wavelength range of a measurement surface of a viscous material. With this setup, a separation between the location and the wavelength in the recorded image is possible. Furthermore, a relationship between the measurement surface and the corresponding spectra in the image can be established. The consequence is that fluorescent and non-fluorescent impurities in the viscous material can be located and evaluated. Further, the quality of the viscous materials can be validated. In addition, the setup is arranged and constructed for a later integration into moving production processes.

2. Materials and Methods

This section covers the fundamentals of the measurement system presented in this work. Firstly, the idea will be explained, followed by the measurement setup, system characterization, calibration and analysis.

2.1. Idea

The setup of an LVF spectrometer is a special type of an optical acquisition unit. Compared to a grating spectrometer, LVF spectrometers feature several differences. LVFs have a huge aperture at high transmissivity. In a hyperspectral configuration, the dispersive element is mounted directly on the top of a CMOS 2D sensor array. Thereby, an optical hyperspectral spectrometer consisting of an LVF and a CMOS 2D sensor array is set up. This low-cost combination is the essential advantage for use as an optical adjustment unit for static fluorescence measurements on viscous materials. Figure 1 describes the idea of the developed method. The setup is explained in Section 2.2. As described in Section 1, similar methods are known from satellite-based remote sensing in the application of the push-broom principle.

Figure 1. Representation of the described idea. The UV light source illuminates the viscous material (not in the figure). The resulting fluorescence of the measurement surface is denoted at an exemplary location. In addition, the spectral fluorescence imaging of the viscous material is denoted for an exemplary line array on the complementary metal oxide semiconductor (CMOS) 2D sensor array. The hyperspectral spectrometer, consisting of a linear variable filter (LVF) and a CMOS 2D sensor array, interrogates the transmitted intensity synchronously to the illumination duration of the measurement surface.

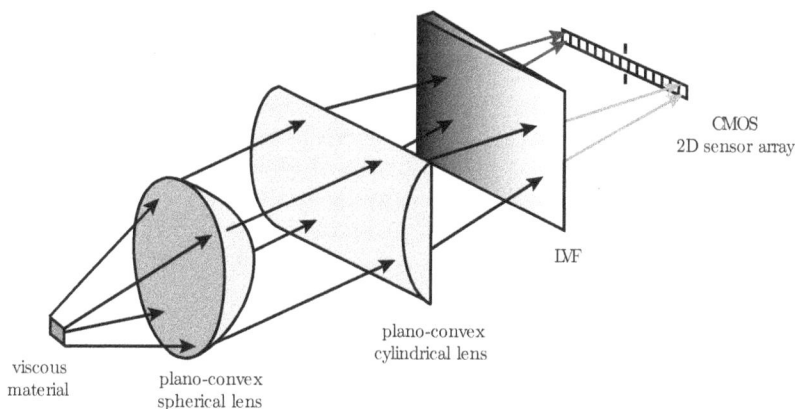

In our setup, a fluorescent viscous material was provided in a notch with a constant thickness of 200 μm. A UV light source illuminates the measurement object. The fluorescence is captured by the LVF/CMOS configuration. In order to give spectral information to the regarded measurement surface, an adjustment of the lenses is used. Due to the spectral characteristic of the detection system, the spectra of a fluorescent measurement object over the location can be obtained. In addition, the existence and location of fluorescent impurities with their corresponding wavelength range can be proven. Further, non-fluorescent impurities can be measured through the decrease of the resulting fluorescence spectra.

The readout of all 1,280 × 1,024 CMOS 2D sensor array elements was synchronized with the illumination duration of a high power UV-LED. An external hardware trigger signal starts the measurement at a manually defined time. If the trigger signal is changing from a high (5 V) to a low (0 V) level, an image with the LVF/CMOS configuration is made. The duration of the recording depends on the adjusted exposure time of the CMOS 2D sensor array. After a successful data acquisition, several spectra of different points on the measurement surfaces are recorded in an image. Each hardware trigger leads to a new spectral capture of the fluorescent viscous material. The recorded image has a complete spectrum in the visible (Vis) wavelength range in each row. Each column represents a narrow-band wavelength range. The wavelength range occurs by the characteristics of LVF, which is in front of the CMOS 2D sensor array. All values in a recorded image are intensity values and are illustrated in an eight-bit gray resolution. The hyperspectral imaging is realized by the spatial resolution over the rows and the wavelength resolution over the columns on the CMOS 2D sensor array. The spot size of the measurement surface on the CMOS 2D sensor array depends on the adjudication of the lenses used in the setup. To find the best lens adjudication for the measurement setup, a simulation with the optical simulation tool Zemax was done before. In the simulation, the spot size was changed by variations of the position of the used plan-convex spherical lens. The criterion for the final lens adjudication was that all rays are still collimated as best as possible for the largest possible measurement area.

Compared to grating spectrometers, hyperspectral sampling can be realized very fast; thus, in combination with the huge aperture, it leads to an enormously reduced measurement time. In relation to the currently available hyperspectral imaging systems, this setup features higher efficiency and little adjustment effort at a low price. Due to the wafer scale production of the LVF and CMOS 2D sensor array, an enormous cost reduction is achieved. The adjustment effort is drastically reduced because of the solid-state nature of the spectral apparatus. In addition, this measurement setup is useable for other applications in which measurement objects are fluorescent. The replacement of optical components, like optical filters and light sources, is very simple, because all components in the setup are constructed modularly.

2.2. Measurement Setup

A high-power UV-LED is used to illuminate the measurement object at an angle of 45°. The measurement object in this case is a fluorescent viscous material. The light-emitting spot on the measurement object is limited by a 750 μm slit. The resulting fluorescence is nearly collimated by the plano-convex spherical lens, L_1, with the focal length $f_1 = 20$ mm. Through a slight shift out of the focal point by $\Delta x_1 = 1$ mm, an imaging of an area on the measurement object is possible. A plano-convex cylindrical lens, L_2, with the focal length $f_2 = 25$ mm is placed at the distance $d = 30$ mm. The curved side of this lens is orientated in the same direction as the slit above the measurement object. In this dimension, the plano-convex cylindrical lens images the parallel parts to the direction of the slit onto the CMOS 2D sensor array. Due to the characteristic of this lens, there is no effect of the light in the other dimension. Thus, all light rays of one point on the measurement surface are distributed in this dimension. The material of the two used lenses is borosilicate glass (BK7). The optical acquisition unit consists of an LVF (wavelength range: 400 nm–700 nm) mounted atop a CMOS 2D sensor array and is orientated in the direction of the slit atop the viscous material. The used CMOS 2D sensor array features $1,024 \times 1,280$ pixels with a sensitive area of 6.66 mm $\times$ 5.32 mm and has a parallel read-out capability. The size of one pixel on the CMOS 2D sensor array is 5.2 μm $\times$ 5.2 μm. The analog digital converter of the sensor has a resolution of eight bits and the exposure time can be chosen between 35 μs and 980 ms. The complete measurement setup in the plan and side view is shown in Figures 2 and 3.

Figure 2. Illustration of the proposed hyperspectral fluorescence spectrometer based on an LVF/CMOS 2D sensor array configuration. In the top view of the measurement setup, the ray path of the fluorescence with the spectral information is shown. The lens adjustment consisting of a plano-convex spherical lens and a plano-convex cylindrical lens enlarges the information of the fluorescent measurement object over the complete CMOS 2D sensor array. Due to the assembly of the plano-convex cylindrical lens, this dimension contains the spectral information to a corresponding location of a considered measurement object.

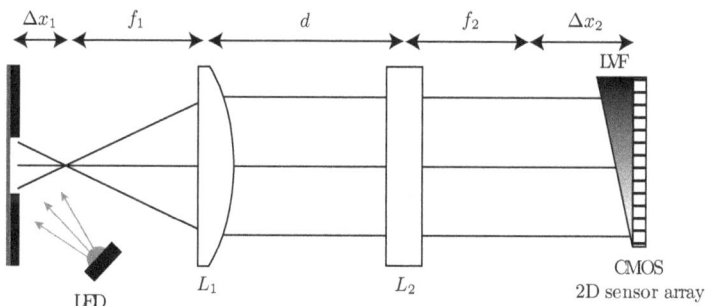

2.3. System Characterization

To adjust the measurement setup, a resolution measurement of the LVF and a measurement of the pixel linearity on the CMOS 2D sensor array had to be carried out. The resolution of the LVF was analyzed in a previous work [12]. Considering the resolution of the LVF and the pixel size of the CMOS 2D sensor array, the system resolution is 1.6% of the LVF center wavelength. Thus, the measurement setup can measure impurities that are larger than ten nanometers. The pixel linearity of the CMOS 2D sensor array was determined with a light-emitting film in the visible wavelength range, which illuminated the complete CMOS 2D sensor array with a constant intensity. This measurement was done with and without the LVF atop the CMOS 2D sensor array. The intensity of the illumination was adjusted by two polarization filters in constant rotation steps of 5°. For each measurement step, the maximum variance of the mean pixel intensity over all pixels was lower than 3.1%. Furthermore, for each pixel, the maximum variance from the ideal linearity over all intensity values was lower than 1.4%. Due to the results of the measurement characterization, it has to be stated that all pixels on the CMOS 2D sensor array show nearly linear characteristics. In addition, the signal-to-noise ratio (SNR) of the measurement setup was calculated as 59.

Figure 3. Illustration of the proposed hyperspectral fluorescence spectrometer based on an LVF/CMOS 2D sensor array configuration. In the side view of the measurement setup, the ray path of the fluorescence with the spatial information is shown. The lens adjustment consisting of a plano-convex spherical lens and a plano-convex cylindrical lens images the fluorescent measurement object onto the CMOS 2D sensor array. Due to the imaging through the plano-convex cylindrical lens, this dimension contains the spatial information of the considered measurement object.

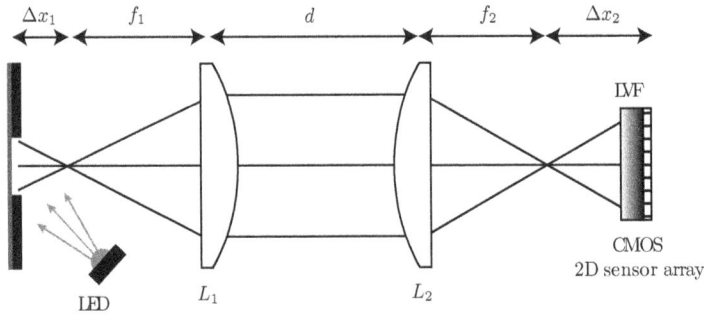

2.4. Calibration and Measurement Process

A wavelength calibration of the complete measurement setup is necessary for an interpretation and evaluation of the fluorescence data. The wavelength range is defined by the used LVF and CMOS 2D sensor array. For the wavelength calibration, a mercury lamp was used as the illumination, emitting four distinctive peaks in the range of the LVF. The four peaks are located at 406 nm, 436 nm, 546 nm, and 578 nm. The detected wavelengths of each pixel on the CMOS 2D sensor array are filtered by the LVF and change with little shifting of the position of the LVF. Hence, the specific combination of the LVF and CMOS 2D sensor array has to be calibrated with respect to its wavelength distribution. Because of the linear wavelength characteristic of the LVF, each row of the CMOS 2D sensor array was calibrated by applying a linear fit algorithm to the measured wavelengths of the characteristic peaks. The wavelength calibration method is described in more detail in [13]. After a successful wavelength calibration, it is possible to allocate for each intensity value the corresponding wavelength, and the LVF/CMOS configuration can be integrated into the measurement setup. In addition, it is possible to create a complete spectrum for each row of the

CMOS 2D sensor array. Due to the eight-bit intensity resolution of the CMOS 2D sensor array, there is no additional intensity calibration for the system necessary.

A measurement process of the fluorescence spectra consists of four phases. First, the parameters for the illumination duration of the UV-LED, the frequency for the external hardware trigger, the exposure time of the CMOS 2D sensor array and the periodic time of one measurement have to be set. These settings are necessary, so that the images have no overexposure and the fluorescence intensity does not decrease through a warm-up of the viscous material. In the second phase, the data acquisition starts by the manually defined hardware trigger. Then, data processing is deployed with respect to the recorded measurement signals. Finally, an analysis and interpretation of the resulting data is realized.

In summary, the described measurement setup is a hyperspectral fluorescence spectrometer where the spatial information along the direction of the slit is detected in one dimension of the CMOS 2D sensor array. The other dimension of the CMOS 2D sensor array shows the corresponding spectral information divided by the LVF.

2.5. Analysis

Due to the different concentrations and properties of the fluorescent components of the viscous materials, it is important that no fluorescence decreasing effects during the measurements exist. One of these effects occurs especially at static measurements, where the measurement surface is illuminated for an extended time and warm-up. Thereby, the molecules in the viscous materials obtain stronger vibrations, and the collision probability increases with rising object temperature. This is known as the quenching effect and leads to a decrease of the fluorescence signal [14]. For our case, the effect can be classified at the dynamic fluorescence extinction and can be described with the Stern–Volmer Equation [15]. This equation can be written as:

$$\frac{F_0}{F} = 1 + K_D[Q] \tag{1}$$

where F_0 and F are the fluorescence intensities in the absence and presence of the quencher and Q is the concentration of the quencher. The dependency of the temperature in the equation is represented by the Stern-Volmer constant, K_D. The value of K_D is inversely proportional to the temperature of the viscous fluorescent materials.

The root-mean-square error (RMSE) between a reference and measured fluorescence spectrum gives the difference of both in the quantity being measured (in this case, % fluorescence) and can be written as:

$$\mathrm{RMSE}(j) = \sqrt{\frac{\sum_{i=1}^{n}(F_{i,ref} - F_{i,j})^2}{n}} \tag{2}$$

where n is the number of sampling points of the spectrum and j is the row number of the CMOS 2D sensor array with a complete spectrum.

3. Results and Discussion

In this section, exemplarily static fluorescence measurements are presented, and a discussion of the obtained results is carried out. The efficiency of the complete system is demonstrated by fluorescence measurements with and without impurities in a viscous material.

In the first measurement, a fluorescent viscous material with no impurity was used. The measurement surface was illuminated with the wavelength of 370 nm (10 nm full width at half maximum, FWHM). The resulting fluorescence spectra included a wavelength range from 380 nm to 550 nm. The sample thickness was in all places 200 μm. A reference measurement was carried out by a commercial UV-Vis-spectrometer. The measured spectra started at a wavelength of 400 nm. Due to the optical properties of the LVF, the wavelength was restricted on a range of 400 nm to 700 nm. Figure 4 shows the results of the measured fluorescence spectra for three exemplary rows on the CMOS 2D sensor array. The locations of the three rows are on both sides and in the middle of the relevant area on the CMOS 2D sensor array. The regarded area on the CMOS 2D sensor array for the hyperspectral imaging was restricted through the dimensions of the LVF. The relevant area comprised row 400 to 700 and all columns (6.66 mm × 1.56 mm). Furthermore, Figure 4 illustrates a reference fluorescence spectrum of a UV-Vis-spectrometer.

Figure 4. Static measured hyperspectral fluorescence spectra of a viscous material without an impurity. Exemplarily, spectra of rows 400, 550 and 700 and a reference spectrum of a UV-Vis-spectrometer are illustrated. The location of the three rows is on both sides and in the middle of the relevant area on the CMOS 2D sensor array. The intensities of all fluorescence spectra are specified in arbitrary units (a.u.).

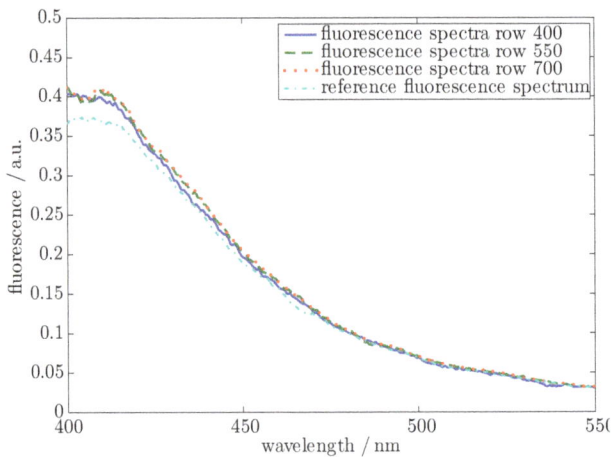

The results of the three exemplary fluorescence spectra show a good accordance between one another. The reference spectrum of the UV-Vis-spectrometer is almost equivalent to the three exemplary rows. The variances between the reference spectrum and the exemplary spectra were calculated with Equation (2). The differences amount to 1.75% (row number: 400), 1.69% (row number: 550) and 1.42% (row number: 700). In order to prove that all rows on the CMOS 2D sensor array have the same fluorescence spectrum, a RMSE trend was calculated. As a reference spectrum, the middle row (row number: 550) on the CMOS 2D sensor array was chosen. In Figure 5, the RMSE trend of all other relevant rows to the reference row is illustrated.

The results agree well with the presumption that the measured viscous material contains no impurities, which leads to a constant fluorescence over the complete measurement surface. The highest RMSE of all relevant rows is 0.65% at row 442. The average RMSE of the considered rows is 0.55%. Minor differences in the measured spectra compared to the reference row can be explained with small irregularities in the used optical components.

After the successful proof of the idea presented in Section 2.1 by references to fluorescent viscous materials, another measurement with a non-fluorescent impurity at a known location was performed. In this measurement, the same area on the CMOS 2D sensor array as in the previous measurement was considered. In Figure 6, the fluorescence spectra of three exemplary rows are shown. The locations of the exemplary rows correspond to the investigated rows of the first measurement.

Figure 5. Root-mean-square error (RMSE) trend between a reference row and all other rows of the CMOS 2D sensor array. The chosen rows (row number 400 to 700) show a maximal RMSE of 0.65% to the reference row (row number: 442).

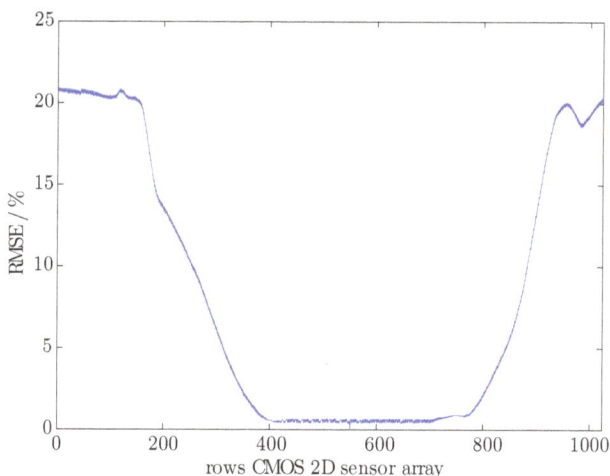

The results in Figure 6 demonstrate two identical fluorescence spectra and a fluorescence spectrum with a lower intensity. This confirms that in the area of row 400, a non-fluorescent impurity is included, thus enabling the measurement and location of non-fluorescent impurities in viscous materials based on the hyperspectral imaging with a small and low-cost LVF-spectrometer.

Besides measurements of a non-fluorescent impurity in viscous materials, it is possible to use the system for measurements in viscous materials with a fluorescent impurity. The expected results of the spectra for the area without an impurity should be similar to the spectra in Figure 4. For the area with the fluorescent impurity in the viscous material, an increase in the fluorescence intensity over the complete wavelength range is expected. This is different from the non-fluorescent impurity, where a decrease of the fluorescence intensity over the complete spectrum exists.

214

Figure 6. Static measured hyperspectral fluorescence spectra of a viscous material with a non-fluorescent impurity. Exemplary fluorescence spectra of rows 400, 550 and 700. The non-fluorescent impurity is in the area of row 400 on the CMOS 2D sensor array.

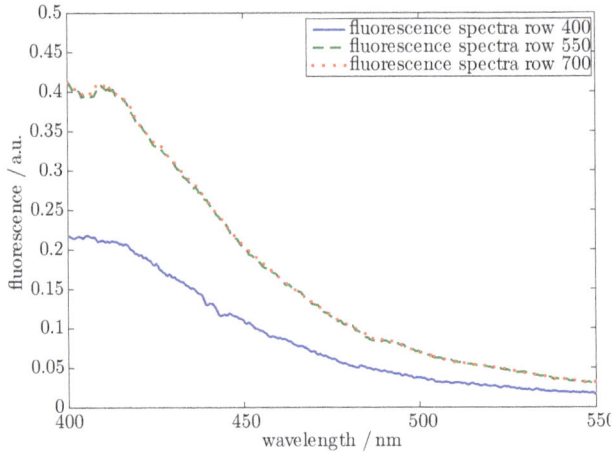

A first step in order to increase the robustness and accuracy of the system is to integrate the LVF directly atop the CMOS 2D sensor array. Further, it is desirable to increase the area of the CMOS 2D sensor array, which is covered with an LVF. Moreover, measurements with higher resolutions of the CMOS 2D sensor array up to twelve bits can be tested. In addition, the setup can be modified for moving measurements of viscous materials or other applications.

Other possibilities are measuring different viscous materials and investigating different impurities in one sample. An aim of such a measurement can be, e.g., the identification and location of several fluorescent impurities in one sample by variations in the fluorescence spectra.

4. Conclusions

A new approach and application for static hyperspectral fluorescence measurements has been presented. The main idea is to build a universally useable, contactless, online, small and low-cost LVF-spectrometer. The acquisition unit with lens adjustment is mounted vertically above the sample. The light source illuminates the viscous material at an angle of 45° at a wavelength of 370 nm. By an appropriate synchronization between the illumination time and the exposure time of the CMOS 2D sensor array, it is possible to get static fluorescence spectra of a measurement surface separated in location and wavelength.

Exemplary static measurements of a fluorescent viscous material with and without a non-fluorescent impurity show the efficiency of this approach. A good agreement with a reference fluorescence spectrum of a UV-Vis-spectrometer was achieved. In addition, the accordance between the relevant fluorescence spectra at the CMOS 2D sensor array in a non-fouled fluorescent sample was illustrated. The possibility to detect and locate a non-fluorescent impurity on a measurement surface of a viscous material through the fluorescence spectra has been proven. Actually, the minimum cross-section dimension for a successful detection of an impurity in the measurement system is about 75 nm. The maximum size for the impurity is limited through the dimension of the slit in the system.

The implementation of this approach for moving measurements of fluorescent samples is conceivable due to small changes in the presented setup and software. The described technology enables a cost-effective and high-speed monitoring of the production processes of viscous materials based on fluorescence. Such an application has not been developed until now.

By replacing the optical components, like LVF and the light source, with identically constructed components with other optical parameters, it is possible to use this setup in other applications where the measurement objects are fluorescent. Further investigations will show the suitability of this setup for the detection and location of different kinds of impurities in fluorescent viscous materials. In addition, the efficiency of the approach for moving fluorescence measurements and other applications will be determined.

Conflicts of Interest

The authors declare no conflict of interest.

References

1. Kim, M.S.; Chen, Y.R.; Mehl, P.M. Hyperspectral reflectance and fluorescence imaging system for food quality and safety. *Trans. ASAE* **2001**, *44*, 721–729.
2. Gowen, A.; Odonnell, C.; Cullen, P.; Downey, G.; Frias, J. Hyperspectral imaging—An emerging process analytical tool for food quality and safety control. *Trends Food Sci. Technol.* **2007**, *18*, 590–598.
3. Kim, I.; Kim, M.S.; Chen, Y.R.; Kong, S.G. Detection of skin tumors on chicken carcasses using hyperspectral fluorescence imaging. *Trans. ASAE* **2004**, *47*, 1785–1792.
4. Kong, S.G.; Martin, M.; Vo-Dinh, T. Hyperspectral fluorescence imaging for mouse skin tumor detection. *ETRI J.* **2006**, *28*, 770–776.
5. Emadi, A.; Wu, H.; de Graaf, G.; Enoksson, P.; Correia, J.H.; Wolffenbuttel, R. Linear variable optical filter-based ultraviolet microspectrometer. *Appl. Opt.* **2012**, *51*, 4308–4315.
6. Schmidt, O.; Bassler, M.; Kiesel, P.; Knollenberg, C.; Johnson, N. Fluorescence spectrometer-on-a-fluidic-chip. *Lab Chip* **2007**, *7*, 626–629.
7. Chang, C.I. *Hyperspectral Imaging: Techniques for Spectral Detection and Classification*; Springer: New York, NY, USA, 2003.
8. Klein, M.E.; Aalderink, B.J.; Padoan, R.; de Bruin, G.; Steemers, T.A.G. Quantitative hyperspectral reflectance imaging. *Sensors* **2008**, *8*, 5576–5618.

9. Stuffler, T.; Kaufmann, C.; Hofer, S.; Förster, K.; Schreier, G.; Mueller, A.; Eckardt, A.; Bach, H.; Penné, B.; Benz, U.; *et al.* The EnMAP hyperspectral imager—An advanced optical payload for future applications in earth observation programmes. *Acta Astronaut.* **2007**, *61*, 115–120.

10. Tauro, F.; Mocio, G.; Rapiti, E.; Grimaldi, S.; Porfiri, M. Assessment of fluorescent particles for surface flow analysis. *Sensors* **2012**, *12*, 15827–15840.

11. Miettinen, J.; Andersson, P. Acoustic emission of rolling bearings lubricated with contaminated grease. *Tribol. Int.* **2000**, *33*, 777–787.

12. Murr, P.J.; Wiesent, B.R.; Hirth, F.; Koch, A.W. Thin film measurement system for moving objects based on a laterally distributed linear variable filter spectrometer. *Rev. Sci. Instrum.* **2012**, *83*, 035110, doi:10.1063/1.3697750.

13. Wiesent, B.R.; Dorigo, D.D.; Koch, A.W. Limits of IR-spectrometers based on linear variable filters and detector arrays. *Proc. SPIE* **2010**, *7767*, 77670L, doi:10.1117/12.860532.

14. Lakowicz, J. *Principles of Fluorescence Spectroscopy*; Springer: New York, NY, USA, 2009.

15. Stern, O.; Volmer, M. Über die abklingungszeit der fluoreszenz. *Phys. Z.* **1919**, *20*, 183–188.

Reprinted from *Sensors*. Cite as: Sun, T.; Xing, F.; You, Z. Optical System Error Analysis and Calibration Method of High-Accuracy Star Trackers. *Sensors* **2013**, *13*, 4598–4623.

Article

Optical System Error Analysis and Calibration Method of High-Accuracy Star Trackers

Ting Sun [1,2], Fei Xing [1,2,]* and Zheng You [1,2]

[1] Department of Precision Instruments and Mechanology, Tsinghua University, Beijing 100084, China; E-Mails: sunting09@mails.tsinghua.edu.cn (T.S.); yz-dpi@mail.tsinghua.edu.cn (Z.Y.)
[2] State Key Laboratory of Precision Measurement Technology and Instruments, Tsinghua University, Beijing 100084, China

* Author to whom correspondence should be addressed; E-Mail: xingfei@mail.tsinghua.edu.cn; Tel.: +86-10-6277-6000; Fax: +86-10-6278-2308.

Received: 29 December 2012; in revised form: 27 March 2013 / Accepted: 3 April 2013 / Published: 8 April 2013

Abstract: The star tracker is a high-accuracy attitude measurement device widely used in spacecraft. Its performance depends largely on the precision of the optical system parameters. Therefore, the analysis of the optical system parameter errors and a precise calibration model are crucial to the accuracy of the star tracker. Research in this field is relatively lacking a systematic and universal analysis up to now. This paper proposes in detail an approach for the synthetic error analysis of the star tracker, without the complicated theoretical derivation. This approach can determine the error propagation relationship of the star tracker, and can build intuitively and systematically an error model. The analysis results can be used as a foundation and a guide for the optical design, calibration, and compensation of the star tracker. A calibration experiment is designed and conducted. Excellent calibration results are achieved based on the calibration model. To summarize, the error analysis approach and the calibration method are proved to be adequate and precise, and could provide an important guarantee for the design, manufacture, and measurement of high-accuracy star trackers.

Keywords: star tracker; error analysis; calibration; parameter estimation

1. Introduction

With the development of Earth-observing satellites and deep-space exploration satellites, requirements for attitude measurement accuracy are increasing. Thus, error analysis of the accuracy and calibration of the star tracker have become particularly important.

At present, research and analysis of the effect factors on the star tracker accuracy are being conducted. References [1] and [2] provides a general overview of the effects of the optical parameters. References [3] and [4] use a geometric method to establish a complicated error model, and obtain variations in accuracy for a certain range of optical parameters, but most of the existing analysis methods discuss the effects of factors separately and qualitatively. Up to now systemic error analysis and accurate error propagation model are inadequate.

Factors such as misalignment, aberration, instrument aging and temperature effects [5] could cause a departure of the star trackers from the ideal pinhole image model, and contribute to the attitude measurement error. Misalignment and aberration are time-independent, or static errors, which need to be calibrated prior to launch, and can be called ground-based calibration. By contrast, instrument aging and temperature effects are time-varying, or dynamic errors, which must be calibrated in real time, and can be called on-orbit calibration. This paper focuses only on the ground-based calibration method.

The ground-based calibration of star trackers generally includes real night sky observation and laboratory calibration. Real night sky observation can take advantage of the characteristics of the star tracker utilizing the star angular distance. This method is relatively easy to apply, whereas the model parameters interact with one another. Obtaining the global maximum is difficult, and this method is greatly influenced by the environment. Laboratory calibration could use a star simulator as the source. However, it is not easy to manufacture a high-accuracy star simulator.

Camera calibration techniques [6–10] can be choices for calibrating the star tracker considering that they are both optical imaging devices. However, there are problems for the star tracker to apply the calibration methods of the camera. First, most of these methods need to establish a complicated calibration model with scores of parameters. Good calibration results depend largely on the initial values and large amounts of calculation are needed for the optimization. Observability and convergence can be problematic. Second, the star tracker focuses more on the accuracy of the position of the image point, while the camera focuses more on the MTF or other image quality. Since the noteworthy parameters of the calibration methods of the camera and the star tracker are not exactly the same, the accuracy of general camera calibration techniques is not enough for the calibration requirement of the star tracker, which is one of the highest precision attitude measurement devices on the satellite. Moreover, majority of the camera calibration techniques have not considered the inclination of the image plane.

Last but not the least, the optical imaging principle and focus matters of the star tracker and the camera are not the same due to their functions. The camera uses a finite distance imaging mode, while the star tracker adopts an infinite distance imaging mode. General camera calibration methods are not suitable for the star tracker. Taking reference [11] as an example, the cubic 3-D calibration object applies to camera calibration as the camera can take a clear photograph of a finite distance object, but the star tracker is used to take pictures of infinite distance stars, so it cannot take a clear photograph of the 3-D calibration object. Even though there are a few reports about how to add another high accuracy lens to make this finite imaging calibration method apply to the star tracker, the accuracy and the position of the added lens, the accuracy of the 3-D cubic object all need to be discussed. These bring new troubles and are not easy to carry out.

Therefore, the calibration method provided in literature [11] or other similar camera calibration methods work better on short focal length, small view field camera. Convenient calibration methods for large FOV and high accuracy star tracker are still problems needed to be figure out. The calibration method using composite mode of high accuracy autocollimator theodolite and the features of the star tracker proposed in the manuscript is a good choice for this topic.

To summarize, the literature on the analysis and evaluation of the error sources of star trackers has not been adequate until now. This paper proposes a systematic method for weight analysis of the error source. Optical parameters that play key roles in the accuracy of the star tracker (such as the principal point deviation, focal length error, imaging plane inclination error and distortion) [1,12] are analyzed by the proposed method. From the analysis results, a calibration method is put forward. The calibration can separate the radial distortion from the image plane inclination, thus the optimization processes are simplified. The calibration result proves that the analysis of the optical systematic error and the calibration method for the high-accuracy star trackers proposed in this paper are reasonable and adequate, and can improve the accuracy of the star tracker.

2. Star Tracker Mesurement Model

The star tracker is a high-accuracy attitude measurement device, which considers the stars as the measuring object. It obtains the direction vector from the celestial inertial coordinate system by detecting the different locations of the stars on the celestial sphere. After many years of astronomical observations, star positions on the celestial sphere are predictable. Stars in the celestial sphere coordinate system can be expressed in the right ascension and declination (α, δ). Based on the relationship between the rectangular coordinate system and the spherical coordinate system, the direction vector of the stars in the rectangular coordinate system is expressed as follows:

$$v = \begin{bmatrix} \cos\alpha\cos\delta \\ \sin\alpha\cos\delta \\ \sin\delta \end{bmatrix} \tag{1}$$

Navigation stars are selected from the star catalog to meet the imaging requirement, and their data are stored in the memory of the star tracker.

When a star tracker with attitude matrix A is in the celestial coordinate system, the ideal measurement model of the star tracker can be considered as a pinhole imaging system. Navigation star S_i with direction vector v_i under the celestial coordinate system can be detected through the lens, whereas the vector of its image can be expressed as w_i in the star tracker coordinate system, as shown in Figure 1.

Figure 1. Star tracker ideal imaging model.

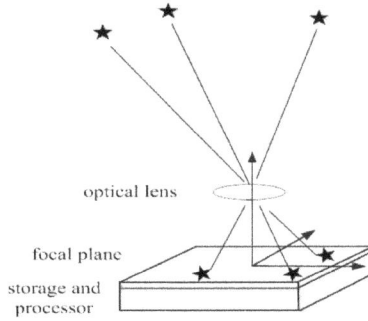

The position of the principal point of the star tracker on the image plane is (x_0, y_0). The position of the image point of navigation star s_i on the image plane is (x_i, y_i). The focal length of the star tracker is f. Vector w_i can be expressed as follows [13]:

$$w_i = \frac{1}{\sqrt{(x_i - x_0)^2 + (y_i - y_0)^2 + f^2}} \begin{bmatrix} -(x_i - x_0) \\ -(y_i - y_0) \\ f \end{bmatrix} \quad (2)$$

The relationship between w_i and v_i under the ideal condition can be expressed as follows, where A is the attitude matrix of the star tracker:

$$w_i = A v_i \quad (3)$$

When the number of navigation stars is more than two, the attitude matrix can be solved by the QUEST algorithm [14]. In this method, the optimal attitude matrix A_q in the inertial space of the star tracker can be calculated.

3. Star Tracker Error Analysis

3.1. Summary of the Error Sources of the Star Tracker

The existence of errors and noise in the system are inevitable. According to the pinhole model shown in Figure 1 and Equation (2), the factors that directly affect the results of the attitude measurement of the star tracker include the extraction error of star point position, principal point error, error of focal length, direction vectors of the navigation stars, and attitude solution algorithm error. The accuracy is also related to the number of stars in the field of view (FOV). Further, the effect factors of the star tracker are classified as follows [1–5]:

3.1.1. Star Vector Measurement Error

Star vector measurement error concerns the accuracy of vector w_i in Equation (3). Star vector measurement error includes:

(1) Extraction error of the star point position

The process in which the star tracker detects the navigation stars includes background radiation, optical systems, photoelectric detectors and signal processing. Each segment affects the extraction quality of the target signal.

Stars are far from the Earth, thus, starlight rays are considered as parallel light rays and can converge to a point on the focal plane. However, most star trackers adopt the defocus form [15] so that the image point can be diffused to cover several pixels. Using the signal energy of multiple pixels enables the star point-extraction accuracy to achieve a sub-pixel level. This process means that after considering the various factors, the extraction error of the star tracker can be considered as 0.1 pixels [16]. This concept lays the foundation of the following analysis.

(2) Star tracker optical parameter errors

The star tracker system cannot achieve the ideal image model because of the principal point deviation, focal length error, inclination of the image plane, distortion in practical use. Therefore, it is necessary to establish a calibration model for above parameter errors, and analyze parameter error and model error.

3.1.2. Star Catalog Error

Star catalog errors concern the accuracy of vector v_i in Equation (3). The number of stars is very large; hence, we must select the appropriate ones for storage in the memory and meet the performance requirements of the star tracker. The different stars selected may influence the star numbers that appear in the FOV. The star catalog set-up time could also contribute small errors in the direction vectors of the stars in the celestial coordinate system. But the influence can be ignored if the star catalog can be corrected every once in a while considering proper motion of stars.

3.1.3. Star Tracker Internal Algorithm Error

Star tracker internal algorithm error concerns the accuracy of the final attitude matrix A of the star tracker. However, algorithm errors such as star pattern recognition methods and attitude solution algorithm are irrelevant to this work. Among the errors enumerated above, those described in Section 3.1.1 exert a relatively larger effect. The error analysis in this paper focuses mainly on this error source.

3.2. Error Propagation Model

In the following analysis, we use the angle measurement error (ξ_A) to represent the star tracker accuracy. According to the pinhole image model, ξ_A is expressed as follows: the change in the angle between the incident ray and the optical axis in the FOV is called the angle of real light change (ξ_{AR}), and the light calculated from the star position and focal length is called the calculated light. The change in the angle between the calculated light and the optical axis is called the angle of calculated light change (ξ_{AC}). The deviation in the ξ_{AR} and ξ_{AC} is called the ξ_A. The ξ_A represents the star tracker attitude measurement accuracy. The optical parameter errors such as the principal point error, error of focal length, inclination of the image plane and distortion influence the ξ_A and cause

222

a difference between the ξ_{AR} and ξ_{AC}. Therefore, analysis of the effect of the different optical parameter errors on ξ_A could identify the key factors that must be calibrated, which is important in improving the star tracker accuracy and can provide optimization guidelines for the star tracker design.

Figure 2 shows the sketch of complete error propagation. β_{ri} is defined as the angle between the incident ray and the optical axis. The initial value of β_{ri} is β_{r0}, $\beta_{r0} = 0°$. The maximum value of β_{ri} is equal to the angle of the FOV. $\Delta\beta_{ri}$ is the ξ_{AR}, and $\Delta\beta_{ri} = \beta_{ri} - \beta_{r0}$. β_{ci} is the angle between the calculated light and the optical axis. $\Delta\beta_{ci}$ is the ξ_{AC}, and $\Delta\beta_{ci} = \beta_{ci} - \beta_{c0}$. We made the following assumptions: axis e_1 is along the direction of the maximum error of the principal point, and the inclination angle of the image plane in this direction is also the maximum of all directions. This can represent the worst case of error conditions. Axis e_3 is the ideal optical axis of the system, and e'_3 represents the actual optical axis. Δs represents the deviation of the optical axis. Δf is the deviation of the focal length. Δx represents the star point-extraction error and Δd is the distortion value. The distortion discussed in this paper concerns radial distortion only.

Figure 2. Sketch of complete error propagation.

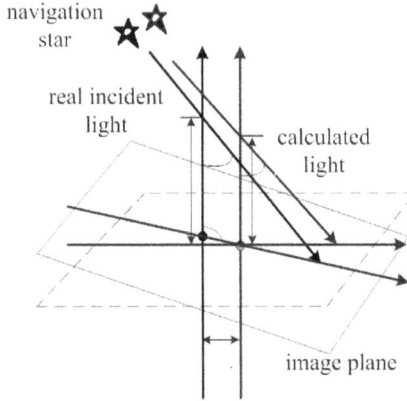

$$O'B' = \frac{f + \Delta f + \Delta s \cdot \tan(\theta)}{\sin(90 - \theta - \beta_n)} \cdot \sin(\beta_n) + \Delta x + \Delta d \tag{4}$$

$$OB = OB' = O'B' - O'O \tag{5}$$

Angle β_{ci} is obtained as:

$$\beta_{cl} = \arctan\left(\frac{OB}{f}\right) \tag{6}$$

The ξ_A is calculated based on the above analysis:

$$\xi_A = \arctan\left(\frac{\left(\frac{f + \Delta f + \Delta s \cdot \tan(\theta)}{\cos(\theta + \beta_n)} \cdot \sin(\beta_n) + \frac{\Delta s}{\cos(\theta)} + \Delta x + \Delta d\right)}{f} - \arctan\left(\frac{\Delta s}{f\cos(\theta)}\right) - \beta_n\right) \tag{7}$$

When the actual optical axis is along the positive direction of the ideal optical axis, Δs is defined as positive. When the inclination angle θ is in the clockwise direction, it is defined as positive. Another ξ_A form is expressed as follows:

$$\xi_P = O'B' - f\tan(\beta_{rl}) = \frac{f + \Delta f + \Delta s \cdot \tan(\theta)}{\cos(\theta + \beta_{rl})} \cdot \sin(\beta_{rl}) + \Delta x + \Delta d - f\tan(\beta_{rl}) \tag{8}$$

We define ξ_P as the position measurement error of the incident light ray. ξ_A and ξ_P describe the measurement error in different views, however, they can be transformed from one form to the other. ξ_A is more concerned with the accuracy of the star tracker, whereas ξ_P is more suitable for using in the calibration.

3.3. Star Tracker Optical Parameter Errors Simulation

We adopt two methods to discuss the error effects. First, we use the Monte Carlo (MC) stochastic modeling method. In this method, it is assumed that the errors after calibration, such as noise, inclination angle, focal length and principal point errors are random errors. These errors are considered to coincide with the normal distribution. In addition, the number of stars that can be captured in the sky is more than 6,000. Therefore, it is reasonable to consider the incident angle in the FOV as uniformly distributed. Based on the two statistical assumptions, we can combine the geometry and MC random models, and develop the complete error effect analysis. Second, we use the maximum error method to prove the simulation result of the MC method. This method can easily identify the error distribution of the different incident angles. Analysis of the position of the maximum error can also provide information for further study. The object analyzed in this paper is a star tracker of 7″ accuracy and the FOV is 17°. The focal length of the system is approximately 49.74 mm. The star tracker adopts the APS CMOS image sensor with 1,024 × 1,024 pixels, and the size of each pixel is 0.015 mm.

3.3.1. MC Error Analysis Method

The MC method [17] used the statistical rule of random numbers for the calculation and simulation. The following subjects analyze the single factor and the combination of factors using the MC method. Using a 1,000,000 times simulation, the statistical results are obtained. The simulation is conducted with an 8.5° incident angle.

Then, combination error analysis using the MC stochastic simulation is conducted. The ξ_A satisfies the distribution rule: $\mu = -0.0225$, $\sigma = 4.4218''$. The accuracy is in the range of $\mu - 3\sigma \sim \mu + 3\sigma = -13.2881 \sim 13.2430''$. Considering that at least 4 stars can be captured in the normal working state, the simulation boresight accuracy can be calculated as: $-13.2881/\sqrt{4} \sim 13.2430/\sqrt{4} = (-6.6440'' \sim 6.6215'')$. Thus, the allocation of permissible errors of the star tracker with a 7″ accuracy is obtained.

Table 1 and Figure 3 show that, for the analyzed system, error of star point extraction, error of principal point, error of focal length, error of inclination of the image plane and distortion are respectively distributed in the range of 0.1 pixels, 4.5 pixels, 0.6 pixels, 0.075° and twenty thousandth, they bring the ξ_A to the same level. Considering their combined effects can ensure the accuracy of the star tracker.

Table 1. Single-error factor analysis using MC stochastic simulation.

Item	Distribution of the Item	Distribution of ξ_A
Error of star point extraction:	$\mu = 0, \sigma = 0.1/3$ pixels (Gaussian distribution)	$\mu = 0.0073, \sigma = 2.0281('')$
Error of principal point:	$\mu = 0, \sigma = 4.5/3$ pixels (Gaussian distribution)	$\mu = -0.0139, \sigma = 2.0400('')$
Error of focal length:	$\mu = 0, \sigma = 0.6/3$ pixels (Gaussian distribution)	$\mu = -0.0079, \sigma = 1.8182('')$
Error of inclination angle:	$\mu = 0, \sigma = 0.075°/3$ (Gaussian distribution)	$\mu = 0.0097, \sigma = 1.9703('')$
Distortion:	$\mu = 0, \sigma = 0.1/3$ pixels (Gaussian distribution)	$\mu = -0.0185, \sigma = 2.0325('')$

Figure 3. Synthetic analysis result using MC stochastic simulation.

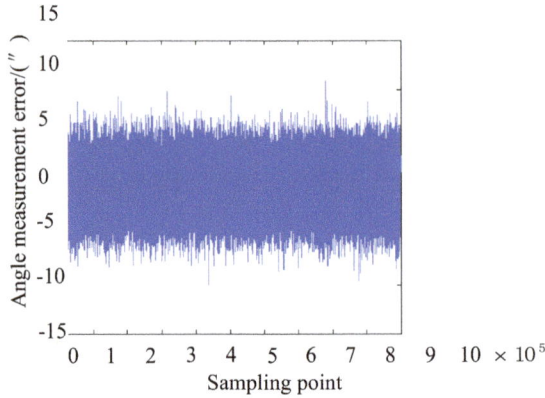

3.3.2. Simulation of the Maximum Error Method

We conduct a simulation using Maximum Error Method to compare with the method in Section 3.3.1 based on the error propagation model in Section 3.2 and the system parameters of the star tracker, as well as the range of the star point extraction error, the principal point error, focal length error, inclination of the image plane and distortion. The incident angle β_{ri} is in the range of 0–8.5° and the focal length is approximately 49.74 mm. Under the simulation conditions, the effect of the error of star point extraction on the star tracker accuracy, along with the incident angle is shown in Figure 4.

Under the simulation conditions, the effect of the error of the principal point on the star tracker accuracy, along with the incident angle is shown in Figure 5.

Under the simulation conditions, the effect of the error of focal length on the star tracker accuracy, along with the incident angle is shown in Figure 6.

225

Figure 4. Influence of the star point extraction error on the star tracker accuracy. (**a**) is obtained when the star point extraction error Δx is among the range from -0.1 pixels to 0.1 pixels; (**b**) is the contour line of (a).

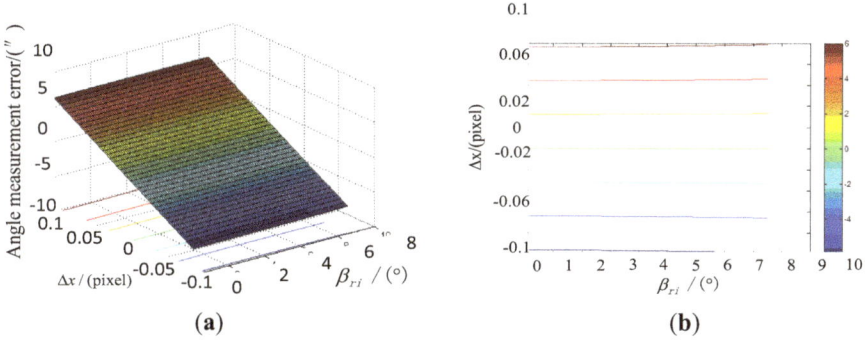

(**a**)

(**b**)

Figure 5. Influence of the principal point error on the star tracker accuracy. (**a**) is obtained when Δs is among the range from -4.5 pixels to 4.5 pixels; (**b**) is the contour line of (a).

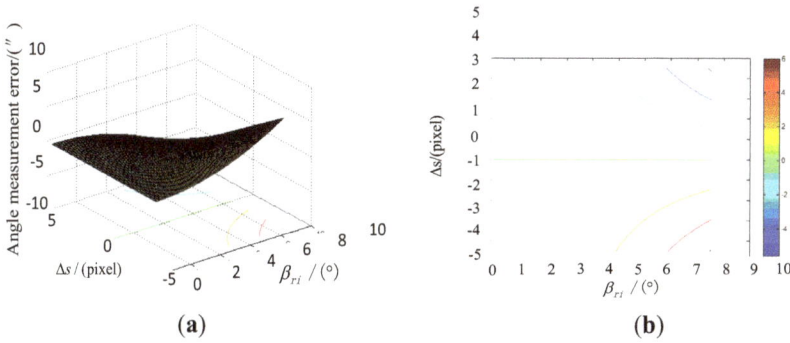

(**a**)

(**b**)

Figure 6. Influence of the focal length error on the star tracker accuracy. (**a**) is obtained when Δf is within the range from -0.6 pixels to 0.6 pixels; (**b**) is the contour line of (a).

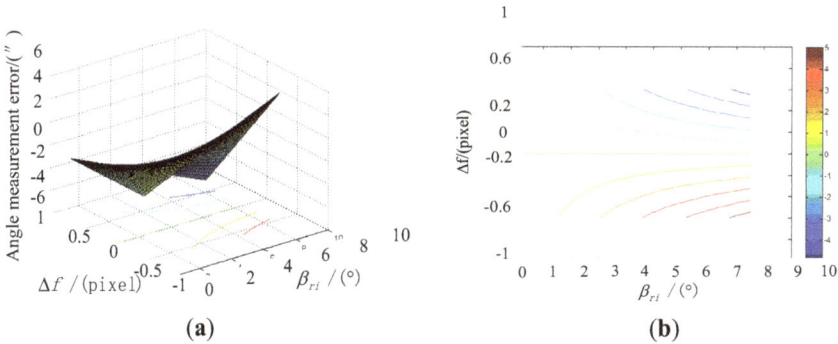

(**a**)

(**b**)

226

Under the simulation conditions, the effect of the error of inclination image plane on the star tracker accuracy, along with the incident angle is shown in Figure 7.

Figure 7. Influence of the inclination angle error on the star tracker accuracy. (a) is obtained when θ is in the range from 0° to 0.075°.(b) is the contour line of (a). (c) is obtained when θ is in the range from −0.075° to 0°, and (d) is the contour line of (c).

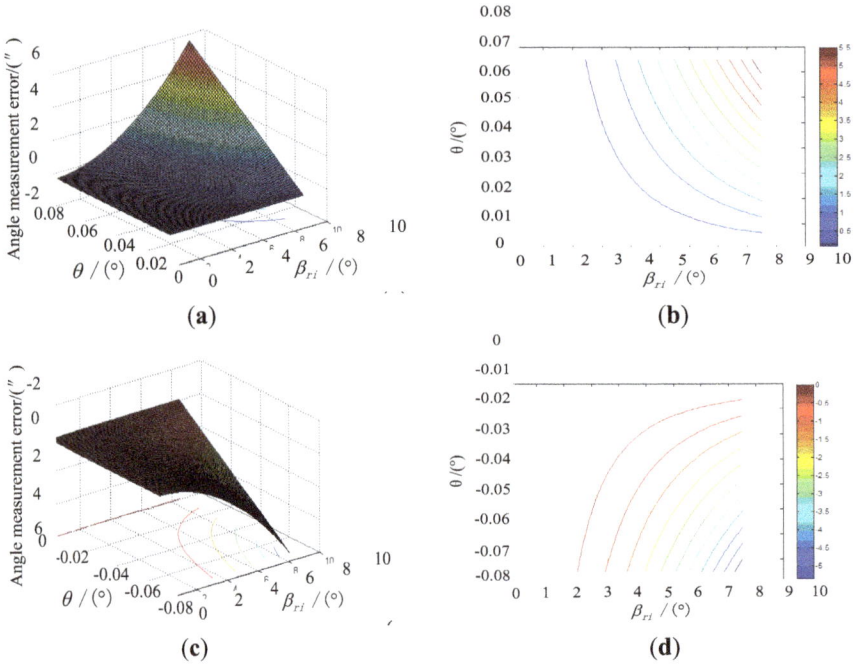

(a)

(b)

(c)

(d)

Under the simulation conditions, the effect of the error due to distortion on the star tracker accuracy, along with the incident angle is shown in Figure 8.

Figure 8. Influence of the distortion on the star sensor accuracy. (a) is obtained when the relative distortion is approximately 2/10,000th; and (b) is obtained when the relative distortion is approximately 1/1,000th.

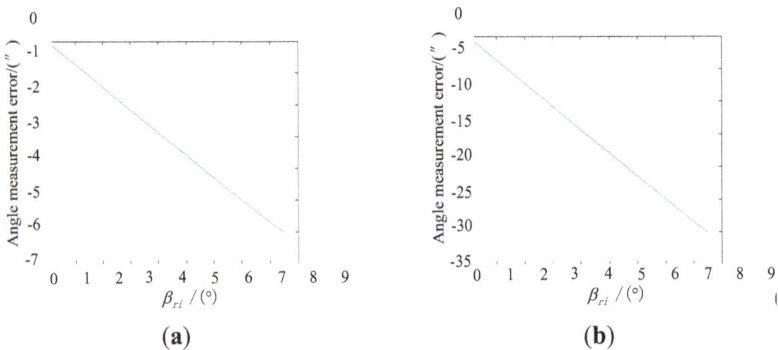

(a)

(b)

The simulation results show that the error effects obtained by the maximum error method agree with the results obtained by the MC simulation method. We can also find that inclination of the image plane and the distortion are two key factors that need to be calibrated. The calibration method will be elaborated in the next section.

3.4. Conclusion of the Star Tracker Error Analysis

Optical systematic error analysis method proposed in this paper can perform analysis on the sensitivity of factors (such as the error of star position extraction error, position error of principal point, error of focal length, inclination of the image plane and the distortion) that may influence the accuracy of the star tracker.

The above analysis of the system error can be applied in the following areas: (1) reference for the calibration target, *i.e.*, if the five indicators are simultaneously satisfied, the calibration results could meet the requirements; (2) analysis for the highest accuracy of the system; and (3) determining the major factors that emphasized the calibration experiment.

For application (1) above, because of the limitations in the calibration method, the effect factors cannot be separated from one another. Thus, determining whether the five indicators are all satisfied is difficult. Therefore, the proposed method is used primarily in the determination and demonstration of the design indicators. For application (3), some of the restrictions in certain error range can easily meet, such as the principal point position error, whereas satisfying the others are more difficult. These error factors need to be calibrated elaborately, such as the inclination angle of the image plane and the distortion. Therefore, emphasis on the calibration method is related to the system parameters. The calibration method and the processes are designed according to the characteristics of the system, so that the ξ_A or ξ_P of the star tracker is within the design range.

4. Laboratory Calibration Method

4.1. Star Tracker Calibration Device

The calibration object of this paper is a star tracker with 7″ accuracy. Based on the result of the above analysis, optical parameters of the system are calibrated. The laboratory calibration of the star tracker can be performed using a three-axis turntable and a collimator or an autocollimator theodolite. In essence, their operating principle is the same. However, because the collimator does not have a self-collimation function, which could introduce trouble to the calibration of the principal point, we adopt the autocollimator theodolite.

The autocollimator theodolite we employ in the experiment is the Leica 6100A. Figure 9(a) shows its external view. It has a small size, high accuracy 0.5″ and simple operation. We can use its auto-collimation eyepiece to determine whether the crosslines coincide, as shown in Figure 9(b). Other experiment devices consist of the optical table and auxiliary fixtures. It is worth noting that the aperture of the autocollimator should be comparable or larger than the aperture of the star tracker in order to avoid vignetting.

228

Figure 9. (a) External view; (b) Internal structure of the autocollimator Theodolite 6100A.

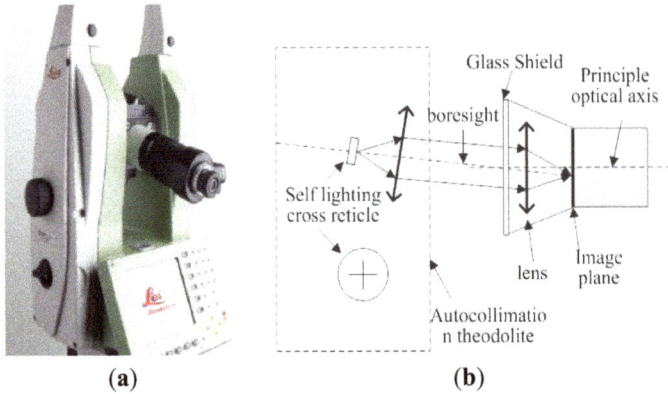

(a)   (b)

4.2. Calibration Algorithm and Experiment

According to the analysis in Section 3, the calibration objective should be focused on the inclination of the image plane and the distortion. The basic block diagram of the calibration process is shown in Figure 10.

Figure 10. Calibration flow diagram.

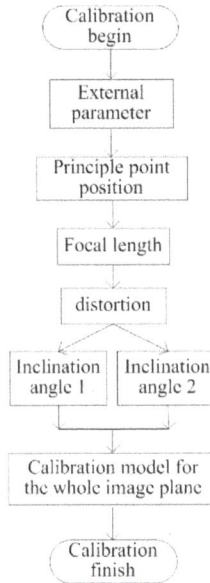

4.2.1. Coordinate System

For better representation of the location and the relationship, we create the coordinate systems as in Figure 11. Axes X_C Y_C in coordinate system CS_C represent the movement directions of the outgoing light rays of the theodolite in the two orthogonal axes, respectively. Z_C, X_C and Y_C comply with the right-hand rule. Coordinate system CS_N represents the normalized coordinate of the focal plane. The positive directions of X_C Y_C are consistent with the directions of the increase in the image pixel value. The coordinate origin is the same as the optical system principal point. The distance between CS_N and the pinhole is normalized to one. Coordinate system CS_{OXY} has the same direction as that of CS_N, and the only difference is that the distance between CS_{OXY} and the pinhole is equal to the focal length f. CS_{OUV} represents the actual coordinate system of the image plane. The positive directions of u, v are consistent with the directions of the increase in the image pixel value. The coordinate origin is at the zero pixel of the image plane, and the optical system principal point can be expressed as (u_0, v_0). The distance between CS_{OUV} and the pinhole is f. The axes of $CS_{O'U'V'}$ and CS_{OUV} have the same direction, with the coordinate origin O' is the same as the principal point. Ideally, $CS_{O'U'V'}$ coincides with CS_{OXY}. Angle α is the longitude value of the outgoing light ray of theodolite in coordinate system CS_C, whereas δ is the latitude value. Angle φ is the angle between the axis X_C and axis X_N. Thus, the coordinates of the ideal pinhole imaging model are established as shown in Figure 11. The relationship of the parameters are presented in Equations (9)–(11).

Figure 11. Coordinate systems in the ideal pinhole imaging model.

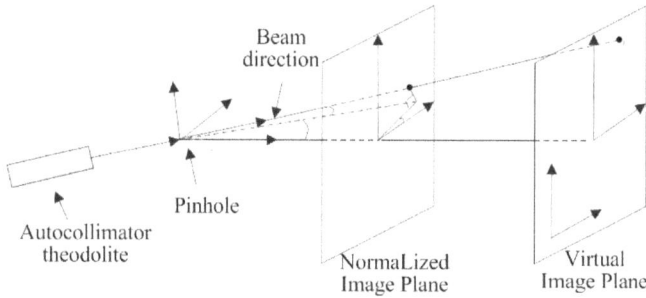

$$\begin{cases} x_N = \tan(\alpha)\cos(\varphi) - \dfrac{\tan(\delta)}{\cos(\alpha)} \cdot \sin(\varphi) \\[2mm] y_N = \tan(\alpha)\sin(\varphi) + \dfrac{\tan(\delta)}{\cos(\alpha)} \cdot \cos(\varphi) \end{cases} \tag{9}$$

$$\begin{cases} x = fx_N = f \cdot (\tan(\alpha)\cos(\varphi) - \dfrac{\tan(\delta)}{\cos(\alpha)} \cdot \sin(\varphi)) \\[2mm] y = fy_N = f \cdot (\tan(\alpha)\sin(\varphi)) + \dfrac{\tan(\delta)}{\cos(\alpha)} \cdot \cos(\varphi) \end{cases} \tag{10}$$

$$\begin{cases} u = x + u_0 \\ v = y + v_0 \end{cases} \tag{11}$$

Angle α is positive along the positive direction of X_C, whereas it is negative along the negative direction of X_C. Angle δ is positive along the positive direction of Y_C, whereas it is negative along the negative direction of Y_C. When the angle from X_N to X_C is counterclockwise, angle φ is considered as positive. The ideal pinhole imaging model cannot be achieved in practical application. There are position errors of principal point, focal length error, inclination of image plane and distortion to cause the $CS_{O'U'V'}$ departure from CS_{OXY} as shown in Figure 12, and the equations above are invalid. It is necessary to estimate the parameters by calibration.

Figure 12. Coordinate systems in the case of errors.

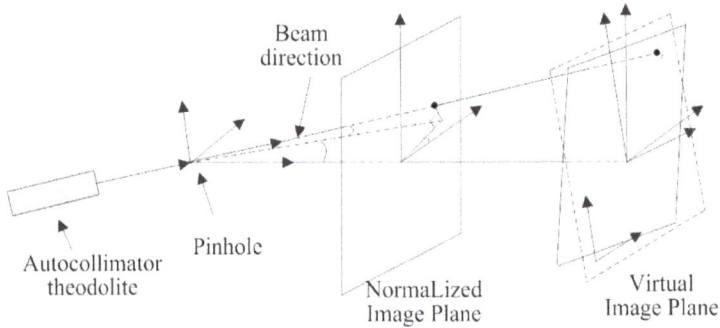

4.2.2. Description of the Calibration Experiment Operating

First, we adjust the theodolite to ensure that its outgoing light travels only along the longitude direction. The latitude value does not change during this process. Imaging conducted at every 0.5° can yield a series of measured values that determines the external parameters.

Then, according to the 17° FOV of the star tracker, we adjust the theodolite so that its light travels along the two orthogonal directions as shown in Figure 13 while conducting imaging at every 0.5°. If any of the two directions satisfies the requirement of the first step, two steps can be combined to determine the internal parameter of the star tracker. Because the outgoing rays of the theodolite are crosslines, the images on the image plane appear similar to that shown in Figure 13(b). It is worth mentioning that the star tracker is supposed to be settled on the same optical table with the auto-collimation theodolite to avoid relative vibration. Considering that our experiment is conducted at the State Key Laboratory of Precision Measurement Technology and Instruments, the floor of the laboratory has been treated with vibration isolation. So the relative vibration could be reduced and ignored in the calibration form in Figure 13(a), and it is easier to adjust the relative position of the theodolite and the star tracker with the tripod.

Figure 13. (a) Calibration experiment device; (b) Imaging method sketch map.

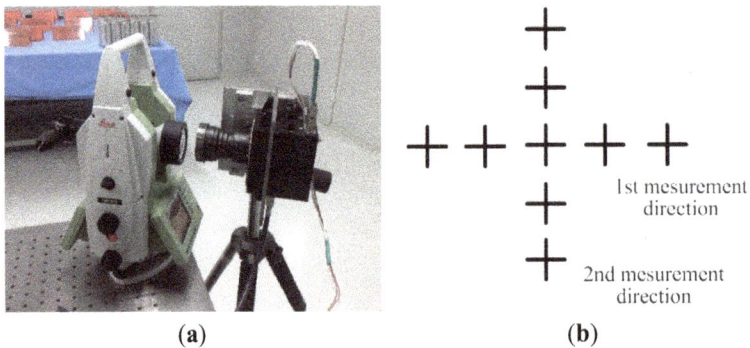

(a)   (b)

4.2.3. Image Processing

The image obtained by the star tracker is shown in Figure 14. An appropriate image processing method should be adopted to obtain the precise center position of the crossline which represents the outgoing ray of the theodolite. For the pixels in the first area, we regard the pixels in the same row as a group, and determine their gray value center of gravity (*i.e.*, weighted average). For the pixels in the second area, we consider pixels in the same column as a group, and also determine their gray value center. The centers of gravity are marked with a circle as shown in Figure 15. Finally, we use the least square method to fit the two straight lines. The point of intersection of the two lines is considered as the center of the crossline. This work provides a basis for further algorithm. Since the light intensity of the theodolite could be adjusted by a knob, obtaining image before experiment and observing whether the image is saturation is also important.

Figure 14. Image of the emergent crossline of the theodolite. (**a**) is original image; (**b**) is partially enlarged view.

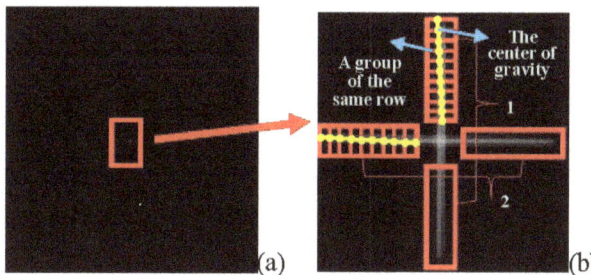

Figure 15. Sketch map of the solution of the gravity center of the cross-line.

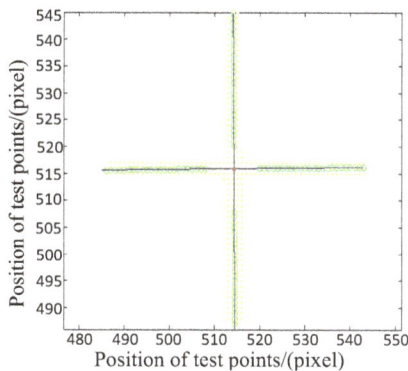

4.2.4. Calibration Process

External Parameter Estimation

The series of point coordinate values obtained in Section 4.2.2 can be used to solve angle φ between the X_C-axis of coordinate system CS_C and the X_N-axis of CS_N. Symbol n represents the number of sampling points:

Linear fitting can also be adopted to solve the value of φ besides Equation (12).

$$\varphi = \sum_{i=1}^{n} \left(\frac{a \cos\left(\dfrac{|u_{Ri} - u_0|}{((u_{Ri} - u_0)^2 + (v_{Ri} - v_0))^{1/2}} \right)}{n} \right) \tag{12}$$

Optical Parameter Estimation

The positions of the series of specific points are obtained according to the above discussion and preparation. The characteristics of these points include the following: the test points are distributed in two orthogonal directions and the test points in the same direction are almost centrosymmetrical. Taking advantage of relationship between the symmetric points, we build a calibration model, shown in Figure 16. In the actual calibration process, we try to make one of α, δ equal to zero for simplicity of the calculation. In reality, no restriction is imposed on the theodolite longitude and latitude angle, and the two orthogonal directions formed by the test points do not necessarily coincide with any of the defined coordinates. For the general situation, we name the two orthogonal directions distributed with the test points as L_1 and L_2. β_i is the synthesis angle of the theodolite longitude and latitude. When there are distortion and inclination of the image plane in the system, the imaging model is shown in Figure 16. β_{i+} and β_{i-} respectively represent the incident angles of the two centrosymmetric points in the same test direction. P_{Ii+} and P_{Ii-} are the ideal image point positions corresponding to the incident light rays when no distortion and inclination of image plane occurs. P_{Di+} and P_{Di-} are the positions of the image points when distortion is present but with no inclination of the image plane. P_{Ti+} and P_{Ti-} are the positions of the image points with the presence of image plane inclination but with no distortion. P_{Ri+} and P_{Ri-} are the positions of the image points when both distortion and inclination of the image plane are present.

Figure 16. Calibration schematic diagram.

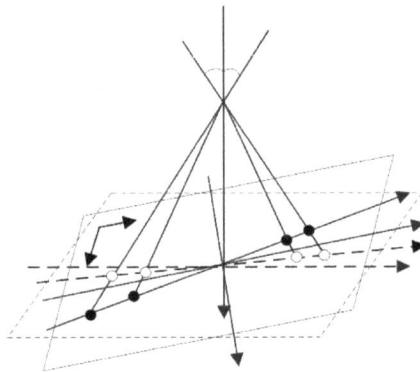

The longitude and latitude values (α_i, δ_i) corresponding to the incident rays and the coordinate positions (u_{Ri}, v_{Ri}) on the image plane can be used to solve the principal point position (u_0, v_0), focal length f, distortion coefficients α_1, α_2, α_3, α_4, α_5 and the inclination angle corresponding to the test direction. Finally, the calibration model for the entire image plane is obtained. This method is elaborated as follows:

(1) Firstly, we adjust the auto-collimation theodolite, and ensure that the crossline in the theodolite eyepiece is coincident with the specular reflection image of the star tracker glass shield, as shown in Figure 17. We consider that the boresight of the theodolite is coincident with the spindle of the star tracker lens. The imaging position of the crossline on the image plane is considered as the principal point (u_0, v_0) of the system.

Figure 17. Principal point measurement principle.

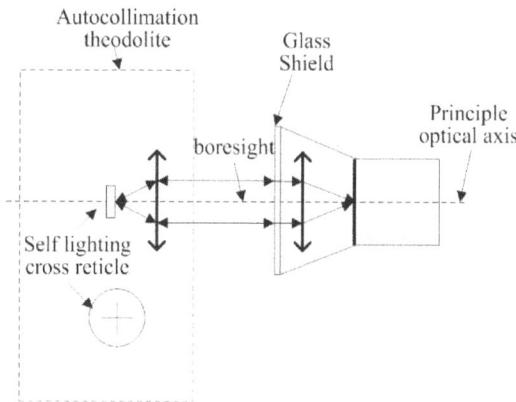

(2) Secondly, we can obtain a series of focal lengths utilizing the incident light in different directions and their image point (u_{Ri}, v_{Ri}). The average focal length is considered as the focal length value of the system. n represents the number of test points except for the principal point:

$$\beta_i = a\cos(\cos(\alpha_i)\cos(\delta_i)) \tag{13}$$

$$f = \frac{\sum_{i=1}^{n} \frac{((u_{Ri} - u_0)^2 + (v_{Ri} - v_0)^2)^{1/2}}{\tan(\beta_i)}}{n} \tag{14}$$

(3) The simulation results shown in Figure 18 prove that when the incident angle $|\beta_i| < 8.5°$, inclination angle $|\theta| < 0.8°$, there are following rules of example points: $|P_{Ti+}P_{Ti-}| = |P_{Ti+} P_{Ti-}|$. As shown in Figure 18, the errors between the two segments are less than 0.1 pixels, which is less than the extraction accuracy of the star tracker. Using the same simulation method, we find that the segments $|P_{Ri-}P_{Ti-}|$, $|P_{Ri+}P_{Ti+}|$, $|P_{Di+}P_{Ii+}|$ and $|P_{Di-}P_{Ii-}|$ are equal to one another, whereas the errors are extremely small as shown in Figure 19. Therefore, we consider $(|P_{Ii+}P_{Ii-}| - |P_{Ri+}P_{Ri-}|)/2$ as the radial distortion value at point P_{Ri+} or P_{Ri-}, and this concept is an important basis of our calibration method.

234

Figure 18. Error between the ideal and real distances of the symmetric points. The Z-axis represents the value $(|P_{Ti+}P_{Ti-}| - |P_{Ii+}P_{Ii-}|)$/pixel.

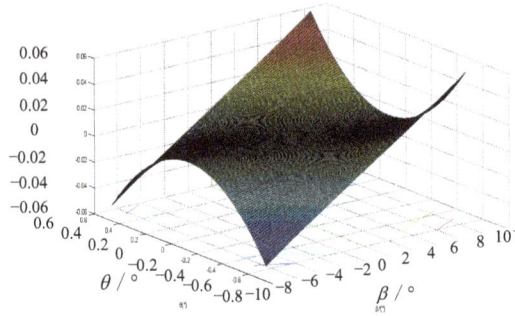

Figure 19. Relationship of the different distortion expressions. Z-axis of (**a**) represents $(|P_{Ti-}P_{Ri-}| - |P_{Ti+}P_{Ri+}|)$/pixel; Z-axis of (**b**) represents $(|P_{Ti-}P_{Ri-}| - |P_{Ii-}P_{Di-}|)$/pixel; Z-axis of (**c**) represents $(|P_{Ti+}P_{Ri+}| - |P_{Ii-}P_{Di-}|)$/pixel.

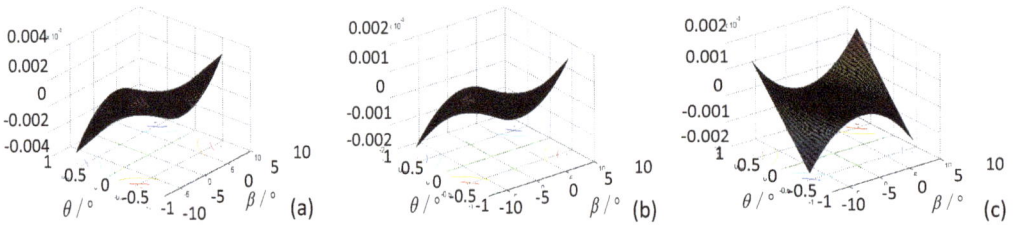

From the multiple groups of symmetrical points, we can obtain optimized distortion coefficients a_1, a_2, a_3, a_4, a_5 by linear least-squares fitting. The form of distortion can be chosen according to the distortion values in different cases. Here we adopt $\sum_{k=1}^{5} a_k R_i^k$:

$$R_i = \frac{|P_{Ri+}P_{Ri-}|}{2} \tag{15}$$

$$\Delta_i = |P_{Di+}P_{Ii+}| = |P_{Di-}P_{Ii-}| = |P_{Ri+}P_{Ti+}| = |P_{Ri-}P_{Ti-}| = \frac{(|P_{Ii+}P_{Ii-}| - |P_{Ri+}P_{Ri-}|)}{2} \tag{16}$$

$$\begin{cases} R_i = \dfrac{((u_{Ri+} - u_{Ri-})^2 + (v_{Ri+} - v_{Ri-})^2)^{1/2}}{2} \\ \sum_{k=1}^{5} a_k R_i^k = (((x_{Ii+} - x_{Ii-})^2 + (y_{Ii+} - y_{Ii-})^2)^{1/2} - ((u_{Ri+} - u_{Ri-})^2 + (v_{Ri+} - v_{Ri-})^2)^{1/2})/2 \end{cases} \tag{17}$$

(4) When the principal point, focal length and distortion coefficients are determined, the inclination angle of the image plane can be obtained using the geometric relationship, and the average value can be calculated using the multiple-set of symmetrical points.

$$|\theta_i| = \left| \arccos(\frac{|OP_{li}|}{|OP_{Ri}|+\Delta_i}) - \beta_i \right| = \left| \arccos\left(\frac{(x_{li}^2 + y_{li}^2)^{1/2}}{((u_{Ri}-u_0)^2 + (v_{Ri}-v_0)^2)^{1/2} + \Delta_i} \cdot \cos(\beta_i) \right) - \beta_i \right| \quad (18)$$

Equation (18) is suitable for the symmetric points. When $|OP_{Ri}| + \Delta_i > |OP_{li}|$, θ_i is defined positive; when $|OP_{Ri}| + \Delta_i < |OP_{li}|$, θ_i is defined negative.

So far, the principal point, the focal length, the radial distortion coefficients and the inclination angle of image plane in one direction are obtained. The principal point, focal length, and radial distortion coefficients are suitable for the entire plane. Inclination angle in the other measurement direction can be obtained in the same manner.

Calibration Model Applied to the Entire Image Plane

The inclination angle in the two measurement directions is not sufficient. The ultimate goal of calibration experiment is to obtain an ideal image position of incident rays from any direction in the FOV. Based on the parameters obtained above, there are many methods to solve this problem. We adopt a coordinate transformation method, and establish a coordinate transformation framework for incident rays. Therefore, we make the following analysis.

We can obtain the position of point P_{li+} as $(x_{li+}, y_{li+}, 0)$ in the coordinate system CS_{OXY} by considering the longitude and latitude of the incident ray, external parameter φ, focal length f and Equations (9)–(11). Then, the straight line $O_C P_{li+}$ can be expressed as:

$$\frac{x_{Ti+}}{x_{li+}} = \frac{y_{Ti+}}{y_{li+}} = \frac{k_{Ti+} - f}{-f} \quad (19)$$

Considering the angle $\angle P_{li} + OP_{Ti+} = |\theta_i|$ and the positive or negative of the angle value, the position of point P_{Ti+} in coordinate system CS_{OXY} can be solved though the following equations:

$$\begin{cases} \dfrac{x_{Ti+}}{x_{li+}} = \dfrac{y_{Ti+}}{y_{li+}} = \dfrac{k_{Ti+} - f}{-f} \\ (x_{Ti+}, y_{Ti+}, k_{Ti+}) \cdot (x_{li+}, y_{li+}, 0) = |(x_{Ti+}, y_{Ti+}, k_{Ti+})\| (x_{li+}, y_{li+}, 0)| \cos(\theta_i) \end{cases} \quad (20)$$

Thus, the position of point P_{Ti+} is obtained $(x_{Ti+}, y_{Ti+}, k_{Ti+})$. The position of point P_{Ti-} can also be obtained in the same way. So far, the positions of the multiple sets of points as P_{Ti+} and P_{Ti-} in both the two measurement direction in CS_{OXY} can be obtained. Therefore, the plane equation of actual image plane with inclination $O'U'V'$ in coordinate system CS_{OXY} can be expressed as $p(O'U'V')$.

The transformation matrix from coordinate system CS_{OXY} to coordinate system $CS_{O'U'V'}$ can be obtained at the same time:

$$CS_{O'U'V'} = CS_{OXY} R_1(-\varepsilon) R_2(-\eta) \quad (21)$$

R_1 and R_2 are the coordinate system transformation matrices.

$$R_1(-\varepsilon) = \begin{pmatrix} 1 & 0 & 0 \\ 0 & \cos(-\varepsilon) & \sin(-\varepsilon) \\ 0 & -\sin(-\varepsilon) & \cos(-\varepsilon) \end{pmatrix}, \ R_2(-\eta) = \begin{pmatrix} \cos(-\eta) & 0 & -\sin(-\eta) \\ 0 & 1 & 0 \\ \sin(-\eta) & 0 & \cos(-\eta) \end{pmatrix}$$

Thus, the whole parameter estimation model of the optical system of the star tracker is completed.

For any point on the image plane of the incident ray, we can determine its position in $CS_{O'U'V'}$ as $(u_{Ri+}-u_0, v_{Ri+}-v_0, 0)$. The position of point P_{Ti+} can be calculated by considering the distortion as $(u_{Ti+}-u_0, v_{Ti+}-v_0, 0)$. According to the coordinate system transformation matrix, the position of point P_{Ti+} in CS_{OXY} can be calculated as:

$$(x_{Ti+}, y_{Ti+}, k_{Ti+})' = R_1(-\varepsilon)R_2(-\eta)(u_{Ti+} - u_0, v_{Ti+} - v_0, 0)' \tag{22}$$

The equation for straight line $O_C P_{Ti+}$ can be obtained. Further, by solving the position of the point of intersection of straight line $O_C P_{Ti+}$ and plane OXY, the point of intersection G_{Ii+} is the ideal position of point P_{Ri+}. Thus, the ideal image position of the incident ray can be obtained. This solution is particularly important for the star pattern recognition and attitude solution of the star tracker.

4.3. Calibration Results and Discussions

4.3.1. Error Analysis

In the calibration process described in Section 4.2.4.2, the principal point position is obtained and considered as its true position. In reality however, due to lens installation, accuracy in manufacture as well as the limitations in the eyepiece alignment when using the theodolite, a deviation error of the optical axis is inevitable. That is to say, there may be an error β_0 between the optical axis obtained in Section 4.2.4.2 and the true optical axis of the lens. Maximum value of β_0 is approximately 40 in this calibration system, consisting of 30" installation error of the lens and 10" eyepiece alignment error. In this situation, the optical axis error will cause the deviation error of principal point, the focal length and inclination angle. The specific effect analysis is described as follows:

As shown in Figure 20(b), assuming the deviation error of the optical axis is 40" as analyzed above, the position error of principal point is about $f \cdot \tan(40")$ (equal to 0.64 pixels); the error of inclination angle is approximately 0.01°. The errors due to the focal length and the distortion are merged and the residual error of the distortion after calibration is within 0.1 pixels. Because the light ray is brighter than a real star in practical use, the signal to noise ratio is higher and the extraction error of the light ray position can reach approximately 0.05 pixels. We can use the MC error analysis method described in Section 3.3.1 to determine the ξ_A and ξ_P.

Figure 20. (a) Principal optic axis deviation error and its effect; **(b)** Its effect on the calibration process.

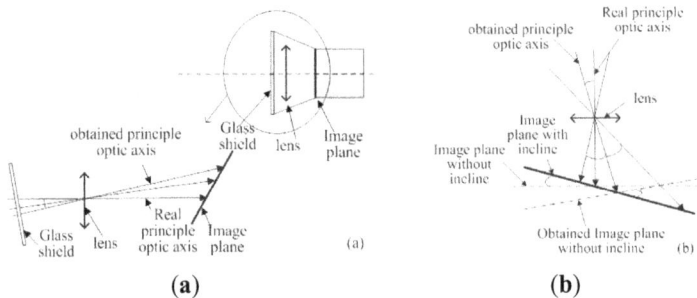

(a)

(b)

The ξ_A meets distribution rule: $\mu = 0.0034$, $\sigma = 2.1192''$. The ξ_P meets distribution rule: $\mu = 0.0007$, $\sigma = 0.0338$ pixels. We adopt ξ_P as the basis of the calibration target. That means the calibration residual error should in the range of $\mu+3\sigma$ distribution of ξ_P.

4.3.2. Calibration Results and Discussion

The calibration experiment is conducted in the laboratory. From the above analysis, we can know the longitude angle α_i and latitude angle δ_i of the incident ray of the theodolite. We also can obtain the position of point as P_{Ri+} or P_{Ri-} by the image processing mentioned in Section 4.2.3. Then we do the same procedure as Section 4.2.4.3, and calculate the position of point G_{Ii+}. The position of point P_{Ii+} can be also obtained through Equations (9)–(11). Comparing the two positions, the calibration model is proved to be effective and adequate if the error between the two positions is less than the ξ_P for points on the image plane at any incident light angle. In the experiment, we conduct two orthogonal direction measurements to get the parameters and the calibration model. Then we conducted measurements in another two independent and orthogonal directions that have an angle of 45° or 135° with the 1st measurement direction L_1 to prove the analysis and calibration model in this paper.

(1) As described in Section 4.2.4.1, we can firstly obtain external parameter φ using a linear fitting method. Measurement points and linear fitting curve are showed in Figure 21.

Figure 21. Measurement points and linear fitting curve (**a**). (**b**) and (**c**) show the two fitting curves separately.

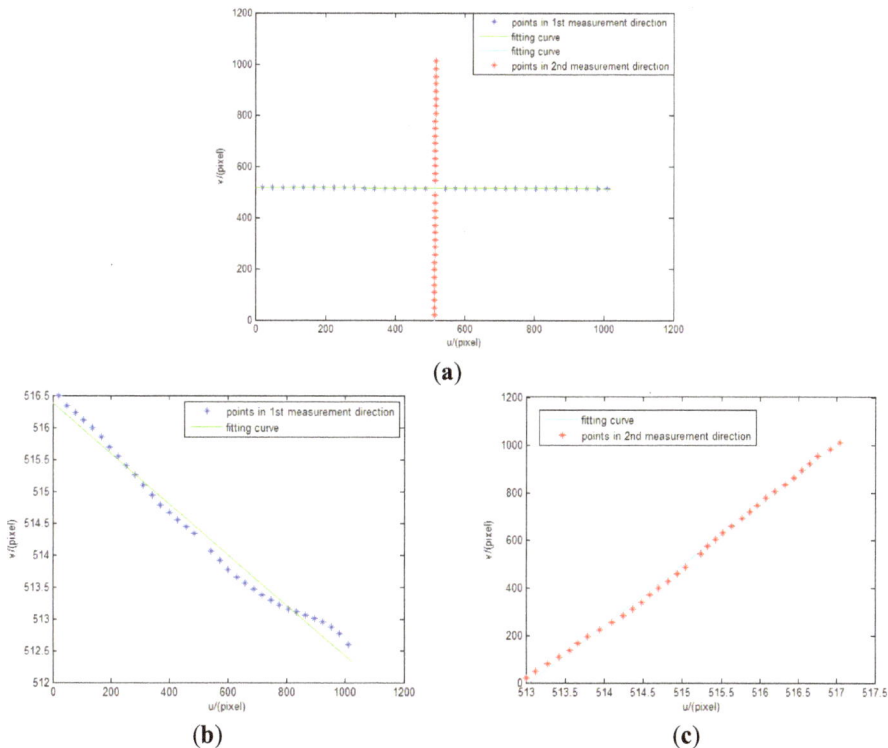

(a)

(b)   (c)

Solving the slope of the line can determine the value of φ. In our experiment, the slope of the two lines are −0.003968 and 0.003949, and root mean squared errors (RMSE) are 0.1614 and 0.0576 pixels. So the values of φ are 0.2273° and 0.2263°. We adopt 0.2268° as the value of φ.

(2) As described in Section 4.2.4.2—(1), we can obtain the position of the principal point as (515.1859, 514.2069) pixels.

(3) As described in Section 4.2.4.2—(2), we can obtain a series of focal lengths utilizing the incident light in different directions and their image point (u_{Ri}, v_{Ri}). The estimations of f is shown in Figure 22.

Figure 22. Estimations of f. (**a**) is the result of 1st measurement direction L_1, and (**b**) is the result of 2nd measurement direction L_2. Points in red and in blue represent centrosymmetry incident light rays separately.

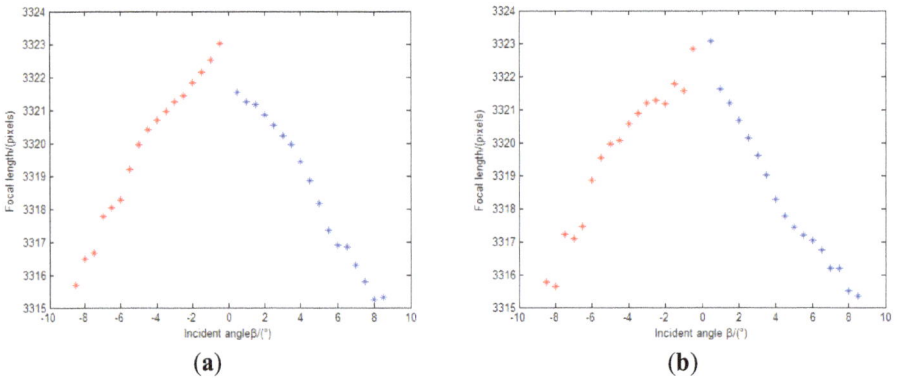

(**a**)   (**b**)

Theoretically, the values of the calculated focal lengths are equal. However, distortion can cause departure of the calculated focal lengths in different incident angles. Inclination of the image plane may cause deviation even when points are in centrosymmetry incident angles. Therefore, the estimations of f in Figure 22 is reasonable. We determine the focal length of the system as 3319.15 pixels using Equation (14). An initial estimate of f is adequate for calibration model as long as the radial distortion and other calibration parameters are matched with it.

(4) As described in Section 4.2.4.2—(3), we can obtain the distortion values Δ_i using Equation (16). The values are shown in Table 2.

Table 2. Distortion values.

Item	1st Measurement Direction		2nd Measurement Direction	
	R_i	Δ_i	R_i	Δ_i
1	29.0039	−0.021951	28.9778	−0.041757
2	58.0162	−0.050355	57.9644	−0.042748
3	87.0237	−0.065911	86.9740	−0.061112
4	116.0016	−0.076456	115.9721	−0.061893
5	145.0044	−0.079762	144.9961	−0.067778
6	174.0553	−0.082395	174.0393	−0.065835
7	203.1132	−0.079229	203.0897	−0.048521

Table 2. *Cont.*

Item	1st Measurement Direction		2nd Measurement Direction	
	R_i	Δ_i	R_i	Δ_i
8	232.1929	−0.063527	232.1630	−0.019044
9	261.2868	−0.037760	261.2486	0.018263
10	290.4139	0.007676	290.4194	0.039629
11	319.5688	0.084626	319.6433	0.061659
12	348.7565	0.164076	348.8699	0.095181
13	378.0469	0.194652	378.0526	0.226332
14	407.3608	0.258506	407.3291	0.330992
15	436.6681	0.383765	436.7842	0.321828
16	466.1053	0.461704	466.1020	0.503454
17	495.6124	0.545419	495.6529	0.536663

(5) We use linear least squares fitting to obtain the distortion curve. The fitting result is shown as follows, and Figure 23(b) shows that the residual error of the distortion after calibration can meet the calibration requirement analyzed in Section 3.3.

Figure 23. Distortion curve. (**a**) is fitting distortion curve and (**b**) is residual errors of distortion after calibration.

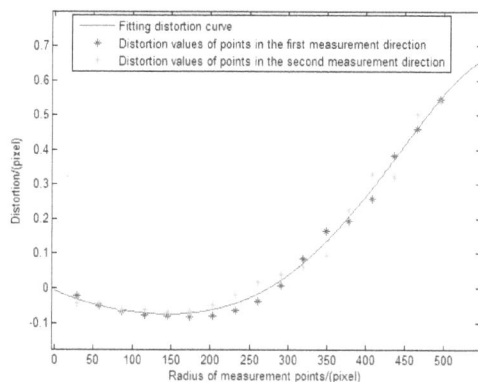

(**a**)

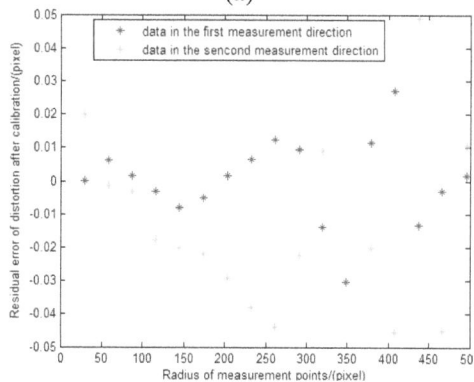

(**b**)

Calibration result is summarized in Table 3:

Table 3. Calibration result.

Item	Value
$\varphi(°)$	0.2268
Principal point (pixels)	(515.1859, 514.2069)
Focal length f (pixels)	3319.15
a_1	$-7.038e-04$
a_2	$-2.344e-06$
a_3	$2.4705e-08$
a_4	$-2.7031e-11$
a_5	$3.1804e-15$
Inclination angle of the image plane(°) (L_1 direction)	$0.1469\ (\sigma = 0.0413)$
Inclination angle of the image plane(°) (L_2 direction)	$0.0524\ (\sigma = 0.0221)$

After calibration, we can obtain the calibrated image point position of the incident light in different directions. The estimations of f can be conducted again to show the effect of the calibration (Figure 24) compared to the initial values in Figure 22. We can see from Figure 24 that the values of focal length do not change with the incident angles any more. They are in a range of 1 pixels (standard deviation≈0.23 pixels). This can also demonstrate that the residual errors are improved after calibration. The position error between points G_{li+} and P_{li+} is shown in Figure 25.

Figure 24. Re-estimations of f. (a) is the result of 1st measurement direction L_1; and (b) is the result of 2nd measurement direction L_2. Points in red and in blue represent centrosymmetry incident light rays separately.

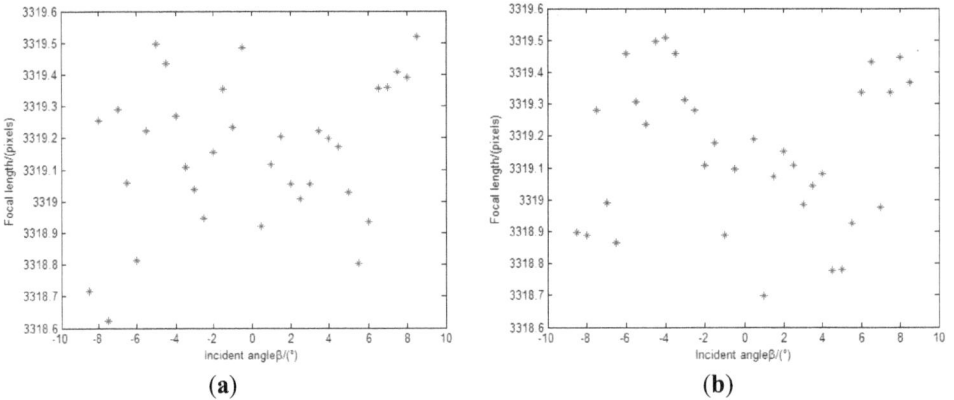

(a)

(b)

We can see from Figure 25 that the position errors in the entire FOV are within the ξ_P and in accordance with analysis in Section 4.3.1. The results demonstrate that the point position errors can be kept within the ξ_P after calibration. Based on MC analysis, the accuracy of this calibration method can reach 7″, which related to the focal length and the FOV of the system. The accuracy of the star tracker is supposed to be better than 4" after calibration, compared with the worse accuracy of more than 10" if no calibration is conducted.

Figure 25. Calibration residual error.

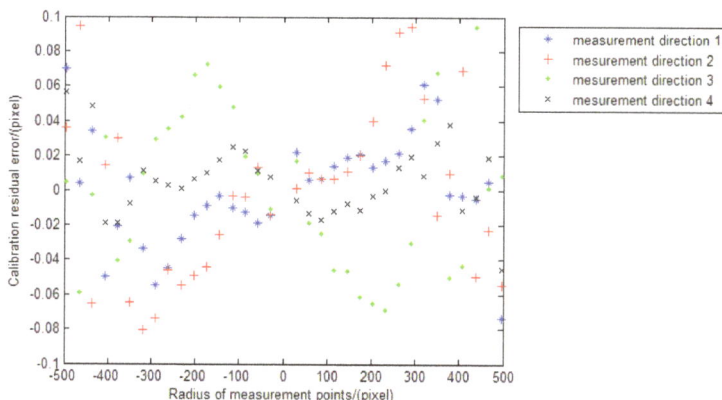

5. Summary and Conclusions

A synthetic error analysis approach for the star tracker has been proposed in detail in this paper. This approach can provide the error propagation relationship of the star tracker. Based on the analysis results, a calibration experiment is designed and conducted. Excellent calibration results are achieved. The calibration experiment can not only guarantee the accuracy to meet the design requirement, but can even improve the accuracy of the star tracker to a higher level. To summarize, the error analysis approach and the calibration method are proved to be adequate and precise, and are very important for the design, manufacture, and measurement of high-accuracy star trackers.

Acknowledgments

This work is partially supported by a grant from the National High Technology Research and Development Program of China (863 Program) (No. 2012AA121503). The laboratory calibration is performed in the State Key Laboratory of Precision Measurement Technology and Instruments at Tsinghua University. Both of them are gratefully acknowledged.

References

1. Liebe, C.C. Accuracy performance of star tracker-a tutorial. *IEEE Trans. Aerosp. Electron. Syst.* **2001**, *38*, 587–599.
2. Griffith, D.T.; Singla, P.; Junkins, J.L. Autonomous on-orbit calibration of approaches for star tracker cameras. *Adv. Astronaut. Sci.* **2002**, *112*, 39–57.
3. Wang, H.T.; Luo, C.Z.; Wang, Y.; Zhao, S.F. Star sensor model parametric analysis and calibration method study. *J. Univ. Electron. Sci. Technol. China* **2010**, *39*, 880–885.
4. Hao, X.T.; Zhang, G.J.; Jiang, J. Star sensor model parameter analysis and calibration method. *Opto-Electron. Eng.* **2005**, *2*, 5–8.

5. Xing, F.; Dong, Y.; You, Z. Laboratory calibration of star tracker with brightness independent star identification strategy. *Opt. Eng.* **2006**, *45*, doi:10.1117/1.2213996.

6. Faig, W. Calibration of close-range photogrammetry systems: Mathematical formulation. *Photogramm. Eng. Remote Sens.* **1975**, *41*, 1479–1486.

7. Abdel-Aziz, Y.I.; Karara, H.M. Direct Linear Transformation into Object Space Coordinates in Close-Range Photogrammetry. In Proceedings of Symposium on Close-Range Photogrammetry, Urbana, IL, USA, January 1971; pp. 1–18.

8. Tsai, R.Y. A versatile camera calibration technique for high accuracy 3D machine vision metrology using off-the-shelf TV cameras and lenses. *IEEE J. Robot. Automon.* **1987**, *3*, 323–344.

9. Weng, J.; Cohen, P.; Herniou, M. Camera calibration with distortion models and accuracy evaluation. *IEEE Trans. Pattern Anal. Mach. Intell.* **1992**, *14*, 965–980.

10. Xiang, Z.Y.; Sun, B.; Dai, X. The camera itself as a calibration pattern: A novel self-calibration method for non-central catadioptric cameras. *Sensors* **2012**, *12*, 7299–7317.

11. Heikkila, J.; Silven, O. A Four-step Camera Calibration Procedure with Implicit Image Correction. In Proceedings of 1997 IEEE Computer Society Conference on Computer Vision and Pattern Recognition, San Juan, Argentina, 17–19 June 1997; pp. 1106–1112.

12. Liu, H.B.; Li, X.J.; Tan, J.C.; Yang, J.K.; Yang, J.; Su, D.Z.; Jia, H. Novel approach for laboratory calibration of star tracker. *Opt. Eng.* **2010**, *49*, doi:10.1117/1.3462042.

13. Gwanghyeok, J. Autonomous Star Sensing, Pattern Identification, and Attitude Determination for Spacecraft: An Analytical and Experiment Study. Ph.D. thesis; Texas A&M University: Texas, TX, USA, 2001.

14. Wahba, G. A least squares estimate of satellite attitude. *SIAM Rev.* **1996**, *8*, 384–386.

15. Liebe, C.C. Star trackers for attitude determination. *IEEE Aerosp. Electron. Syst. Mag.* **1995**, *10*, 10–16.

16. Yang, J.; Liang, B.; Zhang, T.; Song, J.Y. A novel systematic error compensation algorithm based on least squares support vector regression for star sensor image centroid estimation. *Sensors* **2011**, *11*, 7341–7363.

17. Roth, J.P. Diagnosis of automata failures: A calculus and method. *IBM J. Res. Dev.* **1966**, *10*, 278–291.

Reprinted from *Sensors*. Cite as: Fasano, G.; Rufino, G.; Accardo, D.; Grassi, M. Satellite Angular Velocity Estimation Based on Star Images and Optical Flow Techniques. *Sensors* **2013**, *13*, 12771–12793.

Article

Satellite Angular Velocity Estimation Based on Star Images and Optical Flow Techniques

Giancarmine Fasano *, Giancarlo Rufino, Domenico Accardo and Michele Grassi

Department of Industrial Engineering (DII), University of Naples "Federico II", P.le Tecchio 80, Naples I80125, Italy; E-Mails: giancarlo.rufino@unina.it (G.R.); domenico.accardo@unina.it (D.A.); michele.grassi@unina.it (M.G.)

* Author to whom correspondence should be addressed; E-Mail: giancarmine.fasano@unina.it; Tel.: +39-081-7682-361; Fax: +39-081-7682-160.

Received: 20 July 2013; in revised form: 13 September 2013 / Accepted: 17 September 2013 / Published: 25 September 2013

Abstract: An optical flow-based technique is proposed to estimate spacecraft angular velocity based on sequences of star-field images. It does not require star identification and can be thus used to also deliver angular rate information when attitude determination is not possible, as during platform de tumbling or slewing. Region-based optical flow calculation is carried out on successive star images preprocessed to remove background. Sensor calibration parameters, Poisson equation, and a least-squares method are then used to estimate the angular velocity vector components in the sensor rotating frame. A theoretical error budget is developed to estimate the expected angular rate accuracy as a function of camera parameters and star distribution in the field of view. The effectiveness of the proposed technique is tested by using star field scenes generated by a hardware-in-the-loop testing facility and acquired by a commercial-off-the shelf camera sensor. Simulated cases comprise rotations at different rates. Experimental results are presented which are consistent with theoretical estimates. In particular, very accurate angular velocity estimates are generated at lower slew rates, while in all cases the achievable accuracy in the estimation of the angular velocity component along boresight is about one order of magnitude worse than the other two components.

Keywords: spacecraft angular velocity estimation; star field images; optical flow; performance analysis; hardware-in-the-loop simulation

1. Introduction

Spacecraft requiring accurate three-axis attitude control are all equipped with star sensors to support attitude determination with high accuracy. In recent years, star tracker technology has seen a remarkable evolution. In particular, these sensors have gained significant improvements in their autonomy and capabilities [1–3]. Indeed, modern star sensors are expected to offer new advanced functionalities in addition to the assessed capability of high-precision pointing determination during low angular rate mission phases. The ultimate goal in modern star sensor design is achieving performance, functionality, and reliability levels that allow star sensors to be the only attitude sensor on-board the spacecraft [4]. In particular, the following advanced functionalities can be cited as characterizing modern star sensors:

- to produce high-accuracy, high-reliability attitude angle and rate estimates without external support;
- to operate in a wide range of mission conditions;
- to solve the lost-in-space problem autonomously and in a short time;
- to deliver angular rate information also when attitude determination is not feasible, as during platform de tumbling or slewing.

These functionalities should be achieved via additional software routines rather than by hardware enhancements (apart from improved sensitivity of photodetectors), and different operating modes should control sensor operation. As a result, software for system control and management becomes very complex.

Among the cited advanced functionalities, one of the most demanding, in terms of algorithm and software complexity and sensor operation management, is the determination of the satellite inertial angular velocity during slewing and/or de-tumbling phases. Indeed, many existing satellites execute slewing maneuvers at rates lower than 1°/s, at which the star sensor is still able to acquire star field images, so that star centroids can be computed on the focal plane. Instead, higher angular rates (>1°/s) are being proposed for high-agility small satellites and next generation Earth Observation satellites [5]; in this case the stars are typically acquired as strips, thus calling for different algorithms to be used for angular rate computations.

On the other hand, there is a growing interest in systems able to propagate attitude of very small satellites (such as CubeSats) using low cost sensors and optics [6] (no star trackers available), in order to maintain accurate attitude estimates during eclipse avoiding the drift that characterizes gyroscopes.

In this paper a technique for angular rate determination based on optical flow computation is analyzed. Besides being adopted for vision-based guidance and control of Unmanned Aircraft Systems, optical flow techniques have found usage in space applications within the fields of remote sensing and space exploration. Regarding spaceborne remote sensing, optical flow measurements have been used for example to estimate glacier motion from multi-temporal sequences of electro-optical (EO) images [7], to detect sandstorms [8], to estimate atmospheric motion from geostationary meteorological satellites [9]. Within space exploration, optical flow approaches have been widely proposed for planetary landing (see for example [10,11]).

The optical flow technique proposed in the paper relies on the computation of a displacement field between successive star images, then a least squares method is used to find the best estimate of the angular velocity vector components in the rotating frame matching the observed displacement field. The effectiveness of the proposed techniques is tested by using star field scenes reproduced by an indoor testing facility and acquired by a commercial-off-the shelf camera sensor, shortly described in the paper. Specifically, star field scenes relevant to representative satellite slewing maneuvers are simulated. Then the corresponding images are processed with the optical flow algorithm in order to extract the angular rate information. This information is then compared with the one used in input to the testing facility.

Satellite angular rates estimation, independent of star identification and attitude measurement, has been also discussed in [12] and more recently in [6,13,14].

In particular, [13] discusses a technique that (unlike the one presented in this work) is applicable to electronic rolling shutter imaging mechanisms, since it is aimed at compensating distortion effects due to this technology, thus improving centroiding accuracy and attitude measurement performance in nominal conditions.

In [6] the q-method [15] is used to solve the relative attitude problem between successive frames, while [12] refers to Poisson relation as the basic algorithm equation. [14] illustrates an angular velocity technique based on a least squares approach that starts from knowledge of star vectors and the time sampling interval, and focuses on dynamic estimation techniques such as adaptive Kalman filtering. Validation is based on numerical simulations and night sky observations.

With regards to these latter works, the work presented in this paper provides the following original contributions:

- the entire angular velocity measurement process is presented comprising accurate and efficient optical flow computation and relation with algorithm tuning;
- a complete theoretical error budget is developed that allows predicting the expected measurement accuracy as a function of camera and geometric parameters;
- the developed methodology is tested in hardware-in-the-loop simulations of representative satellite slewing maneuvers.

The paper is organized as follows: Section 2 describes the adopted algorithm with a preliminary error budget to estimate the expected angular accuracy, then Sections 3 and 4 describe, respectively, the adopted indoor facility and the simulation scenario, and the results of the algorithm test on star field scenes acquired with the laboratory facility.

2. Algorithm

The developed algorithm is composed of a few basic steps: given a couple of subsequent star field images, first the acquired images are pre-processed to eliminate background noise, and the velocity vector field (which is indeed a displacement field) is calculated in pixels. Then, unit vectors and unit vector derivatives corresponding to the computed velocity vectors are evaluated by exploiting a neural network calibration to estimate at the same time intrinsic and extrinsic parameters relevant to the adopted experimental setup. Once unit vectors and their derivatives are known, the Poisson's equation expressing the time derivative of a unit vector in a rotating reference frame and a least square method are used to find the best estimate of the angular velocity vector components in the rotating frame.

The above mentioned process is summarized in Figure 1. The different blocks are described in details in the following sub-sections, with particular regard to the adopted optical flow methodologies and the equations used for estimating the angular velocity.

Figure 1. Algorithm flow-chart.

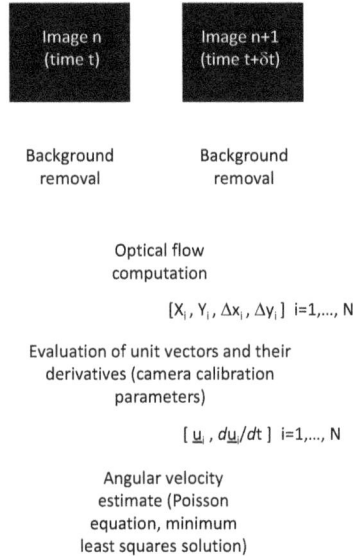

Image n
(time t)

Image n+1
(time t+δt)

Background
removal

Background
removal

Optical flow
computation

$[X_i, Y_i, \Delta x_i, \Delta y_i]$ i=1,..., N

Evaluation of unit vectors and their
derivatives (camera calibration
parameters)

$[\underline{u}_i, d\underline{u}_i/dt]$ i=1,..., N

Angular velocity
estimate (Poisson
equation, minimum
least squares solution)

2.1. Image Processing and Optical Flow Computation

Given a couple of consecutive grey level images, first of all a background noise removal process is carried out separately on both images to eliminate sensor noise which can affect accuracy of optical flow computation. To this end, a global threshold technique [16] is applied in which a $\mu + 3\sigma$ threshold is applied to identify the illuminated pixels, with μ and σ being, respectively, the intensity mean and standard deviation computed over the entire image. All the pixels with intensity below the noise threshold are set equal to zero. This processing may slightly affect centroid accuracy in dynamic conditions when stars are spread over several pixels and the signal-to-noise ratio is degraded, as it will be discussed in the following when dealing with results from high rate simulations.

An example of background noise removal process around a star is reported in Figure 2. After background noise removal, a labeling technique [16] is applied to distinguish the different stars detected on the focal plane. Within this phase, stars whose dimension is smaller than three pixels are discarded to increase algorithm accuracy, as it is better explained in the error budget sectiosn. It is important to underline that all the subsequent calculations are applied only to the detected stars and not to the whole image. This thresholding procedure significantly reduces the computational burden of optical flow techniques, which is very important in view of real time applications. In fact, modern, multifunction star trackers with large-medium size field of view (FOV, e.g., 15° to 20°) and capable of autonomous multi-mode operation have a detection limit up to visible magnitude m_v of

6–6.5. Assuming as reference a 20°-FOV and $m_v = 6.2$ as detection limit, the resulting average number of detectable stars in the sensor FOV is 40 [17].

Figure 2. Background noise removal process (pseudo colors are used for the sake of clarity).

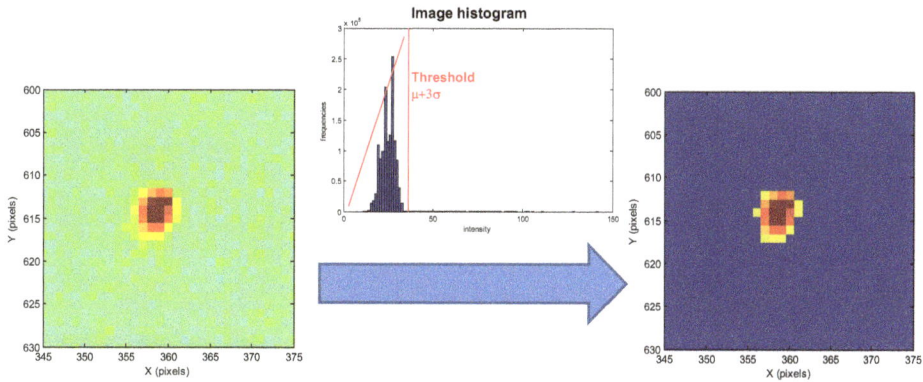

In general, the optical flow is the 2-D motion field, which is the perspective projection onto the image plane of the true 3-D velocity field of moving surface in space [18,19], arising from the relative motion between the surface and the viewer.

The basic assumption in measuring the image motion is that the intensity structures of local time-varying image regions are approximately constant for, at least, a short time duration. The classical "optical flow constraint equation" [20] can be expressed in differential terms as follows:

$$\frac{\partial I}{\partial x}V_x + \frac{\partial I}{\partial y}V_y + \frac{\partial I}{\partial t} = 0 \tag{1}$$

where I represents the image intensity, x and y the two spatial coordinates in the image, V_x and V_y the corresponding apparent velocity components, and t is time.

Different approaches can be adopted to compute optical flow [20–22] such as differential techniques, phase-based and energy based methods, and region-based matching.

Differential techniques compute velocity from spatiotemporal derivatives of image intensity or filtered version of the images (using low pass or band pass filters). In this framework, Equation (1) is an under-constrained equation, since only the motion component in the direction of the local gradient of the image intensity function may be estimated: this is known as "the aperture problem" [20] and one more assumption is necessary.

As an example, Horn and Schunck's method assumes that the motion field is smooth over the entire image domain and tries to maximize a global smoothness term [20], while Lucas and Kanade's method (first introduced in [22] and then developed into the most implemented tracking algorithms [23–25]) divides the original image into smaller sections, assumes a constant velocity in each section, and performs a weighted least-square fit of the optical flow constraint equation, to a constant model for the velocity field in each section.

Differential techniques are not the best solution in the considered case for several reasons. First of all, after background removal, images are very sparse, with a few non zero pixels and a significant departure from the smoothness properties these techniques are based on. Thus, accurate numerical differentiation is typically unachievable. This also happens if background removal is not applied because of the negative impact of noise. Then, it has to be considered that if a very high resolution camera is used, *i.e.*, with a very small Instantaneous FOV (IFOV, *i.e.*, the angle subtended by a single pixel of the imaging system) as it typically happens for a star tracker, apparent star motion can be of several pixels per frame even during medium-rate rotations, while differential techniques typically work well for apparent velocities of the order of 1 pixel/frame, at most. Coarse to fine pyramid representations can be used [24], but with high computational cost since they should be carried out over the entire image, and with degraded performance because of the very sparse image structure.

Since phase-based and energy-based methods work in the Fourier domain, in the star sensor case they also suffer from the same problems of differential techniques.

Region based matching is, instead, an appealing solution because it works well even in noisy images without smooth intensity patterns, and in case of large pixel velocities, such as the ones we have to work with.

The basic principle is to evaluate velocity as the displacement that yields the best fit between image regions at different times. Specifically, in the considered application, a customized two-step method is adopted in which a coarse estimate of the star displacement on the focal plane is computed first and then refined to improve accuracy in the velocity field estimate:

- First of all, the integer shift in pixels ($\underline{d}$) is computed for each star that minimizes over $\underline{\delta}$ the sum of squared differences:

$$SSD(x,y,\underline{\delta}) = \sum_{j=-k}^{k}\sum_{i=-k}^{k}\left[I_n(x+i,y+j)-I_{n+1}(x+i+\delta_x,y+j+\delta_y)\right]^2 \qquad (2)$$

As before, I_n and I_{n+1} indicate two consecutive star images. The sum is calculated on a window whose center is the star centroid calculated in the first image (whose coordinates are x and y) and whose dimensions (*i.e.*, k) depend on the maximum foreseen star dimensions, while δ has to vary in an interval which depends on the maximum measurable star displacement. These are the basic parameters for algorithm tuning, and the computational burden of the algorithm increases for larger angular velocities to be measured;

- The coarse estimate of $\underline{d}$ is then refined by computing in the second image the centroid of a window centered at the coarse estimation, whose size and shape are the same of the considered star, plus a margin of 2 pixels. This margin is used to ensure that all the pixels of the considered star (whose intensity is above the threshold) are used for centroid computation in the second image. In fact, the coarse centroid computation has an intrinsic accuracy of 1 pixel due to the integer nature of the solution, and one more pixel is considered as a "safety margin". This second step is customized to the considered application. It allows a very precise determination of $\underline{d}$ with a very small increase of the computational weight, as it needs very few pixels to be further processed.

The two steps are repeated for each star detected and labeled in the first image. Once star displacements are determined, the information can be easily translated in a velocity information (in pixels) by taking the frame rate into account. Within this framework, it is assumed that accurate image timing is available, thanks to the adoption of proper hardware (camera and shutter technique) and software (real time operating systems and proper coding of image acquisition).

2.2. Angular Velocity Estimation

Once star centroids and vector displacements between two consecutive frames are known, the subsequent step is to convert this information in unit vectors and their derivatives. This has to take camera calibration parameters into account and can be done in different ways.

For example, a classical calibration procedure can be used to estimate, firstly, camera intrinsic parameters to be used in a pinhole camera model plus distortion effects (e.g., focal length, optical center, radial and tangential distortion, *etc.*) [16,26], and, then, the extrinsic parameters relevant to the test facility (*i.e.*, the translation vector from camera optical center to a point on the LCD screen assumed as the origin of the display reference frame, and the rotation matrix that relates camera reference frame to the axes of the display reference frame).

In the considered case, an end-to-end neural-network-based calibration procedure is used, which correctly takes account of all the intrinsic and extrinsic parameters relevant to the camera and the test facility [27,28].

Once unit vectors and their derivatives are known, angular velocity estimation is based on the Poisson equation, that relates the temporal derivatives of the stars unit vectors in the Inertial Reference Frame (IRF) and in the Star sensor Reference Frame (SRF):

$$\frac{\partial \underline{u}}{\partial t}\bigg|_{IRF} = 0 = \frac{\partial \underline{u}}{\partial t}\bigg|_{SRF} + \underline{\omega} \wedge \underline{u} \tag{3}$$

where we take into account that stars are fixed in the IRF, $\underline{u}$ is the star unit vector, and $\underline{\omega}$ represents the angular velocity of the SRF with respect to the IRF.

Equation (3) can be rewritten through the vectors components in the SRF as:

$$\begin{bmatrix} 0 & u_{3s} & -u_{2s} \\ -u_{3s} & 0 & u_{1s} \\ u_{2s} & -u_{1s} & 0 \end{bmatrix} \begin{bmatrix} \omega_{1s} \\ \omega_{2s} \\ \omega_{3s} \end{bmatrix} = - \begin{bmatrix} \dfrac{\partial u_{1s}}{\partial t} \\ \dfrac{\partial u_{2s}}{\partial t} \\ \dfrac{\partial u_{3s}}{\partial t} \end{bmatrix} \tag{4}$$

Thus, three non independent linear equations (in three unknown variables) can be written for each star, leading to $N \times 3$ linear equations if N is the number of stars for which the optical flow has been calculated.

These $N \times 3$ equations can be solved in ω by a classical minimum-least-squares technique based on orthogonal-triangular decomposition, which is computationally light thanks to the sparse structure of the problem matrix. Once the solution for ω is obtained, measurement residuals can be calculated to detect anomalous values and thus to have a first assessment of the method reliability.

250

2.3. Performance Analysis

A theoretical analysis can be carried out to derive a first order error budget for the selected technique. The input parameters for the error budget are: the angular resolution of the considered sensor, the angular velocity to be measured and the consequent velocity field pattern of the stars, the attitude of the SRF with respect to the inertial reference frame (which determines the star distribution within the camera field of view), and the number of detected stars (which depends on star sensor sensitivity and, again, on sensor attitude).

Equation (4) can be rewritten as:

$$
\begin{bmatrix}
\omega_{2s}u_{3s} - \omega_{3s}u_{2s} \\
\omega_{3s}u_{1s} - \omega_{1s}u_{3s} \\
\omega_{1s}u_{2s} - \omega_{2s}u_{1s}
\end{bmatrix}
= -
\begin{bmatrix}
\dfrac{\partial u_{1s}}{\partial t} \\
\dfrac{\partial u_{2s}}{\partial t} \\
\dfrac{\partial u_{3s}}{\partial t}
\end{bmatrix}
\tag{5}
$$

With reference to Figure 3 let us introduce the angles ϕ and θ that define the star line of sight orientation in SRF: θ is the elevation angle over the X_s,Z_s plane of the star line-of sight, and ϕ is the angular separation from the sensor boresight Z_s of its projection on X_s,Z_s. In addition, we define χ as the angle of the generic star line-of-sight with respect to the sensor boresight axis.

Figure 3. Definition of the generic star angles in SRF: the star line-of-sight is in red, Z_s is the sensor boresight axis.

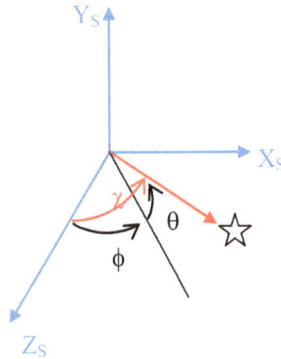

The error analysis can be carried out separately for the different components of the angular velocity in SRF (ω_{1s}, ω_{2s}, ω_{3s}). Let us first consider ω_{1s}, *i.e.*, the case under analysis is $\omega_{2s} = \omega_{3s} = 0$, $\omega_{1s} \neq 0$. In this case Equation (5) reduces to:

$$
\begin{bmatrix}
0 \\
-\omega_{1s}u_{3s} \\
\omega_{1s}u_{2s}
\end{bmatrix}
= -
\begin{bmatrix}
\dfrac{\partial u_{1s}}{\partial t} \\
\dfrac{\partial u_{2s}}{\partial t} \\
\dfrac{\partial u_{3s}}{\partial t}
\end{bmatrix}
\tag{6}
$$

The unit vector components can be written in terms of the ϕ and θ angles. Since star sensors typically have small FOVs, we can apply the small angle approximation thus getting:

$$\begin{cases} u_{1s} = \cos\theta\sin\phi \approx \phi \\ u_{2s} = \sin\theta \approx \theta \\ u_{3s} = \cos\theta\cos\phi \approx 1 \end{cases} \tag{7}$$

And from Equation (6):

$$\begin{bmatrix} 0 \\ -\omega_{1s} \\ \omega_{1s}\theta \end{bmatrix} \cong - \begin{bmatrix} \dot{\phi} \\ \dot{\theta} \\ \dfrac{\partial u_{3s}}{\partial t} \end{bmatrix} \tag{8}$$

Then, we can relate the ϕ and θ rate of change directly to the star displacement on the focal plane:

$$\dot{\phi} \cong \frac{x_c}{f} \tag{9}$$

$$\dot{\theta} \cong \frac{y_c}{f} \tag{10}$$

where f is the sensor focal length and x_c and y_c are the coordinates on the focal plane of the generic star centroid.

Thus, we get the final approximate relation in which the first component of the inertial angular velocity vector is directly related to the velocity component along the y_s axis computed by means of the optical flow techniques and expressed as an angular velocity:

$$\omega_{1s} \cong \dot{\theta} \cong \frac{\dot{y_c}}{f} \cong V_y \tag{11}$$

Equation (11) allows us to derive the error budget for ω_{1s}. In what follows, we use x and y as non-dimensional coordinates, $i.e.$, they are calculated as $x = \dfrac{x_c}{f}$ and $y = \dfrac{y_c}{f}$.

From a numerical point of view:

$$V_y \cong \frac{y_{n+1} - y_n}{\Delta t} \tag{12}$$

where n and $n + 1$ refer to two generic successive frames, Δt is the time elapsed which is inversely proportional to sensor frame rate. Thus we have, for a single star:

$$\sigma_{V_y} \cong \frac{\sqrt{2}}{\Delta t}\sigma_y \cong \frac{\sqrt{2}}{\Delta t} \cdot \left(\frac{IFOV}{\sqrt{N_{starpixels}}} \right) \tag{13}$$

where $N_{starpixels}$ is the number of pixels of the focal plane collecting the radiation from the generic star. The term between brackets in Equation (13) approximates the actual accuracy of the centroiding operation.

Since ω_{1s} represents a rotation around an axis perpendicular to the sensor boresight, the corresponding velocity field measured on the focal plane is uniform, *i.e.*, it does not depend on the distance from the boresight axis. Thus, if N is the number of detected stars, since the number of pixels of the different stars is more or less the same, we can produce an estimate of ω_{1s} by combining N identical, and identically distributed, measurements of Vy. Thus the uncertainty in ω_{1s} does not depend on the star position in the FOV and it can be estimated as:

$$\sigma_{\omega_{1s}} \cong \frac{\sigma_{V_y}}{\sqrt{N}} \cong \frac{\sqrt{2}}{\Delta t} \cdot \frac{IFOV}{\sqrt{N}\sqrt{N_{starpixels}}} \tag{14}$$

Assuming realistic values for the frame rate (10 Hz), the number of pixels per star (10), and the number of detected stars (40), we get the uncertainty in ω_{1s} as a function of camera IFOV presented in Figure 4. It can be seen that within the considered range for camera IFOV, the uncertainty in ω_{1s} goes from about 0.0035°/s to about 0.035°/s.

Figure 4. Approximate theoretical uncertainty in ω_{1s} estimate as a function of sensor IFOV.

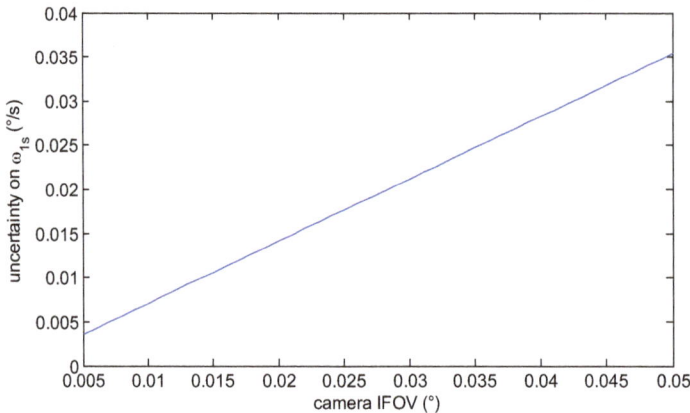

Uncertainty in ω_{2s} can be estimated exactly in the same way, and the error budget is identical since azimuth and elevation IFOVs usually coincide. It is worth noting that the estimated uncertainty does not depend on the angular rotation value which produced the observed velocity field. Of course this conclusion relies on the validity of the proposed model, depending on the assumption that the slew rate is small enough so that a star-field image can be imaged on the focal plane in the considered subsequent images.

The error budget in ω_{3s} is somewhat different. Combining Equations (5) and (7) in the case $\omega_{1s} = \omega_{2s} = 0$, $\omega_{3s} \neq 0$, and with the small angles assumption, we get:

$$-\omega_{3s}\theta \cong -\frac{\partial u_{1s}}{\partial t} \tag{15}$$

$$\omega_{3s}\phi \cong -\frac{\partial u_{2s}}{\partial t} \qquad (16)$$

$$\frac{\partial u_{3s}}{\partial t} \cong 0 \qquad (17)$$

Combining Equations (15) and (16), and taking Equations (7) into account, we get:

$$\omega_{3s}\sqrt{\phi^2 + \theta^2} \cong \sqrt{\dot{\theta}^2 + \dot{\phi}^2} \qquad (18)$$

The first term can be further developed by using spherical trigonometry. Indeed, with reference to Figure 3 we have:

$$\cos \chi = \cos \phi \cos \theta \qquad (19)$$

where χ is the angle between the direction to the generic star and the sensor boresight axis. Then:

$$\sin^2 \chi = \sin^2 \theta + \sin^2 \phi - \sin^2 \phi \sin^2 \theta \qquad (20)$$

From the small angle assumption we get:

$$\chi^2 \cong \theta^2 + \phi^2 \qquad (21)$$

Thus, from Equation (18) we get:

$$\omega_{3s}\chi \cong \sqrt{\dot{\theta}^2 + \dot{\phi}^2} \qquad (22)$$

The χ angle obviously depends on the observed star, and its maximum value depends on the FOV size.

Equation (18) can be rewritten by using the non-dimensional coordinates x and y as follows:

$$\omega_{3s} \cong \frac{\sqrt{V_x^2 + V_y^2}}{\sqrt{x^2 + y^2}} \qquad (23)$$

and the uncertainty in ω_{3s} can be then calculated at a first order, and for a single star, as:

$$\sigma^2_{\omega_{3s}} \cong \left(\frac{\partial \omega_{3s}}{\partial x}\right)^2 \sigma^2_x + \left(\frac{\partial \omega_{3s}}{\partial y}\right)^2 \sigma^2_y + \left(\frac{\partial \omega_{3s}}{\partial V_x}\right)^2 \sigma^2_{V_x} + \left(\frac{\partial \omega_{3s}}{\partial V_y}\right)^2 \sigma^2_{V_y} \qquad (24)$$

By developing the different terms we get:

$$\sigma^2_{\omega_{3s}} \cong \left(-\frac{xV}{\chi^3}\right)^2 \sigma^2_x + \left(-\frac{yV}{\chi^3}\right)^2 \sigma^2_y + \left(\frac{V_x}{V\chi}\right)^2 \sigma^2_{V_x} + \left(\frac{V_y}{V\chi}\right)^2 \sigma^2_{V_y} \cong$$

$$\cong \left(-\frac{V}{\chi^2}\right)^2 \sigma^2_x + \left(\frac{1}{\chi}\right)^2 \sigma^2_{V_x} \qquad (25)$$

where:

$$V = \sqrt{V_x^2 + V_y^2}$$

and it has been assumed that:

$$\sigma^2{}_x = \sigma^2{}_y$$
$$\sigma^2{}_{V_x} = \sigma^2{}_{V_y}$$

By using Equations (12) and (13) we finally have:

$$\sigma^2{}_{\omega_{3s}} \cong \left(-\frac{V}{\chi^2}\right)^2 \frac{IFOV^2}{N_{starpixels}} + \left(\frac{1}{\chi}\right)^2 \frac{2 \cdot IFOV^2}{\Delta t^2 N_{starpixels}} = \left(-\frac{\omega_{3s}}{\chi}\right)^2 \frac{IFOV^2}{N_{starpixels}} + \left(\frac{1}{\chi}\right)^2 \frac{2 \cdot IFOV^2}{\Delta t^2 N_{starpixels}} \qquad (26)$$

For the typically encountered angular velocities and high frame rates (10 Hz or more), the second term in the above equation is larger than the first one, which yields the following approximate form of the uncertainty in ω_{3s} for a single star :

$$\sigma_{\omega_{3s}} \cong \left(\frac{1}{\chi}\right) \frac{\sqrt{2} \cdot IFOV}{\Delta t \sqrt{N_{starpixels}}} = \left(\frac{1}{\chi}\right) \sigma_{V_x} \qquad (27)$$

Equation (27) shows the very intuitive result that the uncertainty in the estimation of the apparent velocity affects the estimation of angular velocity in a way which depends on the star position in the field-of-view: the farther the star line-of-sight is from the boresight, the more accurate the angular velocity estimate will be for a given optical flow uncertainty.

The final ω_{3s} estimate is obtained by combining star measurements having different error distribution. However, a preliminary estimate of the ω_{3s} uncertainty can be obtained by taking an average value of χ and using again the factor $\dfrac{1}{\sqrt{N}}$. Thus we get:

$$\sigma_{\omega_{3s}} \cong \left(\frac{1}{\overline{\chi}}\right) \frac{\sqrt{2} \cdot IFOV}{\Delta t \sqrt{N} \sqrt{N_{starpixels}}} = \left(\frac{1}{\overline{\chi}}\right) \sigma_{\omega_{1s}} \qquad (28)$$

Considering an average value of 5° for χ (realistic considering typical medium-large size FOVs) we get that the achievable accuracy is about one order of magnitude worse than the one attainable for ω_{1s}. This is also consistent with the usual difference existing between the attitude measurement uncertainties across and along the boresight axis of a star sensor [17]. Assuming again a frame rate of 10 Hz, an average number of 10 pixels per star, and 40 detected stars, in Figure 5 we get the uncertainty in ω_{3s} as a function of camera IFOV. Of course, the actual estimation uncertainty depends on the distribution of detected stars within the sensor FOV, and thus also on the actual attitude of the satellite.

Figure 5. Theoretical uncertainty in ω_{3s} as a function of sensor IFOV.

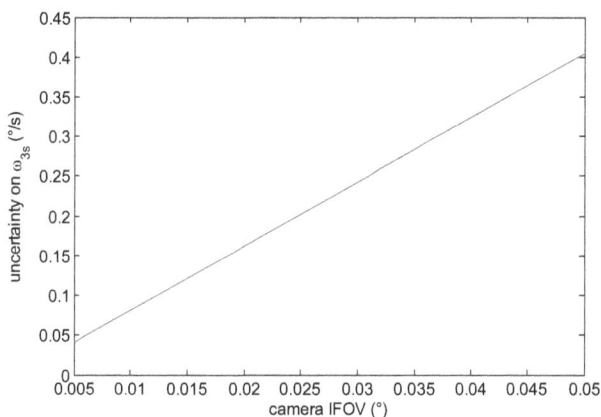

3. Hardware-in-the-Loop Facility

Tests for performance assessment of the discussed procedure were carried out by means of a functional, hardware prototype of star sensor operated in a laboratory facility for star field scene simulation.

The star sensor prototype was designed to implement the operational modes suggested by the European Space Agency [29]: autonomous operation, initial acquisition from lost-in-space state, attitude tracking, cartography mode for in-depth operation monitoring. It was realized by using COTS hardware: MATROX IRIS P1200HR System [30] is the hardware basis while sensor algorithms were developed in-house. The IRIS P1200HR is composed of two separate units: camera head and a compact embedded CPU which makes this camera fully programmable (it is a so called "smart sensor"). The former exploits SONY CCD detector and focal plane electronics, the latter is based on a 400-MHz Intel Celeron processor equipped with 128-MB RAM, 128-MB flash disk, Microsoft Windows CE 5.0 operating system. Camera head is connected to the processor unit by means of a standard Camera Link[TM] cabling. Main sensor specifications are in Table 1. Sensor algorithms and the relevant performance are discussed in the literature [31,32].

Table 1. Star sensor prototype specifications.

Field Of View	$22.48° \times 17.02°$
Effective Focal Length	16 mm
F-number	1.4
Star Sensitivity	<visible magnitude 7
Image Sensor	½" CCD Progressive Scan
Image Size	$1,280 \times 1,024$ pixel
Instantaneous Field Of View	$0.017° \times 0.017°$

The laboratory test facility (Figure 6) consists of a dark room where a high-resolution, computer-controlled LCD display produces star field scenes as computed on the basis of a star catalog and of assigned star sensor orientation [27,28]:

- a single pixel of the LCD screen is exploited to simulate a single star of a star field if a static pointing is considered or in the case of a low-rate dynamics of the orbiting platform. Differently, when high-rate attitude dynamics are accounted for in the simulation, a single star is represented by the strip of pixels reproducing its apparent trajectory in the sensor FOV during the update time of the displayed star field scene. Pixel brightness control is used to reproduce star apparent brightness. Approximations result in this simulation approach as a consequence of spatial, temporal, and pixel brightness digital discretization of the synthetic star field scenes and relevant sequences. However a theoretical, worst-case analysis [27] showed that, for high rate dynamic rotation simulation, approximation on large velocity components is at most of the order of 0.01°/s, taking into account the typical number of simulated stars. As it is shown in the following, this does not represent a significant artificial contribution to the estimated algorithm accuracy;
- a collimating lens allows for simulating the large distance of the star sensor from light source;
- a high-performance video processor is adopted for LCD display control by an embedded computer, to carry out static but also dynamical simulations. The former ones simply consist of sequences of star field scenes, as resulting from assigned sensor attitude. The latter ones reproduce the evolution of the star field observed by the sensor during assigned maneuvers (orbit and/or attitude dynamics), with accurate timing;
- sensor position within the darkroom and collimating lens selection guarantee matching of instrument FOV and LCD apparent angular size. Micro translators and rotators are used for fine regulation and alignment of sensor orientation and facility intrinsic reference frame, *i.e.*, the display;
- finally, precise matching is software-based. In particular, it is realized by means of a neural calibration function used to compensate for residual misalignment after installation in the darkroom, and to adjust sensor output to LCD star angular position finely [27,28,33]. This neural network is trained on the basis of a preliminary set of acquisitions to obtain accordance between input star field and sensor position measurements.

Figure 6. Laboratory facility set-up for star field simulation and star sensor tests.

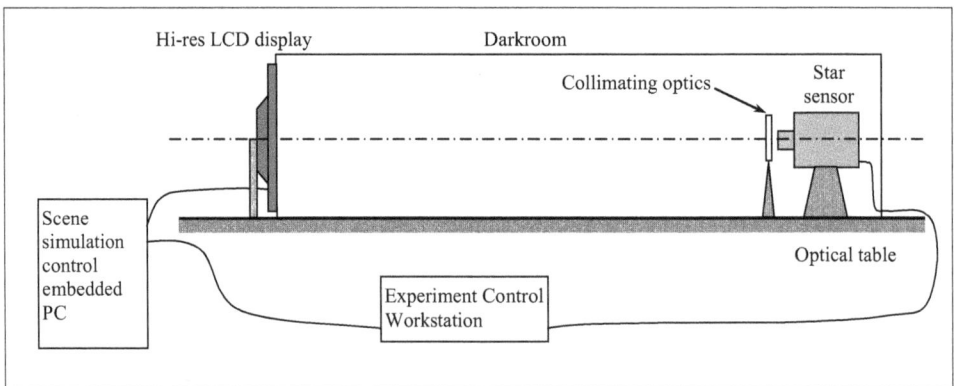

The above hardware is completed by the Experiment-Control Workstation that coordinates simulation and sensor operation during test, and it also generates the needed simulated star field data, off-line before star sensor testing.

Table 2 shows the main feature of the system when it is specialized to be coupled to the star sensor prototype in use.

Table 2. Test facility features relevant to sensor FOV match.

Display active area H × V (m)	0.641 × 0.401
Display resolution H × V (pixel)	2,560 × 1,600
Collimating lens focal length (m)	1.3
Collimator diameter (mm)	50
Display apparent angular size (deg)	27.6 (H) × 17.5 (V)
Display pixel apparent angular size at screen centre (deg)	0.011 × 0.011 (H × V)
Overall magnification ratio (with 16-mm-focal sensor optics)	1.23×10^{-2}

4. Simulation Results

Accuracy and reliability of the proposed method can be evaluated by exploiting the described hardware-in-the-loop facility. In all the simulated cases, a circular equatorial Low Earth Orbit (LEO) at altitude of 500 km is considered. This choice does not compromise the general validity of the results since a wide range of attitude maneuvers is simulated to evaluate the effect of different star image patterns on method accuracy. Initially, the satellite body reference frame (BRF) is supposed to coincide with the classically defined orbital reference frame (ORF), *i.e.*, the axis 1 is along the orbital velocity direction, the axis 2 is anti-parallel to the orbital angular momentum vector, and the axis 3 is in the nadir direction. In all the considered cases, the SRF also initially coincides with the BRF apart from sign conventions. In fact, the axis Ys coincides with the axis 2, whereas the other two axes have opposite directions.

Figure 7. Reference frames for the considered simulations.

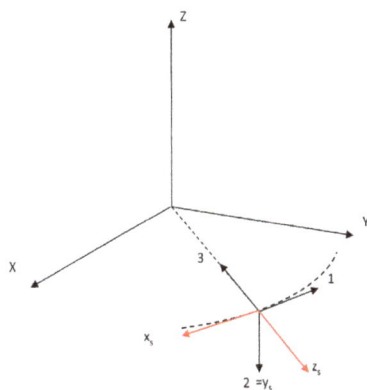

The SRF is thus obtained from the BRF by a 180°-rotation around the axis 1. As a consequence, the star sensor boresight axis initially points in zenith direction in the equatorial plane. The reference frames used for the simulations are depicted in Figure 7, with IRF origin at the Earth's centre.

The simulated cases differ for the considered attitude maneuvers. In the first two cases (case 1 and case 2) a satellite rotation around the 1 axis with constant angular velocity (1 deg/s in case 1, 5 deg/s in case 2) is superimposed to the constant angular velocity of the keplerian orbit (6.2430^{-2} deg/s along the negative 2 axis initially) so that the star sensor boresight axis moves outside the equatorial plane towards the North pole while the satellite rotates around Earth. This condition allows evaluating method performance with a varying number of detected stars and an almost uniform apparent velocity field on the focal plane (pure translation).

In the other two cases (case 3 and case 4), the satellite is supposed to rotate around the star sensor boresight axis, again with constant angular velocity (1 deg/s in case 3, 5 deg/s in case 4). This condition is representative of the case in which the velocity field on the focal plane is not uniform (pure rotation). Actually, a small translational component due to the orbital angular velocity is present in the acquired images.

The simulated angular rates are relevant to the slew maneuvers of many existing satellites, which are typically executed at rates lower than 1°/s. In this condition, the star sensor is able to acquire star field images, and star centroids can be computed on the focal plane. Higher angular rates (>1°/s) are instead proposed for high-agility, small satellites, and next generation Earth Observation satellites [4]. In this case, the stars are typically acquired as strips. This condition can affect the accuracy of the proposed technique.

For reader convenience, all the simulated cases are summarized in Table 3. It is worth recalling that the reported "true" angular velocity components (ω_{1s}, ω_{2s}, and ω_{3s}) represent the components along the SRF axes of the inertial angular velocity vector of the SRF.

Table 3. Summary of simulated test cases: initial conditions.

	Out of Plane Rotation		Radial Rotation	
	Case 1	Case 2	Case 3	Case 4
ω_{1s} (°/s)	1	5	0	0
ω_{2s} (°/s)	$-6.243 \cdot 10^{-2}$	$-6.243 \cdot 10^{-2}$	$-6.243 \cdot 10^{-2}$	$-6.243 \cdot 10^{-2}$
ω_{3s} (°/s)	0	0	-1	-5

4.1. Out of Plane Rotation Results

In this case, initially the true angular velocity vector has non-zero components only along the x_s and y_s axes of SRF. As a consequence, the velocity field pattern represents a pure translation with a larger components along the y_s axis. This condition is evident in Figure 8, where the velocity vectors calculated from a couple of consecutive frames in case 1 are depicted (magnified for the sake of clarity). In spite of some noise affecting more the (smaller) horizontal velocity component, the uniformity of the velocity field can be clearly appreciated. In the considered case, pixel displacements are of the order of 0.4 pixels for the horizontal component and 5.9 pixels for the vertical component.

As a result of the relatively large number of detected stars, and velocity vectors, both the larger x_s component (1 deg/s) and the smaller y_s component (0.06 deg/s) are measured with good accuracy, as shown in Figure 9. The described algorithm was run on a sequence of about 100 images, corresponding to a simulation time span of about 10 s. It can be seen that the measurements are unbiased on average, and the measurement noise is very small. The third component estimate is also unbiased, but, in accordance with the error budget analysis, a larger noise is observed in this solution. Slight variations of the number of detected stars (due to stars moving inside or outside camera FOV) are the main cause of small oscillations of measurement noise.

Figure 8. Velocity field as estimated by the optical flow algorithm from a couple of consecutive images (case 1).

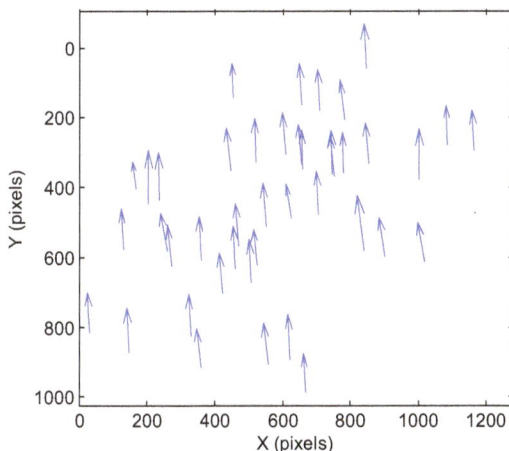

Figure 9. Estimated angular velocity components against "true" values (case 1, 10 frames per second).

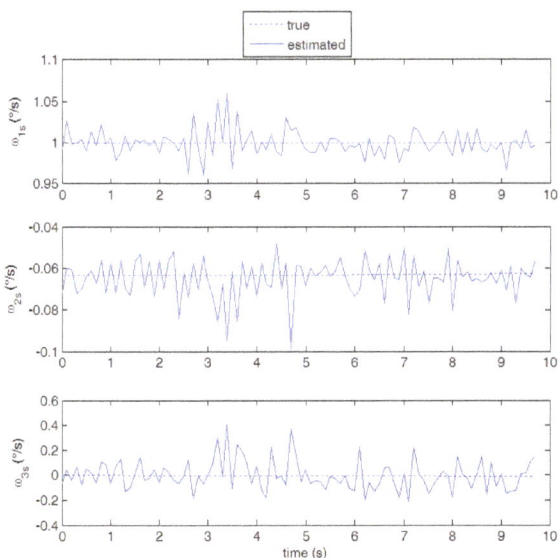

Although the proposed technique is specifically tuned to work with star images, it is of great interest investigating its application to cases with higher angular velocities, where stripes rather than stars are imaged on the focal plane and a large displacement in pixels is measured among consecutive frames. Case 2 is representative of this condition (see Figure 10, where the original star image has been significantly modified in brightness and contrast to enhance clarity). In this case, the computational load of the proposed technique increases since large windows have to be used for effective region-based matching. Moreover, the signal to noise ratio in each frame is reduced, thus reducing the number of valid star measurements, and degrading accuracy in estimating star centroids and their displacement. As it is derived from the theoretical error budget, these phenomena increase the uncertainty in the angular velocity estimates. Nevertheless, as shown in Figure 11, the average performance is still satisfying, with the smaller component ω_0 measured with slightly worse accuracy compared with case 1. Instead, the estimate of ω_{1s} shows a small negative bias (due to a slight under-estimation of stars displacement) and a larger error standard deviation, which is also found in the third component estimate.

Figure 10. Sample image of star stripes relevant to case 2, significantly modified for the sake of clarity (large angular velocity).

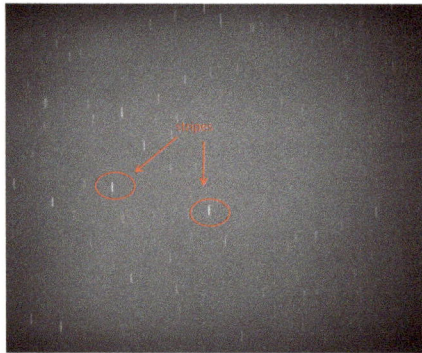

Figure 11. Estimated angular velocity components against "true" values (case 2, 10 frames per second).

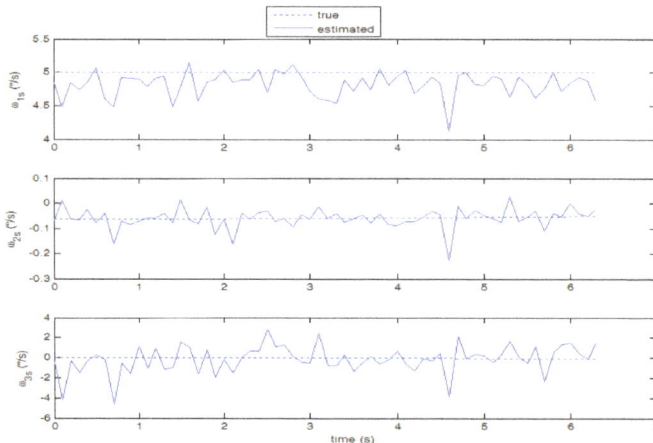

4.2. Radial Rotation Results

Considering now the first radial rotation case (case 3), the velocity field pattern is of course very different from the one detected in cases 1 and 2, with a rotation around the boresight axis superimposed to the horizontal translation due to the orbital angular velocity. Notwithstanding the large variation of the velocity modules on the focal plane, the optical flow is able to capture the motion field (shown in Figure 12) and to measure the angular velocity components with good accuracy (see Figure 13). Again, as foreseen by the error budget analysis, a larger noise is found in the estimate of the third velocity component. In the high rotation case (case 4) satisfying performance is maintained and it is in any case better than case 2 in all velocity components (see Figure 14).

Figure 12. Vector field as estimated by the optical flow algorithm from a couple of consecutive images (case 3).

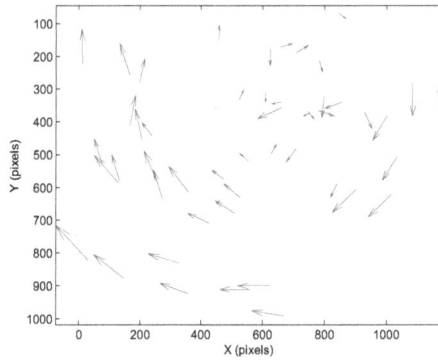

Figure 13. Estimated angular velocity components against "true" values (case 3, 10 frames per second).

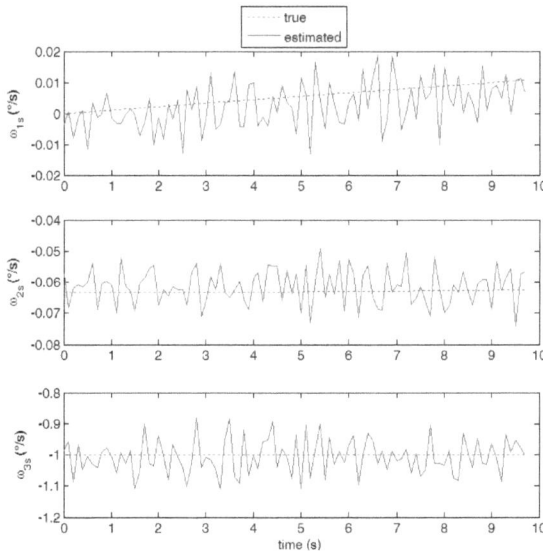

Figure 14. Estimated angular velocity against "true values" (case 4, 10 frames per second).

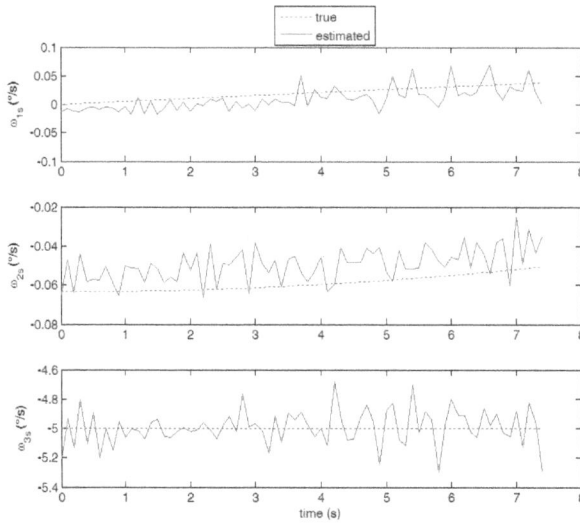

Performance in terms of mean and standard deviation of errors with respect to assigned values is summarized in Tables 4 and 5. Specifically, Table 4 shows statistics relevant low slew rates (cases 1 and 3): as foreseen by the error budget analysis, in the out-of-plane case, the standard deviation in ω_{1s} is of order of 10^{-2} deg/s (about 1% of the "true" value), whereas the noise in the boresight axis component is always about one order of magnitude higher. In absolute terms, a slightly better performance is measured in the radial rotation case, which is still in agreement with the theoretical error budget taking into account that the number of detected stars, and the average off-boresight angle (of the order of 55 and 7°, respectively) were larger than the reference values assumed in deriving Figure 4 and 5.

Table 4. Synthetic statistics relevant to low slew rates.

	Out-of-plane (Case 1)		Radial (Case 3)	
	Mean	**Std**	**Mean**	**Std**
Error on ω_{1s} (°/s)	$-1.20 \cdot 10^{-4}$	$1.64 \cdot 10^{-2}$	$-2.88 \cdot 10^{-3}$	$6.72 \cdot 10^{-3}$
Error on ω_{2s} (°/s)	$-1.90 \cdot 10^{-3}$	$9.20 \cdot 10^{-3}$	$1.61 \cdot 10^{-3}$	$5.71 \cdot 10^{-3}$
Error on ω_{3s} (°/s)	$-4.96 \cdot 10^{-3}$	$1.22 \cdot 10^{-1}$	$-6.66 \cdot 10^{-3}$	$5.57 \cdot 10^{-2}$

Table 5. Synthetic statistics relevant to high slew rates.

	Out-of-plane (Case 2)		Radial (Case 4)	
	Mean	**Std**	**Mean**	**Std**
Error on ω_{1s} (°/s)	$-1.79 \cdot 10^{-1}$	$1.81 \cdot 10^{-1}$	$-9.57 \cdot 10^{-3}$	$1.52 \cdot 10^{-2}$
Error on ω_{2s} (°/s)	$4.26 \cdot 10^{-4}$	$3.90 \cdot 10^{-2}$	$9.61 \cdot 10^{-3}$	$7.47 \cdot 10^{-3}$
Error on ω_{3s} (°/s)	$-1.28 \cdot 10^{-1}$	1.38	$7.95 \cdot 10^{-3}$	$1.21 \cdot 10^{-1}$

Table 5 shows statistics relevant to case 2 and case 4, which, as previously underlined, represent limiting conditions characterized by high slew rates. Although performance is globally worse, a satisfying accuracy is maintained especially in the radial case. This is mostly due to the fact that strips generated by the fast star movement during sensor acquisition time are shorter in the radial case, as it can be seen from Equations (11) and (23). Since in these high-rate conditions strip length is inversely proportional to the signal-to-noise ratio, this implies a better signal to noise ratio for each star, and thus a larger number of detected stars as well as better accuracy in estimating optical flow between consecutive frames. The standard deviation in third component is always one order of magnitude higher with respect to the first component.

5. Conclusions

This paper focused on an optical flow-based technique to estimate spacecraft angular velocity based on successive images of star fields. The main steps of the developed algorithms are image pre-processing for background removal, region-based optical flow computation, and least-squares solution of a linear system obtained expressing the time derivative of a unit vector in a rotating reference frame for each detected star.

Algorithm performance was evaluated on a set of star images generated with different rates and geometries (1°/s and 5°/s out-of-plane or radial rotations) by a hardware-in-the-loop testing facility and acquired by a commercial-off-the shelf camera sensor.

The method showed good performance in terms of accuracy and reliability, and experimental results were consistent with the developed theoretical error budget taking account of star fields and camera parameters. In the case of the out-of-plane rotation at 1°/s, unbiased angular rate estimates were generated and the measurement noise was of the order of 10^{-2} deg/s for the off-boresight components, while the achievable accuracy for the angular velocity component along boresight was of about one order of magnitude worse. A slightly better performance was estimated in the 1°/s radial rotation case due to the number and the average off-boresight angle of detected stars.

Rotation at 5°/s represents a very challenging situation for angular velocity measurement, with star strips on the image plane and a significant reduction of signal-to-noise ratio. Nevertheless, the developed algorithm was able to measure with satisfying accuracy these velocities, especially in the radial rotation case.

Future work is aimed at optimizing algorithm tuning in view of real-time implementation. In fact, the computational burden dramatically depends on settings related to the maximum angular velocity that has to be measured. From this point of view, a feedback control scheme, where the current algorithm settings depend on the latest angular velocity estimate and the measurement residual, seems to be a promising solution. Furthermore, measurement residual can also be used to generate a real-time estimate of measurement covariance, which allows generated output to be effectively integrated in dynamic filtering schemes, possibly also comprising estimates from other sensors.

Conflicts of Interest

The authors declare no conflict of interest.

References

1. Birnbaum, M.M. Spacecraft attitude control using star field trackers. *Acta Astronaut.* **1996**, *39*, 763–773.
2. Liebe, C.C.; Alkalai, L.; Domingo, G.; Hancock, B.; Hunter, D.; Mellstrom, J.; Ruiz, I.; Sepulveda, C.; Pain, B. Micro APS based star tracker. *Proc. IEEE Aeroconf.* **2002**, *5*, 2285–2300.
3. Sun, T.; Xing, F.; You, Z. Optical system error analysis and calibration method of high-accuracy star trackers. *Sensors* **2013**, *13*, 4598–4623.
4. Liebe, C.C.; Gromo, K.V.; Meller, D.M. Toward a stellar gyroscope for spacecraft attitude determination. *J. Guid. Control Dyn.* **2004**, *27*, 91–99.
5. Lappas, V.J.; Steyn, W.H.; Underwood, C.I. Attitude control for small satellites using control moment gyros. *Acta Astronaut.* **2002**, *51*, 101–111.
6. Rawashdeh, S.; Lumpp, J.E.; Barrington-Brown, J.; Pastena, M. A Stellar Gyroscope for Small Satellite Attitude Determination. In Proceedings of the 26th AIAA/USU Conference on Small Satellites, Logan, UT, USA, 13–16 August 2012
7. Vogel, C.; Bauder, A.; Schindler, K. Optical Flow for Glacier Motion Estimation. In Proceedings of the 22nd ISPRS Congress, Melbourne, Australia, 25 August–1 September 2012.
8. Cassisa, C.; Simoens, S.; Prinet, V.; Shao, L. Sub-grid Phycisal Optical Flow for Remote Sensing of Sandstorm. In Proceedings of the 2010 IEEE International Geoscience and Remote Sensing Symposium (IGARSS), Honolulu, HI, USA, 25–30 July 2010; pp. 2230–2233.
9. Bresky, W.; Daniels, J. The Feasibility of an Optical Flow Algorithm for Estimating Atmospheric Motion. In Proceedings of the Eighth International Winds Workshop, Beijing, China, 24–28 April 2006.
10. Janschek, K.; Tchernykh, V.; Beck, M. Performance Analysis for Visual Planetary Landing Navigation Using Optical Flow and DEM Matching. In Proceedings of the AIAA GNC Conference 2006, Keystone, CO, USA, 21–24 August 2006.
11. Izzo, D.; Weiss, N.; Seidl, T. Constant-optic-flow lunar landing: Optimality and guidance. *J. Guid. Control Dyn.* **2011**, *34*, 1383–1395.
12. Crassidis, J.L. Angular velocity determination directly from star tracker measurements. *AIAA J. Guid. Control Dyn.* **2002**, *25*, 1165–1168.
13. Enright, J.; Dzamba, T. Rolling Shutter Compensation for Star Trackers. In Proceedings of the AIAA Guidance, Navigation, and Control Conference, Minneapolis, MN, USA, 13–16 August 2012; doi:10.2514/6.2012-4839.
14. Liu, H.B.; Yang, J.C.; Yi, W.J.; Wang, J.Q.; Yang, J.K.; Li, X.J.; Tan, J.C. Angular velocity estimation from measurement vectors of star tracker. *Appl. Opt.* **2012**, *51*, 3590–3598.
15. Wertz, J.R., Ed. *Spacecraft Attitude Determination and Control*; D. Reidel Publishing Company: Boston, MA, USA, 1978.
16. Pratt, W.K. *Digital Image Processing*; John Wiley & Sons: Hoboken, NJ, USA, 2007.
17. Liebe, C.C. Accuracy performance of star trackers: A tutorial. *IEEE Trans. Aerosp. Elect. Syst.* **2002**, *38*, 587–599.

18. Girosi, F.; Verri, A.; Torre, V. Constraints for the Computation of Optical Flow. In Proceedings of the IEEE Workshop on Visual Motion, Irvine, CA, USA, 20–22 March 1989; pp. 116–124.
19. Beauchemin, S.S.; Barron, J.L. The computation of optical flow. *ACM Comput. Surv.* **1995**, *27*, 433–461.
20. Barron, J.L.; Fleet, D.J.; Beauchemin, S.S. Performance of optical flow techniques. *Int. J. Comput. Vis.* **1993**, *12*, 43–77.
21. Horn, B.K.P.; Schunc, B.G.K. Determining optical flow. *Artif. Intell.* **1981**, *17*, 185–203.
22. Lucas, B.D.; Kanade, T. An Iterative Image Registration Technique with an Application to Stereo Vision. In Proceedings of the Imaging Understanding Workshop, Washington DC, USA, April 1981; pp. 121–130.
23. Shi, J.; Tomasi, C. Good Features to Track. In Proceedings of the IEEE Conference on Computer Vision and Pattern Recognition, Seattle, WA, USA, 21–23 June 1994; pp. 593–600.
24. Bouguet, J.-Y. Pyramidal Implementation of the Lucas Kanade Feature Tracker: Description of the Algorithm. Intel Corporation, Microprocessor Research Labs, 1999.
25. Baker, S.; Matthews, I. Lucas-kanade 20 years on: A unifying framework. *Int. J. Comput. Vis.* **2004**, *56*, 221–255.
26. Camera Calibration Toolbox for Matlab. Available online: http://www.vision.caltech.edu/bouguetj/calib_doc/ (accessed on 10 October 2011).
27. Rufino, G.; Accardo, D.; Grassi, M.; Fasano, G.; Renga, A.; Tancredi, U. Real-time hardware-in-the-loop tests of star tracker algorithms. *Int. J. Aerosp. Eng.* **2013**, *2013*, doi: 10.1155/2013/505720.
28. Rufino, G.; Moccia, A.A. Laboratory test system for performance evaluation of advanced star sensors. *AIAA J. Guid. Control Dyn.* **2002**, *25*, 200–208.
29. European Space Agency. *Stars Sensors Terminology and Performance Specification*; ECSS-E-ST-60-20C; European Cooperation for Space Standardization, ESA-ESTEC, Noordwijk, The Netherlands, 31 July 2008.
30. Datasheet Matrox IRIS, Matrox Inc.: Datasheet of Matrox IRIS P Series. Available online: http://www.matrox.com/imaging/media/pdf/products/iris_pseries/b_iris_pseries.pdf (accessed on 10 October 2011).
31. Accardo, D.; Rufino, G. Brightness-independent start-up routine for star trackers. *IEEE Trans. Aerosp. Electron. Syst.* **2002**, *38*, 813–823.
32. Rufino, G.; Accardo, D. Enhancement of the centroiding algorithm for star tracker measure refinement. *Acta Astronaut.* **2003**, *53*, 135–147.
33. Rufino G.; Accardo D. An Effective Procedure to Test Star Tracker Software Routines Using a Sensor Model. In Proceedings of the 4th International Conference on Space Optics ICSO 2000, Centre National d'Etudes Spatiales, Toulouse, France, 5–7 December 2000; pp. 703–712.

MDPI AG
Klybeckstrasse 64
4057 Basel, Switzerland
Tel. +41 61 683 77 34
Fax +41 61 302 89 18
http://www.mdpi.com/

Sensors Editorial Office
E-mail: sensors@mdpi.com
http://www.mdpi.com/journal/sensors

www.ingramcontent.com/pod-product-compliance
Lightning Source LLC
Chambersburg PA
CBHW051923190326

41458CB00026B/6382